Frank Schmiedchen
Klaus Peter Kratzer
Jasmin S. A. Link
Heinz Stapf-Finé
(eds.)

The World
We Want to Live In

Compendium of Digitalisation,
Digital Networks, and Artificial Intelligence

Bibliographic information published by the Deutsche Nationalbibliothek

The Deutsche Nationalbibliothek lists this publication in the Deutsche Nationalbibliografie; detailed bibliographic data are available on the Internet at http://dnb.d-nb.de.

Editors: Klaus Peter Kratzer, Jasmin S. A. Link, Frank Schmiedchen, and Heinz Stapf-Finé for the Federation of German Scientists (VDW e.V.), in cooperation with the University of Applied Sciences (HTW) Berlin

Scientific Director: Frank Schmiedchen

This compendium is a joint effort of the VDW Study Group on Technology Assessment of Digitalisation; in cooperation with the VDW Study Groups "Education and Digitalisation" and "Health as Self-Determined Participation".

Contact:

Federation of German Scientists e.V. (VDW)
Marienstraße 19/20
10117 Berlin

www.vdw-ev.de
info@vdw-ev.de

Typesetting: Florian Hawemann (satz+layout, Berlin)

ISBN 978-3-8325-5365-4
DOI 10.30819/5365

Logos Verlag Berlin GmbH
Georg-Knorr-Str. 4, Geb. 10
D-12681 Berlin
Tel.: +49 (0)30 42 85 10 90
Fax: +49 (0)30 42 85 10 92
INTERNET: http://www.logos-verlag.de

Foreword

The scientists who discovered semiconductors in laboratory studies in the first half of the twentieth century, which gave rise to microelectronics in the second half of the twentieth century, could hardly have imagined that by using semiconductors in computers they would trigger a worldwide technology revolution, now called digitalisation. Now, some 80 years after the discovery of semiconductors, at the end of a pandemic in which the state had to restrict personal freedoms in almost every country to prevent the health care system from collapsing, the use of digitalisation has not only accelerated, but has also helped to some extent to limit the economic disadvantages of such restrictions. This book, "The World We Want to Live In – Compendium of Digitalisation, Digital Networks, and Artificial Intelligence" (original title "Wie wir leben wollen") is comprehensive and spans from the first part on "Humans and Digital Technology" to the second part on the "Necessity of Legal Design" (with a sub-chapter on autonomous weapons), and finally to the third part on the "Political Design of Digitalisation". All chapters are written by scientists with a high level of experience, and the internal discussion has also led to initial recommendations in the fourth and last part on the "Responsibility of Science". One sub-aspect would have deserved more attention: technical innovation, which is also progressing faster and faster through digitalisation, is widening the gap between unrestricted use, which is often detrimental to at least parts of society, and the limitation of misuse through standards as well as laws. Social debate, and therefore legislation, is increasingly lagging behind technical progress. How can this gap be narrowed? Another central problem is the global inequality that is growing even further with digitalisation, because fewer and fewer countries are contributing to progress in digitalisation and so-called artificial intelligence. Some industrialised countries, but especially emerging economies and developing countries in general, are left behind. The problem mentioned is addressed in the book, but not yet in sufficient detail considering its importance. My recommendation: Please read this book and – stimulated by it – intensify the debate with scientific colleagues, interested citizens, and decision-makers on all subtopics!

Prof. Dr. Dr. h.c. mult. Hartmut Graßl,
Chairman of the Federation of German Scientists e.V. (VDW)

Counting People? People Are Counting!

A Preface

What is actually the opposite of digitalisation? Analogisation?
The opposite of networking? Isolation?
The opposite of artificial intelligence? Human stupidity?

At least about the latter, we have the evidence-supported statement of a respected scientist called Albert Einstein: "Two things are infinite, the universe and human stupidity, but I am not quite sure about the universe yet."[1]

But is human stupidity really the opposite of artificial intelligence (AI)? And if so, would artificial intelligence have to be infinitely small to pass as the opposite of infinitely large human stupidity? Are networking and isolation opposites? Or does not networking by means of modern communication technologies almost automatically lead to isolation – or, conversely, have some Tekkis who feel isolated invented networking technologies in order to no longer be so alone? And is there really a hard difference between digital and analogue? Are they just two ends of a continuum?

Personally, I am not as sure about the answers. But in science, the questions often enough put us on the right track – even if it still takes some hard work to arrive at the answers that are considered valid for a particular time or culture. In this respect, I am very pleased that the volume presented here, "The World We Want to Live In – Compendium of Digitalisation, Digital Networks, and Artificial Intelligence" (original title "Wie wir leben wollen – Kompendium zu Technikfolgen von Digitalisierung, Vernetzung und Künstlicher Intelligenz"), raises many questions and that the authors do not shy away from one or the other answer! The annual conference of the Federation of German Scientists, which took place in October 2019 at the University of Applied Sciences in Berlin, already made the need for these topics and the discussion worthiness of many questions more than clear. So clear, in fact, that now almost two years later – and despite the constraints of the Covid 19 pandemic – an impressive compendium has emerged. I would like to sincerely thank all those involved and wish the articles in this book a broad, ready, and attentive readership!

Digitalisation in its modern manifestations has become one of the defining issues of our time and of global importance, not just because of the Corona pandemic. It only goes through various cycles of attention – analogous to most of our other "humanity issues", such as the finite nature of natural resources, man-made climate change, globalisation, health care, or even the global mobility of people, things, and ideas. In fact, digitalisation is centuries older than the extraordinarily stupid dictum of a German

---

[1] Incidentally, the authorship of this quotation is not entirely certain: although it is usually attributed to Albert Einstein, it is not really documented.

chancellor about the alleged "new territory of the Internet" from 2013 and also the justified mockery of it suggest. And even if it sometimes seems to us as if digitalisation with all its technologies, devices, global networks, etc. is rolling over us like a mighty tsunami of unknown origin, the historical truth is quite different: All elements of digitalisation, including networking and artificial intelligence, were invented by people, are driven by them, or even slowed down once in a while. And their beginnings go back to the dawn of humankind. The term itself provides the first clue: "Digitus" is Latin and was the finger in ancient Rome. Because many people over the centuries have counted and calculated with their fingers, the English dictionary now provides the translations for "digit": digit, number, finger, digit and toe.

The central basic principle of digitalisation, the abstraction and description of physical phenomena with the help of signs, is similarly old. While the first signs and scripts were still characterised by the need to depict the described objects pictorially, as can still be seen today in Egyptian hieroglyphics or Chinese character writing, for example, at some point mankind invented writing systems based on a relatively small set of characters. Our current alphabet is an example of this. With only 26 letters and various special characters such as full stops, commas, etc., we can form an infinite number of concepts, sentences, and descriptions of anything and even put non-physical phenomena such as love, thoughts, or antimatter into words. The core of this perhaps most important human invention is to give things a name.

This makes it possible to speak or write about things and phenomena without them having to be present themselves. The power of this is immediately obvious when one imagines, for example, how good it was to warn children or tribal members of sabre-toothed tigers without them having to stand in front of them and baring their teeth. But crafts and technology have also benefited from this possibility for thousands of years, which ultimately forms the core of all media technology, from cave paintings to books and television to augmented and virtual reality applications. Conceptually, the detachment from the physical presence of objects in conjunction with their media "doubling" is the core of all "tele-technology" and modern networking technologies.

Sign and writing systems became much more economical when people began to express the names and descriptions of things with abstracted signs that are completely detached from the appearance of things. This means that far fewer characters are needed, e.g. fewer than a hundred characters in our current writing system. Seen in this light, digitalisation is a time-honoured phenomenon originating tens of thousands of years ago when humans learned to count and spell. The fact that we could count, calculate, read, and write could well be placed alongside humour among the peculiarities of humans that distinguish us from other living beings.

The special thing about abstracted sign systems like our alphabet is that the number of signs is finite (preferably even quite small), that the signs are clearly distinguishable, do not occur simultaneously, and must not contradict each other. All sign systems that fulfil these conditions are equally powerful and can be translated into each other quasi-automatically with the help of simple rules. Our everyday 100-character system is therefore equivalent, for example, to a system of two characters, such as 0 and 1. This is why it has been possible since the 1940s to develop machines that are specifically designed to help us count and calculate. Most of these so-called computers (= Latin and English: "counters") can now count and calculate better than most people, and with their help we can capture, process, store, or distribute writing much better than with all earlier technologies. Since the 1990s, the WorldWideWeb has added a steadily growing international infrastructure that makes it possible to send and exchange numbers, calculations, and all kinds of content around the globe in fractions of a second.

The origins of these mediumistic concepts and techniques are thus almost as old as humanity and are perhaps even an essential feature of our becoming human. In any case, they are intimately connected with the history of humanity. This in turn means that we have also been familiar with virtually all the challenges of digitalisation and media use for millennials: The danger of misunderstanding when authors and recipients of information are distant from each other; the possibilities of deception; misuse for defaming other people or societies; use for power interests and violence; spread of pornography, voyeurism, and sexual assault; dangers of addiction to the content or even some forms of media technology; exploitation for economic enrichment at the expense of others etc.

The good part is that despite countless excesses, most societies and humanity as a whole have succeeded time and again over the millennia in using digital concepts and techniques to contain the dangers and make their benefits widely available.

That is why it is essential, on the one hand, to constantly face up to the challenges and possible negative consequences of media and digitalisation technologies – as the articles in this volume do – and, on the other hand, to regularly look at what solutions previous societies or individuals have already developed.

For me personally, in addition to looking at history, it often helps to also consider fiction. Therefore, to conclude this little preface, I will take the liberty of referring to two authors to whom I, as a graduate computer scientist from the last century, always like to return, Isaac Asimov and Stanislav Lem. Both have dealt with questions of digitalisation (which was not yet called that at the time), artificial intelligence, and networking in their own way, fortunately often with some humour and certainly sometimes with more depth than many current contributions.

As an example, I would like to highlight Asimov's robot cycle, whose individual stories partly come across as criminal science fiction but at their core almost always revolve around the question of how humans as inventors relate to robots as the invented. This is encapsulated and seemingly resolved in three so-called "robot laws", which superficially establish the primacy of humans, but in their interpretation and delimitation anticipate many questions, for example, about autonomous driving or the limits of artificial intelligence.

If you don't feel like rereading such old stories, you can see a surprisingly intelligent summary in Hollywood's "I Robot" from 2004 with Bridget Moynahan and Will Smith. Key question: If a robot / artificial intelligence concludes that humans tend to harm themselves, wouldn't it be legitimate – and compatible with robot laws – to barrack humans to protect themselves from themselves?

From Lem, I would like to highlight "The Washing Machine Tragedy", a grotesque about washing machines that are at some point capable of many other activities in addition to their washing duties, can look like humans and are finally "bred up" so intelligently and emotionally that they sue for their "civil" rights before the Supreme Court. Again, at its core, is the question of the relationship between humans and their intelligent machines, spiced up with the economic interests of two large corporations, and sharpened to the limit of "Should I laugh or cry about this? Or should I rather think about it?"

Ultimately, digitalisation is man-made, and the dangers or abuses result very much from technology, but even more so from human needs. And we know all about those!

To return to the beginning, part of the message of the old sense of the word digitus to modern "digitalisation" could be not to abstract things so much that they become incomprehensible, or vice versa: the more we digitise, the more we have to make sure that things remain manageable and humanly controllable.

This is because digitalisation only works when people count. In more ways than one.

Or to take up the title of this volume: If we don't care about "the world we want to live in", others will – possibly machines....

Prof. Dr. Carsten Busch,
President of the University of Applied Sciences, Berlin

Table of Contents

The World We Want to Live In – An Introduction

Frank Schmiedchen

We need to understand profoundly, where we stand and what the next steps will be of digitalisation, networks (or interconnectedness), and machine learning (artificial intelligence; AI) to make informed and wise decisions. Our most important questions maybe: What are the effects of the latest digital technologies in various applications and what are the consequences for me, for my country, and for humankind?

Complex digital networks and artificial intelligence are fundamental innovations that permeate the economy and society in many fields, thus becoming the engine of a comprehensive, disruptive structural change. These innovations, being new technical approaches, are also the technological basis for a vast amount of present and future follow-up innovations that are changing our lives profoundly and at high speed.

Despite numerous publications, the profound and multifaceted effects of advanced digital development are usually considered only in isolated areas, i.e. for specific socioeconomic, sociocultural, or sociotechnical areas. In a long-term study on media discourse in Germany, Fischer and Puschmann find that there is a one-sided focus on economic and technology-euphoric aspects of current digital developments. In contrast, other social issues are still discussed too rarely (cf. Puschmann/Fischer, 2020; pp. 29ff). There is still a lack of diversity in terms of perspectives and actors represented (ibid.). The discourse is thus dominated by the assumption of a generally valid and broadly accepted approval of technology-optimistic concepts for the future, which usually remain vague and are not critically questioned.

Given the exponential increase in knowledge, the lack of scientific (e.g. neurological, [social] psychological, legal, or economic) analysis means that fewer and fewer people or institutions have a somewhat rudimentary understanding of recent digital developments. As a result, these few persons or institutions can increasingly determine further developments in an uncontrolled manner, in the sense of "the winner takes it all". However, the greater challenge by far is the fact that a majority of publications to date explicitly or implicitly suggest that the continued pursuit of the chosen technological path is an inevitable fate for humanity. This culminates in the claim that the chosen technological path determines humankind's evolutionary future, i.e. we are already in a so-called lock-in today.

The advanced digital development poses societal challenges, but also existential risks for our human future. There are not many like the US documentary filmmaker and author James Barrat, who dig deeper on possible further developments. He concludes that a broad, social discussion about the fundamental connections and dangers of further digital development, especially of AI, is essential for humanity's survival. Barrat therefore calls for comprehensive technology assessments as a necessity that he describes dramatically as "nor does this alter the fact that we will have just one chance to establish a positive coexistence with beings whose intelligence is greater than our own" (Barrat, 2013, p. 267). Barrat has had an unusual impact with his 2013 book 'The Last Invention', which was one of the triggers for Elon Musk's decision to found Neuralink in 2016, which is expected to build human-machine interface prototypes this year (2021) after a successful trial with a macaque monkey (Musk, 2021; Kelly, 2021). According to Elon Musk, one goal of founding Neuralink is to arm humanity through a symbiotic relationship with digital technology to counter future successful attacks by hostile and powerful AI (Hamilton, 2019).

But even if this is classified as science fiction, since reaching the innovation threshold of the technological singularity is highly controversial in terms of the likelihood of its occurrence (from "in a few years" to "never"), there are numerous challenges that the advanced digital developments pose for our daily life. Especially since the outbreak of the Covid-19 pandemic most voices suggest that there is no alternative to the recent technology path.

An example of the danger that has been growing for the past few years is the consideration of "The Great Reset": a digital "Singaporeisation" of western democracies, i.e. the destruction of open societies through rampant prohibition policies that pretend to be necessary to implement the UN Sustainable Development Goals or the Paris Climate Agreement (SDG 2030, Schwab/Mallert, 2020, WEF, 2020). Good intentions are suggested here (e.g. closing the digital divide, climate protection, diversity). However, the policies proposed are a nightmare smoothie of "1984" and "Brave New World" that would lead to digitally controlled autocratic regimes. These attempts are flanked by illiberal identity policies (cf. Fourest, 2020, Kastner/Susemichel, 2020, Wagenknecht, 2021), that are supposed to define "right action, right speech and right thinking" even in democratic, open societies. Compliance with these is to be digitally monitored and sanctioned in a variety of ways (e.g., higher health insurance contributions for people without smart fitness bracelets). However, such a system would mean that "transparent" people would potentially be constantly monitored in a digital panopticon to see whether they are behaving correctly and appropriately (cf. Chapter 1).

On the other hand, digital technologies have become indispensable tools of everyday life for most people around the world and are extremely useful in many ways: they simplify and speed up many tasks, connect people, reduce dangers, create convenience, are fun, and save lives. This makes them an enrichment of human life, especially when they are used for the common good. Fundamental criticism is counterproductive in view of the broad enthusiasm of most people worldwide for their digital access and the vehement demands of those who do not have (sufficient) access and want it. It is us, as citizens, workers, learners, consumers, or patients, who demand barrier-free access to more and more digitalisation and make extensive use of the digital media available. At least in the three technologically leading regions of the world, North America, East Asia, and Europe, people are largely convinced that the human future will be a digital one, whatever that may mean. The Covid-19 pandemic has once again dramatically accelerated this development.

Here is an excursus by Oliver Ponsold who discusses personal interaction with digital media:

A Personal Approach to the Digital World

Oliver Ponsold

For years, a new reality of life has been taking shape in our personal interaction with the digital: Through the convergence of advanced technical performance and available energy storage in pocket format, a digital personal environment is created where high-performance connectivity is reliably available, which permanently supports and influences our perceptions and evaluations of the environment, and based on this, enables or channels short-term decisions and actions.

As an example, service companies that arrange and deliver meals from third-party providers to customers offer a link to the actual service provider. Permanent use leads to service providers becoming more customer-oriented and learning from them, especially when extra data from the digital customer ego is also included. This reduces their entrepreneurial risk and tends to result in higher customer satisfaction, and thus loyalty. A dedicated, modern digitised manufacturing and distribution environment supports speed and consideration of customer-specific special requests, as well as iterative purchasing models and long-term customer loyalty in service delivery. Service providers earn extremely well from platforms that are ostensibly free of charge for the customer through enormous scaling effects and intelligent evaluation of collected usage data, the actual means of payment of the Internet.

Finally, the natural person converges with their own virtual images on the Internet and the digital assistants they carry with them during long-term intensive use. The digital footprints left behind are evaluated by AI algorithms and used for targeted selection preferences and (product) recommendations, so that an enhanced positive and binding user experience is made possible. This includes, for example, the combination of e-commerce platforms with social networks, the circle of friends, and acquaintances. The resulting psychological and practical path dependencies and lock-in mechanisms are presented and discussed in detail in Chapter 3 of this book.

Four blocks of questions can shed light on this in a very personal way:

1. Is looking at the smartphone a constant daily ritual? How often and on what occasions do I use it? Do I often use different media at the same time and are they smartly connected?
2. From which source or medium do I obtain information and do I double-check it? How quickly do I expect answers? What personal imprint profile do I leave behind in my research?

3. How often and what kind of life moments do I share on social networks and which and how many people do I follow on social networks, blogs, or podcasts?
4. What would be the consequences if my virtual personality with access to email and all social networks would be stolen and used for criminal purposes?

Such an analysis shows how deeply humans and digital technology are already closely interwoven. Digital technology enables us to access information in real time, which we validate and evaluate through personal use in the mirror of our knowledge and preferences.

Since 2016, the Federation of German Scientists (VDW) has been intensively addressing the consequences of networked digitalisation and artificial intelligence and has established a working group for this purpose. The group has published a "Policy Paper on the Asilomar Principles on Artificial Intelligence" in 2018 and in 2019 organised a conference on 'The Ambivalences of the Digital', together with various German universities, the German Trade Union Federation (DGB), and civil society, at the Berlin University of Applied Sciences. This compendium is another milestone in the VDW's deliberations on digitalisation topics and forms the basis for further work.

In view of the almost religious euphoria regarding the advancing use of digitalisation, networking, and AI in practically all areas of life, it is a necessary and natural task of the VDW to point out underestimated or ignored scientifically and socially relevant and existential problems of this development, to provide approaches for technology assessments, and to make well-founded proposals for an ethically justifiable approach.

- Who determines what is "good" when technology becomes increasingly pervasive and affects all people, directly or at least indirectly, not just those who have consciously chosen to use it?
- For example, do we have a consensus in the EU, USA, or ASEAN on what risks we are willing to accept in order to make our lives more and more comfortable?
- How can such a consensus be achieved at the global level?

We see the enormous potential for liberation that digitalisation already means for individuals and societies if these possibilities of individual freedom and socio-ecological development are used.

We also see the dangers of advanced digital networks and AI, which lie in the fact that they create new types of long-term, profound, and unpredictable dependencies for individuals, institutions, and states, which only a few can escape. In this context, we distinguish between opportunities and risks that arise from different social embeddings of technology use and technology-immanent risks that always arise from the development and use of technology regardless of this embedding, and these technics-intrinsic risks must also be comprehensively addressed.

The latter includes, among other things, consequences of the daily amount, we are using digital technology to improve our fitness and health, in our social relations, or in education. But there are also technique-intrinsic risks for our national security. Technics-intrinsic risks include, for example, hidden manipulations through the processing and utilisation of unmanageably large amounts of data (big data), which are used by companies to control consumption, by political groups for disinformation, and by authoritarian states to control their citizens, but also by (il-)liberal groups in western democracies for social control.

Further digitalisation will only lead to humane and socially desirable outcomes if future decisions serve our common interests as humankind to a greater extent than in the past.

This introduction already shows the necessity of a scientifically sound value framework for meaningful discussions on how possible influences of digitalisation (on people, societies, and the environment) can be made visible and assessable and, above all, what standards we apply. Measuring and evaluating is not meant in purely quantitative terms but includes, over all, a qualitatively accurate approach to determine the object. Measuring and evaluating, however, includes the exact determination of which parameters determine the result of digital computing processes as well, and which different "real-world" effects the feeding of certain parameters has (cf. Becker, 2015, pp. 91–97).

This compendium aims at contributing to a rational discourse by providing basic knowledge about the latest trends, impacts, and possible consequences in relevant areas of digital interconnectedness and artificial intelligence.

To this end, the first part, "Humans and Digital Technology", provides basic knowledge and poses important philosophical questions about the relationship between humans and technology. The core question is whether digital technology still has the character of a man-made tool or whether machines and algorithms are increasingly being awarded an intrinsic value, and people are granting digital technology strong influence to the point of de facto domination over their everyday lives as well as essential areas of society. The amount and intensity of smartphone and social network use initially suggest the latter. The second part of the book therefore deals with the need for legal regulation, be it "hard law" (e.g. laws and sanction-proof international agreements) or "soft law" (e.g. norms and technical standards, voluntary commitments). The third part of the book justifies for six central social areas why future steps of digital networks and AI development must be subjected to a comprehensive technology assessment, for which we formulate requirements in the fourth part.

Thus, the book is divided into four parts:

Part I: Humans and Digital Technology,
Part II: Necessity of Legal Design,
Part III: Political Shaping of Digitalisation, and
Part IV: Responsibility of Science

The first three parts each provide different perspectives on the respective thematic focus:

- In Part I, we look at the nature and perception of data, as well as the current state of developments in information technology, mathematics, and physics (Chapters 1 and 2). Building on this, we discuss path dependencies in a mathematical-sociological analysis (Chapter 3) and deal with technology philosophy, the image of humanity, human enhancement, and Trans-/Posthumanism (Chapters 4 and 5).
- Part II shows the need for and the possibilities of regulation with a view to emerging demands for machine rights (Chapter 6); questions of liability (Chapter 7); technical standardisation (Chapter 8) and intellectual property rights (Chapter 9). Of particular importance are the questions of international regulation of lethal autonomous weapon systems, as an example of future digital warfare (Chapter 10).
- Part III highlights socio-economic, cultural, and political issues of application and focuses on those areas of application that we consider particularly important: education (Chapter 11), health (Chapter 12), sustainability (Chapter 13), economy (Chapter 14), labour (Chapter 15), and social systems (Chapter 16).

Throughout the 16 chapters, a chain of reasoning emerges, which we condense and conclude in the fourth part: "Responsibility of Science" with our demand for comprehensive technology assessments of digitalisation, networking and artificial intelligence.

I owe my thanks to Fatih Birol, Dr. Miriam Engel, Dr. Ginelle Greene-Dewasmes, Astrid Jekat, Dr. Peter Michael Link and Dr. Kuma Sumathipala for helping us in translating the book into English!

The end of our introduction is the beginning of our hope that this will be an exciting reading experience for you and we invite you to reflect with each other and with us, the Federation of German Scientists VDW, on the best way forward!

Part I

Humans and Digital Technology

Introduction

Klaus Peter Kratzer

"We can only see a short distance ahead, but we can see plenty there that needs to be done" (Turing 1950) – once again, we are confronted with this last sentence taken from Alan Turing's fundamental publication on the question of whether machines can think. The difference to Turing's situation more than 70 years ago, however, is that we are not only cultivating remote visions, but that we are in the midst of a transformation of our living environment, in which machines exert an increasing influence on us in every way – we communicate with machines, use machines as mediators (in form and content) of our communication with fellow human beings and allow machines to classify, evaluate, and make judgments on us. The fact that the latter usually happens outside of our perception and often without our direct knowledge is irrelevant – we often see ourselves in the grip of an inevitability that is restricting our ability to choose and, driven by a desire towards conformity, is impairing a rational understanding of the situation.

What is required of us is a desperate, blind faith in technological progress, because only very few of those affected can assess the state of the art, the reliability of this technology, and even fewer can grasp and evaluate the projected state of our society in future. In addition, of course, there is the question of who invests in such a technology: Which business model of which organisation, which company will benefit, and in which role does each individual find himself or herself? Is it inevitable that the manufacturer of my car is recording and analyzing my movement patterns? Do I have to put up with being schematically bossed around by a "chatbot" when making a complaint? Is it permissible for my employer to access data on my body functions via a fitness wristband (provided by the company, of course)? ... Of course, I had already agreed in advance to the inevitable: to wear this wristband on the company premises at all times.

The chapters in the first part of this book confront us all with the state of the art and help us to give substance to the reflection and debate that is still ahead of us. Gaining deeper insights, more questions will be raised than answered, as, lamentably, there are no simple solutions to complex problems. Nevertheless, we will all have to resolve these questions in a timely manner, so that as many options as possible for our future world remain open.

The first chapter, *Datafication, Disciplining, Demystification* by Stefan Ullrich, tells the story of the centuries-long tradition of recording, coding, and structuring data, based on technical development, but also, and especially, on the complex relationship between data, facts, and the real world.

In the subsequent chapter, Alexander von Gernler and Klaus Peter Kratzer introduce the technology underlying networking, digitalisation, and artificial intelligence in the chapter *Technical Foundations* and *Mathematical-Physical Limits*. They show where the possibilities and limits of this technology lie today, and, in consequence, what the foreseeable future is holding in store for our society as a whole, but also for each of its members.

In her contribution on *Path Dependency and Lock-in,* Jasmin S. A. Link shows how the socioeconomic dependency on digitalisation and networking, which is already widely observable, can lead to complete dependency and loss of freedom within the system logic by exponential amplifying effects, which can only be countered by an extensive system break. She argues for situational identification and development of alternatives and diversity as a counterweight to homogenisation based on algorithms.

Stefan Bauberger looks at the relationship between technology and humans in his contribution *Questions in the Philosophy of Technology.* Humanity is in danger of being debased, even degraded, into the role of an information-processing system. Technology must not be an end in itself: a differentiated, value-based consideration of technology must be achieved; what is technically possible is not necessarily what is permitted or required.

Following on from this, Frank Schmiedchen takes a look at interfaces between *man* and machine and utopian/dystopian visions of the future of the anthropopocene in his chapter on *Digital Extensions of* Man, *Transhumanism, and Posthumanism.* While the evolutionary advancement or replacement of Homo Sapiens by algorithms seems to be a quasireligious (nightmare) dream in the distant future, digital "enhancements" in the guise of clothing, jewellery, and implants are already commonplace – the symbiotic relationship of many people to their mobile phones, which is nowadays quite commonplace. supports this vision.

These chapters will lead to reflexion, thought, and discourse, surely also to friction. But when we enter such a discourse and when we can feel the heat of friction, an important goal of this section, better even, of the entire book, has already been achieved.

Chapter 1

Datafication, Disciplining, Demystification

Stefan Ullrich

Data, that what is given – if Francis Bacon has his way (Klein and Giglioni, 2020). Data, that what has to be regulated – if the European Commission has its way (EUCOM, 2020a). In between are 400 years of collecting, storing, processing, and disseminating data. While for Bacon it was self-evident that data primarily served science, today we speak of data as a commodity, which is said to be worth over 325 billion euros in Europe alone (EUCOM, 2020b, p. 31). Etymologically, then, the Latin *datum* and *commoditas*, the given here, and the useful, suitable, advantageous there, are approaching each other. This should make the scientific mind sit up and take notice, because whenever something is too suitable for one's own theory, it is necessary to take a closer look. Data should be uncomfortable, should be able to challenge one's own theory (and, if we cheat a little like Mendel with pea cultivation, sometimes also confirm theories).

This year we celebrate the 75th birthday of the Electronic Numerical Integrator and Computer, or ENIAC for short, the first freely programmable, electronic universal computer. Sure, in Germany we could also celebrate the 80th birthday of Konrad Zuse's Z3 and in the UK the 85th birthday of the Turing machine – in short: we have been living in an age of computation using universal computers for several generations. Data have been seen for decades as machine-readable and, above all, computable information. The word information, that which is put into form, reveals the process of transforming the pure notation of numerical values based on observation or deliberation into structured formats. Data are more than numbers or symbols; they have a scheme, have been modelled and prepared in a machine-readable way.

A very simple – and yet ingenious – scheme is the table. The first row contains designations, such as measured quantities and units, while the other rows contain symbols, written down in pictures, letters, and numbers. Leibniz described the power of the table to his sovereign in flowery words. The busy mind of the ruling person could not possibly know how much woolen cloth was being manufactured in which factories and in what quantity was demanded by whom in the population. However, since knowledge of this "connexion of things" was indispensable for good government, Leibniz proposed so-called state tables ("Staatstafeln"), which make complex facts comprehensible at a glance and thus governable and controllable (Leibniz, 1685).

Data serve to control the human being. Initially, this is only to be understood as genetivus subjectivus, i.e. humans use data to control their environment. Recently, however, the meaning in genetivus objectivus has also been debated: Data are used to control people. In this paper, we start with the building blocks of data as we understand it in the modern context, so we begin with the machines that process data. Then we look at the two dimensions of data for human control.

1.1 Building Block: Digital Number

For modern data understanding, the number is probably the most important building block. With its mantra "Everything is digital number", computer science, in direct succession to the Pythagorean school of thought, advocates the view that every gesture, every speech, every image, every writing, in short, all codified human actions can be written down with the help of a number (cf. Ullrich, 2019). This is not true, of course; the most important things cannot be recorded, such as what constitutes a good-night kiss, in essence. Poetry comes closest to this, but even the best poems fail at a high level to capture the innermost part of the human being. But whereof computer science cannot speak, thereof it does not keep silent about, but collects data. The number of goodnight kisses correlates with the size of the family or shared apartment, an important data point for software rights holders who want to sell licenses for use.

The number, or more precisely, the discrete number breaks down the intangible continuum of the environment around us into measurable and countable objects; the measurements and numbers give us a sense of control. It must have been extremely reassuring for the first communities to unravel the mystery of the seasons. That winter just doesn't last forever, but is replaced by spring! Eight hours of daylight at the winter solstice, a good six lunar phases later, it is already sixteen hours – with the help of the calendar and a look at the date, man at the mercy of the weather has little control. Even today, in the age of man-made climate change, it is data that support our climate models and tell us about our future.

The discrete number served above all the empirical sciences, produced with the help of observation or with the help of instruments and tools. In the mechanical age, tools such as the telescope provided analogue signals that first had to be schematised or immediately discretised – Galileo drew the moon with its less than perfect surface schematically, and to this day mushroom pickers and physicians consider the schematic representation of fruiting bodies or nerve cells more didactically valuable than high-resolution photography.

To convert an analogue signal into a discrete one, an analogue-to-digital converter is needed. The continuous signal, such as a sound wave, is measured 44,100 times per second, or "sampled" in technical jargon. The best way to imagine how this works is with the help of a grid. Imagine drawing a wave on a piece of graph paper. Then take a pencil of a different colour, for example red, and mark the intersections of the checks that are closest to the wave. These red points then mark the discrete values of the analogue signal (Figure 1.1).

The resulting discrete, digital signal is, of course, only an approximation of the analogue signal, which is all the more congruent the higher the sampling frequency and the finer the quantification. The advantage is that we now have a machine-processable datum, a digital datum that we can store or communicate.

The first analogue-digital converters were developed by Konrad Zuse between 1943 and 1944 to mechanise the reading of the analogue dial gauges of the Henschel glide

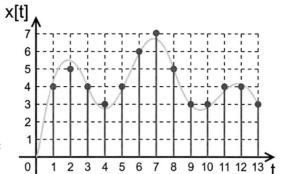

Fig. 1.1: The red digital signal is the sampled and quantised representation of the grey analogue signal. (PD) Public Domain from Wikimedia Commons, the free media repository.

bomb Hs 293. Once the bomb had been dropped, it could be steered by radio thanks to ailerons and elevators to safely deliver its 300 kilograms of explosives to its target. It was the world's most successful sea target missile, with "successful" translating to being responsible for the greatest US human loss in World War II.

This drastic but quite typical example is intended to illustrate the role of data. Data are a demon that can be as subservient as it devours, as this book comprehensively shows. In the real world, there is no such thing as harmless data. Data have left the innocent sphere of mathematics and, since the invention of the punch card, determine the wealth and woe of people.

In the punched card, data are encoded with the help of holes, and you don't even need a machine to decipher them. In view of the enormous investment costs, the first programmers would not have been able to afford the luxury of using valuable machine time just for reading out the data, and even the execution of simple filter algorithms does not require a machine if the holes are suitably designed. Let us take an edge punch card, which, unlike other punch cards, is also suitable for manual processing. An edge punch card has holes all around the edge in the uncoded state. Now a key, a coding is designed, and the cards are notched so that a slot is created at certain points. In Figure 1.2 we see both holes (as created with a hole punch) and slits (notches). All the cards are now placed on a stack and set up so that the holes are on top of each other.

If you now push in a knitting needle and lift the stack of cards, those cards that have a slit where the needle is will fall down. The time saved in searching, in contrast to normal index cards, is enormous: the selection speed is between 30,000 and 40,000 cards per hour.

It is not the data analysis alone, but the very collection and categorisation of the data that promotes both its use and potentially misuse. With the help of an edge-hole map system, Leibniz's state tables would have made a version leap, and who knows, perhaps they would have ensured that the sovereign would have been informed not only about the amount of wool, but about political adversaries, their personal data, habits, and meeting places.

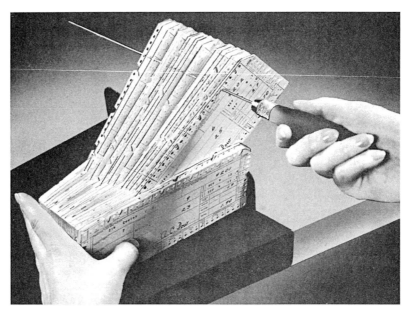

Fig. 1.2: Manual selection of an edge punch card. Illustration taken from: Bourne, Charles: Methods of Information Handling, John Wiley & Sons, New York, 1963, p. 81.

1.2 Data for Human Control

What Leibniz's state tables were to the sovereign, the smart fitness bracelets, and the even smarter universal computers in the trouser pocket are to the health-conscious person, which we still call a "phone" for historical reasons. We count steps, calories, or CO_2 emissions to better discipline ourselves.

We want to control ourselves or our environment with the help of recorded data – but what does that actually mean? The *contre-rôle* is the counter-register to confirm an assertion proven by data. Trust is good, control may be better, but to be effective it needs monitoring: we need to collect data again, this time independent from the data that went into the assertion to be proven (especially if the data come from others), and we need these data preferably in real time.

This is the weak point of Leibniz's state tables: the meaningfulness of the data naturally decreases over time. Sure, for monitoring one's own actions, i.e. testing the effectiveness of political decisions, for example, annually collected data are sufficient, or, as Leibniz wrote: suitable for self-government. However, if we want to rule over someone, we need monitoring and data. We cannot meaningfully talk about current data-based business models or address data markets without explicitly mentioning surveillance, which is an instrument of human control.

The icon of surveillance is, of course, Jeremy Bentham's Panopticon, already briefly mentioned in the introduction to the book, his 1791 design for an "Inspection-House". Bentham (2013) planned this for a variety of institutions, from schools to hospitals, but the first and most famous example is the prison. At the centre of the institution is a tower that allows the cells radiating out to be seen, but protects the observer in the tower from being seen. Thus, the inmate of such a cell does not know if and when she is being observed, but she knows that she can be observed at any time. This possibility of surveillance leads to a change in behaviour; the inmate permanently behaves as if she were actually being watched at all times. She has internalised the surveillance, this is what has become known as the "panoptic principle".

The pan-optical all-seeing eye becomes an all-seeing-for-ever-stored data based eye with the help of surveillance technologies. Just as the modern concept of privacy ("the right to be let alone", Warren/Brandeis, 1890) only emerged ex negativo with the advent of the camera, the fundamental right to informational self-determination was only established with the advent of big data processing systems (BVerfG, 1983). Data protection, which is actually a very inappropriate term, picks up on the information flow direction of the panoptic principle. The informationally more powerful person watches over the informationally inferior occupant in the tower. Data protection, or more precisely: data protection law, is now supposed to ensure that this power is not abused.

But who monitors the monitors? The public, of course, the "body of the curious at large – the great open committee of the tribunal of the world"[2], using publicly available data (Bentham, 1787). Protect private data, use public data, as stated in the Chaos Computer Club's Hacker Ethics, means to recognise the inherent nature of data and use them for the good of society (CCC, 1998). Data also serve to control people, sometimes understood as genitivus objectivus in the case of the Panopticon, sometimes understood as genitivus subjectivus in the case of the Court of Justice of the World.

In biometric recognition systems, all the above-mentioned explanations come together in a complex sociotechnical system. Biometrics, i.e. the measurement of life, is an instrument of statistics. Mortality tables, age structure of the population, and average life expectancy are of interest to state leaders when it comes to taxes, participation, and distributive justice. In one of the first scientific works on biometry, the Swiss natural scientist C. Bernoulli first describes how a table of life expectancy should be structured and what advantages arise from this clear connection of things, before pointing out in a somewhat hidden insertion that it was life insurance institutions that made the collection of these data "a necessity" (Bernoulli, 1841, pp. 398–399). Once the transdisciplinary cultural technology researcher has found this technological history trail, she discovers the true motivations behind biometric systems everywhere. Since Francis Galton, dactyloscopy has not only served law enforcement purposes but also, like all other biometric measurement systems to this day, voluntarily or involuntarily, feeds racist mindsets and practices.

Before we take this thought to its conclusion, the "black box" should be opened a little at this point. To do this, we will look at the typical structure of a system for the automated recognition of fingerprints (after Knaut, 2017, p. 44):

---

2   Quoted after original source https://en.wikisource.org/wiki/Panopticon_or_the_Inspection-House

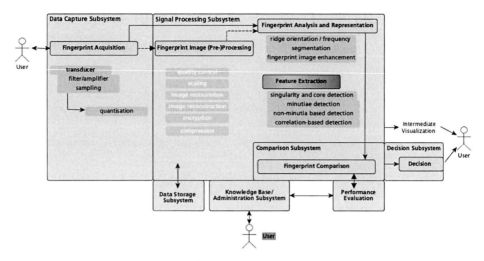

Fig. 1.3: Structure of a biometric system for the automated recognition of fingerprints. (cf. Knaut, 2017, p. 44)

This diagram is already a considerable simplification of the actual architecture of a typical system, which we can see from the "Fingerprint Acquisition" component alone. Before the development of the corresponding sensors, the fingerprint was captured in analogue form, as the name suggests. The print, typically after placing the finger first on an ink pad and then on a piece of cardboard, only leaves colour pigments where the papillary ridge of the finger is located, precisely these typical lines that we also see with the naked eye.

Biometric recognition systems are used for verification and identification and are usually marketed as access systems (verification) or generally as official security technology (identification). The introduction of biometric passports and ID cards in Germany was also presented from this point of view. However, in background discussions and when asked directly, it is clear to all involved that this is a business promotion, as the corresponding readers have to be licensed. However, the data-based business models of biometric recognition systems have a catch: They technically fall under the General Data Protection Regulation (Article 9 (1) GDPR), which makes exploitation so challenging. Biometric data are also the most intimate and visible data: Unless there is a pandemic, we are constantly showing our face. And even in Corona times, our walk in a crowd of people can be quite unique. Finally, there are our fingerprints, which are emblematic of identity, although technicians and scientists have been pointing out for decades that identity constructions and attributions are at stake. However, data can also be used to question these attributions, which is what the last section is about.

1.3 Data for Demystification

Data are the key to knowledge; they are the basis of the empirical sciences and provide a view of the world not only for quantitative but also for qualitative researchers. Data are not facts, which was important to Francis Bacon, and we should always bear that in mind. Data can generate, confirm, or question facts in the scientific mind. Data can also obscure facts. Data science is slowly maturing into the basic cultural technique of the responsible member of the networked society. Data scientist Hans Rosling demonstrated to a large audience (and thanks to audiovisual data also on Youtube, Vimeo, and others) how data can be used to bridge cultural differences, break down prejudices, and ensure common understanding. In a very humorous and exposing way, Rosling holds up the mirror to us that we rely on data, numbers, and facts that we learned in school and that are now reproduced on all media channels. Our conceptions of countries in the global South, for example, are closer to myth than to the present (Rosling, 2006). Demystifying false, perhaps even harmful assumptions with the help of data was the main drive of the humanist Rosling.

But for this to happen, these data must also be available. There is a fundamental bias when it comes to data: we can only measure what is measurable. So, on the one hand, this depends on instruments and tools, but on the other hand it also depends on culture and customs. It is not due to a lack of tools that Caroline Criado Perez (2019) observed a gender data gap, but due to culture of a male dominated society. Data are collected for a purpose, and the more effort put into data collection, the more likely one is to expect a dividend: Data thus becomes a means of payment.

Data are the central element of digitalisation because it comes from both the old world of automated data processing and the new world of heuristic data techniques such as machine learning, big data and artificial intelligence. It is therefore not surprising that the development of data literacy is repeatedly insisted on, without, of course, saying what exactly it should look like. Demystification also includes a sobering look at current data processing practices. The majority of people usually choose not to deal extensively with data, not even with the data they generate themselves, which becomes usable for whatever purpose by consenting to unread declarations of use by "someone". In a free society based on the division of labour, we should also accept this, but then legislators must hold computer scientists and companies with data-based business models more accountable, for example by demanding that data-based business models not be subject to any secrecy obligation or that the data-processing systems be precisely labelled.

Another step towards demystification could be didactic systems like MENACE (cf. Ullrich 2019c). MENACE was the name of a didactic machine to teach machine learning principles that was conceived and described by Donald Michie in the 1960s. His machine could play Noughts and Crosses (also known as Tic-Tac-Toe, Three in a Line, or Tatetí) against a human player (cf. Michie, 1961). The Machine Educable Noughts And Crosses Engine was a machine learning system, but with a special feature: the machine was not made of computer components, but of matchboxes filled with coloured beads. Each colour represents one of the nine possible positions that an X or an O can take on the playing field. The construction was simple and impressive at the same time; no less than 304 boxes were needed for it, one box for each possible configuration in the game. The operator now randomly draws a coloured pearl from the respective box with the corresponding configuration. In the course of the first few games, the matchbox machine will probably lose, as there is no strategy at all, since the beads are drawn randomly. But then machine learning kicks in: If MENACE loses, all the drawn beads that led to the defeat are removed. If MENACE wins, three beads of that colour are added to the boxes used. This means that the chance of losing is reduced, while on the other hand good moves are significantly rewarded. If MENACE is trained long enough, it "learns" a winning strategy (by improving the chances of making good moves) and, therefore, "plays" quite well.

The interesting thing is that no human would attribute any intention to a stack of boxes, unlike machine learning systems implemented with software on a computer hardware. Especially in the case of machine learning or artificial intelligence, as a critical observer of the information society, one can still marvel at "the enormously exaggerated attributions an even well-educated audience is capable of making, even strives to make, to a technology it does not understand" (Weizenbaum 1976, p. 7).

Everything starts with the will to understand, in order to be able to use the power of data accordingly for the benefit of the general public. The Federation of German Scientists is aware of this special responsibility that the influence of technological-scientific progress has on people's mindsets, and that is why we are committed to regaining informational sovereignty as a networked society as a whole.

Chapter 2

Technical Foundations and Mathematical-Physical Limits

Alexander von Gernler and Klaus Peter Kratzer

Acknowledgements: Astrid Jekat helped us a lot when translating our chapter to English by giving us many valuable comments and insights as native speaker that greatly improved the quality our our text.

To discuss the basics used in this book, we chose a slightly different order than in the title of the book. The three terms networking, digitalisation, and artificial intelligence covered here build on each other and are more easily introduced that way.

2.1 Networking

Networking is the foundation of all topics discussed in this chapter. It describes the interconnection of computing or storage units, as well as sensors or actuators, by means of any transmission medium for the purpose of information transfer. The distances bridged can be as short as a few centimetres (near field communication, NFC) or as long as several thousand kilometres (transatlantic Internet cable).

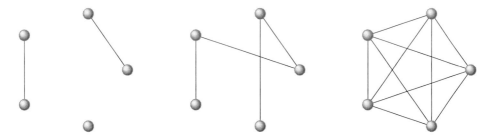

Fig. 2.1: Examples of connectivity using general graphs: A weakly connected and disconnected graph, a connected graph, a fully connected graph. (Source: The authors)

The degree of interconnection can be explained using Figure 2.1: The more edges there are between the nodes in a networking scenario, the more possible paths can be taken between two specific nodes. Accordingly, the availability of individual nodes also increases – improved interconnections make the network robust against failures.

2.1.1 Technical Aspects

2.1.1.1 History

Beginning with the legendary and world's first connection of the first four computers via long-distance links in the so-called ARPANET in 1969 (Figure 2.2 shows a more developed version from 1973) in the USA, the development of ever faster and more robust transmission technologies for information transmission continues unabated. Important parameters for characterising a networking technology are bandwidth[3], latency, carrier medium, and range.

To this day, the transmission media used are almost exclusively copper wires, electromagnetic waves (radio technology), or fibre optic media[4].

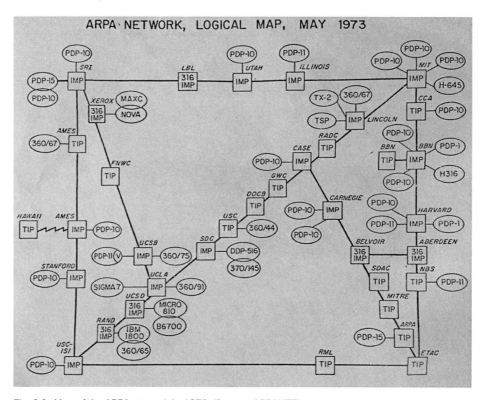

Fig. 2.2: Map of the ARPA network in 1973. (Source: ARPANET)

---

3 Often loosely denoted as *line speed*.
4 In exceptional cases such as hacking or activity of intelligence agencies, alternative media types like local sonic waves or significant variations of computer power intake can be employed to stealthily modulate and transfer information between otherwise non-connected devices. When used in this manner, they are often referred to as side channels. For practical everyday information transfer, however, their significance is negligible.

2.1.1.2 Modelling

To apply information to an otherwise raw transmission medium, so-called logical layer models are used in both telecommunication and information technology. By modelling in different layers, it is possible to realise, among other things, different technical requirements such as a) protection against data loss on the transmission path, b) security properties such as encryption of the communication, c) customer needs such as providing several virtual channels on a single physical line, or d) performance properties such as guaranteeing a certain bandwidth or latency.

An essential model in computer science in the context of networking is the ISO/OSI layer model (Tanenbaum (2012), Figure 2.3). It allows experts to discuss different aspects of data transmission at respectively assigned levels of meaning. In this text, the layer model is intended only as an overview of the terminology used in computer science.

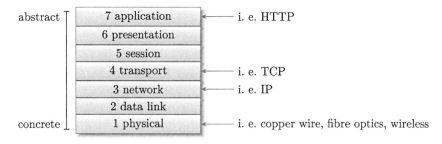

Fig. 2.3: The ISO/OSI layer model used in computer science with a few examples of selected abstraction layers. By means of such models, complex communication issues can be individually discussed based on different concerns. (Source: The authors)

2.1.1.3 Unit of Information: Bit

The information content of a message is measured in bits[5]. A bit can only have the value 0 (negative, false) or 1 (positive, true). The unambiguous answer to a yes/no question can be encoded by means of a single bit. A bit is therefore the smallest possible unit of information.

For the unambiguous coding of the four cardinal directions north, south, east, and west, on the other hand, two bits are necessary: One bit can be used, for example, to state whether the cardinal direction lies on the north-south or on the east-west axis, the other to state whether one or the other direction is meant from the selected axis. To extend the representation to north-east, north-west, south-east, and south-west, a third bit is necessary (Figure 2.4).

---

[5] This definition goes back to Claude Shannon and his fundamentally important Theory of Information (Shannon 1948).

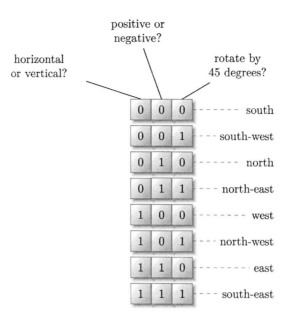

Fig. 2.4: Table of the eight cardinal directions with coding as a sequence of 3 bits. (Source: The authors)

A systematic approach is recognisable in these examples: with a given number of n bits, a certain amount of different N of different possibilities can be coded. The connection here is $N = 2^n$. The number of possibilities thus increases exponentially with the number of bits of information, virtually exploding! 256 bits already allow such an enormous number of possibilities that the resulting power of 2^{256} exceeds the number of atoms on earth assumed by science of about 10^{50} by a multitude of orders of magnitude[6].

2.1.1.4 Entropy and Compression

However, not every data stream of the same length carries the same amount of information. Information also corresponds to the degree of surprise that viewers are exposed to when the next fragment of information arrives (Figure 2.5).

Data with low entropy hardly cause any surprises for the observer. The lowest possible entropy thus is expressed by chain of bits (a so-called bitstring) whose bits all have the same value. A bitstring with very high entropy is hardly distinguishable from so-called white noise, i.e. total randomness: Each additional bit has either the value 0 or 1 with the probability $p = 0,5$.

Data packets with low entropy, such as text files, can be compressed into smaller packets with high entropy using compression algorithms. In Figures 2.6, 2.7, and 2.8, the image

---

6   Consider the story of the chessboard: Place one grain of rice on the first square, two on the second one, then four, then eight, and so on. The amount of rice in the entire world will not be sufficient to fill the chessboard up to the last square when applying this method.

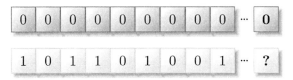

Fig. 2.5: Example bit chains with minimum or high entropy, as well as expected continuations by observers. (Source: The authors)

Figs. 2.6, 2.7, and 2.8: Example images with minimum (left), low (middle), and high entropy (right). (Source: The authors)

with minimal entropy will hardly surprise the viewer: It is entirely filled in one colour and can thus be described exactly in a few words. The middle picture, however, already is so irregular from a human point of view that it can only be roughly, but not exactly, described by natural language. And the third image shows so-called white noise, i.e. the presence of the highest entropy: in contrast to the middle image, each pixel is chosen completely at random.

The degree of compression depends on the algorithm used and corresponds to the achieved approximation of the entropy of the final product to white noise. Optimal compression can be achieved by using an ideal compression algorithm such as Huffman coding (Huffman, 1952). In this case, the data are reduced as best as possible to the size of its actual information content. Compression is often applied to large amounts of low entropy data to reduce the transmission time required within a communication network. Examples include image and video signals (for which even more specialised algorithms exist), but also software updates.

2.1.1.5 Development of Networking

- Bandwidth

Transmission bandwidth has been subject to rapid growth since the 1970s (Figure 2.9), bridging several orders of magnitude in just a few decades. While initially the reliable transmission of a few kilobits per second (kbps) was considered a great success, bandwidths soon rose into the range of megabits (Mbps), gigabits (Gbps), or in the backbone area even terabits (Tbps).

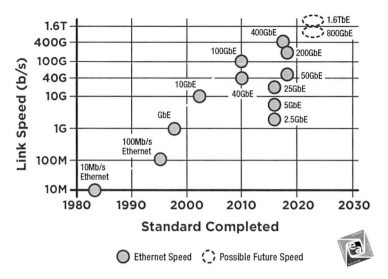

Fig. 2.9: Chart showing the development of Ethernet transmission speeds. While time progresses linearly in the x-direction, a logarithmic scale in the y-direction is necessary for the simultaneous representation of the speeds of all technologies since 1980. (Source: Ethernet Alliance, Ethernet Roadmap 2020)

At the time of printing of this book, up to 100 GBit/s is a still expensive but feasible type of networking in a professional environment, for example to connect parts of a building on a company or university campus. However, networking with 1 GBit/s is much more common, for example for office networks but also for private applications within one's own household.

Wide-area network uplinks, such as the home Internet connection for private users, currently range between 50 and 200 MBit/s, with exceptions at the bottom and at the top. They often represent a bottleneck in the population's access to the Internet. Due to the social importance of Internet access (participation in social life), there is a growing view that the state is obliged to provide the population with fast connectivity nationwide. One keyword here is the German Federal Government's broadband initiative.

- Latency

But pure bandwidth is not everything: If, for example, the transmission of data packets between two computers were to take several seconds, the affected line would be completely unusable for Internet surfing. Also, the so-called real-time capability of sensors and actuators requires them to be connected via a low-latency connection. Usually, acceptable latencies of today's Internet services or news sites are in the range of 20 to 100 milliseconds (ms), i.e. $20 \cdot 10^{-3}$ "s".

The latency of transmission has continuously dropped since the 1980s. In the meantime, however, this variable has largely reached optimised values, since the speed of information is practically always[7] limited by the maximum propagation speed of waves in the respective carrier medium. This in turn is limited by the speed of light in the medium, and ultimately by the speed of light in a vacuum, a natural constant called *c*.

- Energy Efficiency

With each development step towards more bandwidth, the efficiency of energy transmission has also increased: The energy required to transmit one bit of information over the same distance has been continuously reduced. Here too, there is an absolute lower limit based on Planck's constant: Information processing and transmission cannot be done for free, but are always accompanied by a change of energy levels in the media or transmitters involved. And these energies are quantised, i.e. they follow discrete, indivisible steps. Therefore, when a bit of information is transmitted, it is theoretically impossible to fall below a certain minimum quantum of energy expended. In practice, however, mankind is still very far from this lower limit as a large number of electrons or photons are used for reliable transmission instead of only a single one.

But even for this extent, quantisation still applies. This is one of the reasons why IT can have a paradoxical effect: On the one hand, it can be ostensibly energy-saving and sustainable (paperless office, avoidance of travel). On the other hand, IT builds up a foundation of primary energy consumption that stems solely from the provision of fast transmission paths. The video conferences that have become so popular in the Covid Age, but also the streaming of films, have a strong negative climate effect that also depends on the transmission medium employed (BMU, 2020).

- Bridged Distance

Depending on the type of connection purpose, the literature refers to certain classes of communication networks. Therefore, different technologies are used for their respective implementation (Table 2.1).

Tab. 2.1: Transmission distances of different media

Medium	Typical distance
Near Field Communication (NFC)	10 cm
Bluetooth	10 m
Wireless LAN	38 m
Ethernet (copper)	100 m
Fibre optics	up to 100 km, with repeaters even thousands of km

---

[7] By exploiting certain physical effects using quantum entanglement, transferring information in zero-time appears to be possible. However, this kind of zero-time transfer is currently only available in a lab environment, and will thus be of no practical relevance for the conceivable future.

2.1.2 Social Aspects

2.1.2.1 Widening of the Focus

Through networking, people with the same interests are finding together who would not have been able to do so without digital networking, for example because they live on different continents or because it is difficult to communicate their interests. For many unusual activities, hobbies, interests, and ways of thinking, communities are now being formed that are no longer tied to a specific location, as was the case in human associations mere decades ago.

The other side of the coin is that the filtering levels of the old world that were considered meaningful, such as editorial offices, proof readers, and publishers (and many other such instances in our society), are now losing or have completely lost their importance. This has not only resulted in the advance of so-called transparency and immediacy, or even the disappearance of what is perceived as censorship, but also ensures that potentially all news and messages can reach all recipients in an unfiltered manner. In reality, the filtering of messages has only shifted to the recipient. On the recipient side, however, there are now several possibilities: Filtering a) can ideally be carried out in full, but costs additional, unplanned time, b) cannot be carried out due to lack of competence, c) is omitted because of excessive demand.

2.1.2.2 Creation of Parallel Worlds

The relevant social media platforms with business models based on maximising attention have once again reinforced this trend and brought it to life in a new quality. The filtering mentioned above, which was previously shifted to the user, is now taking place again. However, this time the filtering is not carried out in the interest of the user, but following the business interests of the platform. Primarily news that generates attention is disseminated, regardless of its actual truth content (fake news) or its social appropriateness (incitement, defamation, hate speech). The generation of rapidly spreading so-called Internet memes is also observed, i.e. minimised information such as slogans, images, or short videos. Politically, these platforms have also contributed to the rise of populist movements in many countries around the world.

2.1.2.3 Social Isolation

The business model is also based on the man-in-the-middle principle: the platforms can only be successful if they control, evaluate, and influence the information flowing between the participants. It follows that the platform operators have an inherent interest in the isolation of the participants from each other: All communication – even between acquaintances who are networked locally – should exclusively take place via the platform. This isolation is presented to the platform users as attractive through psychological stimuli (such as instant gratification or fear of missing out).

In his book "Ten Arguments for Deleting Your Social Media Accounts Right Now", Jaron Lanier (Lanier, 2018) describes the social consequences of this intention to isolate. The core message here is: hate sells, i.e. sensational news (hate and catastrophes, exaggerated or untrue news) spreads much better than "boring" (but true or good) news (Dizikes, 2018). The business models of the platforms mentioned above aim to maximise the attention of visitors to their own site. This can be achieved by displaying news items that generate a high response from the audience, regardless of the content.

2.1.2.4 Shifting of Metrics Previously Assumed to Be Stable

Networking is also shaking up previous certainties. Such shifts are not new and happen in the course of every innovation. In the 1980s, for example, a grandmother would ask her grandchildren on the phone to keep the conversation short, because long-distance calls were so expensive. That is long over.

However, with digital networking, certainties are being overturned at a much faster rate than before. One such belief has always been that *local* is synonymous with *fast* or *cost-effective*. Since the existence of broadband connections, this no longer unrestrictedly applies: for example, most people would now rather type a search term into their tablet on their couch than look for the answer in a book on their bookshelf, even if the local proximity of the book is only two metres as opposed to several hundred kilometres. This shift in metrics makes new business models attractive that previously were not realistic. The cloud in its entirety, but also the streaming of music and video are among them.

2.2 Digitalisation

2.2.1 Definition

Digitalisation is the progressing spread of various types of digital technology and its increasing penetration of all aspects of social, economic, and political life, combined with an increasing digital representation of analogue processes in the real world and the evaluation and use of the data this generates.

2.2.2 Motivation

This effort is not an end to itself. Especially in private economy, digitalisation is the result of permanent optimisation efforts: The more transparent an organisation's processes, inventories, orders, customer relationships and many other such parameters are, the more efficiently it can plan, and the more informed strategic decisions it can consequently make. Digitalisation also makes new business models possible in the first place.

Of course, the digitalising organisation is in competition with other organisations that are also trying to leverage these advantages.

2.2.3 Interaction of Digital and Real World

As a partial aspect of digitalisation in industry, the German National Academy of Science and Engineering (acatech) created the concept of *cyber-physical systems* (Geisberger and Broy, 2012). It refers to Norbert Wiener's *cybernetics* and describes an important point in digitalisation: computing systems no longer operate only on information in data storage but also exchange data with the real world by measuring and manipulating it using sensors and actuators.

Science fiction literature had already invented the word *cyberspace* for the global digital space by the 1980s, i.e. without knowing about today's Internet (Gibson, 1984). Today, the term is also occasionally used synonymously for the Internet, especially in a popular context.

2.2.4 Pervasion

The digital world is thus increasingly penetrating the analogue. In the early days of information technology, the focus was on well-structured application fields with clearly defined data collection and processing rules (e.g., payroll accounting, merchandise management systems). Now digitalisation is penetrating application scenarios with fuzzy and multidimensional assessments and forecasts as well as complex action spaces with a high number of degrees of freedom. A brief example for the latter: While the guidance of an underground railway with its stringent rail guidance and relatively simple operational organisation is quite controllable algorithmically, and therefore driverless trains are already in operation in many places, the driverless, autonomous guidance of a motor vehicle in open road traffic is still technically extremely difficult and dangerously error prone. This is due to the many times higher number of degrees of freedom in guiding a motor vehicle in road traffic, which cannot be safely controlled with current technology. Nevertheless, vehicles from a certain manufacturer with a beta version of a Level 5 autonomous driving programme are currently driving in the USA in an unregulated manner. The lack of humility visible here when facing the complexity of reality combined with naïve faith in technology is highly questionable and needs to be fully investigated and discussed.

It turns a previously highly structured application environment into an increasingly unstructured one with many parameters. New environments invaded by digitalisation are increasingly complex. Computers are given more "responsibility" but, being machines, cannot logically gauge the effect of their actions, especially the ethical dimension.

As increasing measurement data are now available due to the penetration of digital technology into previously untouched areas, more and more everyday and previously insignificant actions and facts can be linked and analysed in new contexts. For example, it is now possible to count the steps of users by means of the movement sensors in smartwatches or smartphones, or even to obtain vital data via other sensors. From a technical point of view, nothing stands in the way of correlating data obtained this way with data from completely different areas of life (such as account balances, shopping behaviour, GPS movement patterns, call behaviour). Very easily available and apparently irrelevant features such as first names, addresses, and shopping habits can be used to assess the creditworthiness of the persons concerned (Leeb and Steinlechner, 2014).

Another example: There appear to be apps that can detect Parkinson's disease by collecting data on smartphones (EU COM, 2020).

2.2.5 Causality and Correlation

However, as not every coincidentally discovered correlation of data points must point to real causality[8], the users observed in this way are always in danger of having a certain behaviour or causality attributed to them by the data collecting parties without their knowledge.

This also includes the danger of being treated according to categorisations made (e.g., not creditworthy, alcoholic, irregular lifestyle) in the next interactions with digital systems. In contrast to the principle of the rule of law, which would make the accused aware of the accusation made against them and hear them out on the matter presented, here the persons concerned are not informed about the modelling, i.e. the making of assumptions, that takes place in the background. And this, of course, automatically eliminates the opportunity for those affected to object or make a counter statement. Despite all the advantages, digitalisation also makes it possible to charge previously insignificant or meaningless facts and actions with (sometimes only seeming) significance simply because they can now be measured.

2.2.6 Digital Persona

The representation of the individual through the totality of his or her data is often referred to as the digital footprint. The danger outlined above that individuals are judged by their digital representation rather than by their real person or their real actions has a social impact that fits well into the techno-optimistic-utilitarian zeitgeist, aka the self-optimisation of individuals. Just as the stigmatisation of people by algorithms is possible through the Digital Persona, so is their preferential treatment through the portrayal of a best possible digital profile optimised according to criteria of social desirability.

2.2.7 Social Consequences

This influence of the digital world on the decisions of individuals also results in social pressure: for example, in the terror hysteria that followed the attacks of September 11, 2001, page visits of "suspicious" articles with keywords such as *bomb* on Wikipedia fell significantly and disproportionately. Science explains these so-called *chilling effects* with

---

[8]   An excellent visualisation of this principle is offered by the Website ‚Spurious Correlations' (Vigen, n.d.)

peoples' feeling of being watched. This feeling promotes the display of socially desirable behaviour, or behaviour that the individuals concerned consider to be socially desirable (Assion, 2014).

Closely related to chilling effects is the concept of the *panopticon* brought into play by Jeremy Bentham (Bentham, 1995). Originally conceived as an innovation for the penal system in Great Britain, the *panopticon* is now also used synonymously for the possibility of being observed at any time.

Going further, the knowledge of this omnipresence can be elegantly exploited with the help of a so-called *social scoring system* (somewhat more euphemistically also: *social credit system*) (Everling, 2020). Such systems are widely used in the private sector vis-à-vis business partners and employees and attempt to measure integration and the willingness of those concerned to conform by linking a multitude of parameters and to generate corresponding pressure. However, as such systems usually operate in a non-transparent manner, they spur users on to even higher performance in accordance with the principle of intermittent reward. Examples of their use in the private sector include frequent flyer programmes, supplier relations in the automotive industry, variable pricing in online shops, or customer loyalty systems such as Payback or DeutschlandCard in Germany. Social scoring is especially questionable and violates dignity when it is used by the state as a sovereign instrument and is intended to induce fear-driven good behaviour among citizens by means of incentives or sanctions.

2.2.8 Internet of Things

2.2.8.1 Security Implications

The fact that more and more "things" are now suddenly connected to the Internet and can be remotely read or even controlled is also putting previously ignored products and manufacturers in the digital spotlight: Many manufacturers are suddenly IT companies without realising it. Their expertise may be in mechanical engineering, automotive production, production process control, or other areas. These companies have no experience in software development and no history of it. They usually have no technically qualified staff in the IT area, and in view of the shortage of skilled workers in this area, they will only get them at high cost. The lack of a corresponding software-compatible company culture is another problem. Networking and digitalisation have given these companies additional status as software producers – without, however, being prepared for this task. In addition they are unable to create specific additional resources in the short term to cope with these new challenges and with the competition in their respective sectors. The result is usually that such new products from the Internet of Things have a pathetic security record, right up to renowned corporations such as BMW which have already had to deal with several hacks of their car models (BBC, n.d.).

2.2.8.2 Data Protection Implications

In addition to extremely serious security issues, the connection of previously offline devices to the Internet also threatens the privacy of users. Basically, all usage patterns can now leave the device and be uploaded to the Internet, where they can be further processed in a variety of ways. Depending on the jurisdiction and the manufacturer's compliance with the applicable data protection rules, linking such data with data from other areas of the user's life can give the evaluating institutions (companies, governmental agencies) information about lifestyle, political position, personal or financial circumstances, creditworthiness, and much more. And even if the linkage created by the manufacturer does not tally with the real world, the users still have to live with the prejudice of the machine, i.e. with a *Digital Persona* that bears the attributed characteristics in contrast to the real person.

Interesting examples of usage patterns include the following:

- Use of private coffee machines: Is the user potentially addicted to caffeine? How long does she work on weekdays and weekends? Does the user tend to drink coffee at certain times of the day? Is the user currently traveling? Does the user have visitors at the moment?
- Data from eBook readers: What is the reading speed of the reader? What conclusions can be drawn about the user's IQ based on the titles read and the reading speed? Is the user reading the sex scene of the book for the third time?
- Smartphone charging history: Does the user keep the charge level always above 70 %, or does she regularly drain the battery? What conclusions can be drawn about the user's creditworthiness (King and Hope, 2016)?

As already mentioned earlier in the text, the legal process, i.e. the application of the principles of the rule of law, is excluded.

2.2.9 Transformation of the World of Work

Not only has digitalisation created an enormous demand for IT staff – digitalisation is also beginning to transform certain professions or make them completely superfluous. It is not always only supposedly monotonous or simple activities that fall victim to automation, but also specialised activities that can either be significantly rationalised or even eliminated altogether by systems with AI components. (cf. Chapters 14 and 15).

2.2.10 Lemon Market IT Security

With the penetration of the real world by digital mechanisms, it must not be forgotten that not only pure functionality is replaced, but also the trust of users in the correct functioning of the goods or services they use. Especially in the product aspect of IT security, digitalisation initially brings with it a great many problems. Customers expect the digitalisation of their business processes to at least not lower their security level as a result of this change. However, as shown above, many companies are not original software producers and therefore have not learned or understood even the basics of secure software engineering.

From the customer's point of view, security is a feature that is difficult to understand because it is non-functional. It is not explicitly demanded by customers, but rather ignored or tacitly assumed. In any case, it cannot be practically assessed by this target group, because it would require both considerable IT expertise and insight into the internals of the specific products. In the case of the IT security market, we can therefore speak of a so-called *lemon market*, i.e., a market in which customers are unable to judge the primary features of the products offered for sale and are therefore dependent on clues provided by secondary features, or, in the worst case, on marketing statements or popularity scores (supposed customer ratings) of the manufacturing companies. Even IT experts usually fail in the assessment, as the internals of products are almost never voluntarily disclosed by the manufacturer with reference to intellectual property and company secrets. However, as customers cannot judge the safety of a product anyway due to the lemon market property, security is often neglected or omitted to save costs or even increase performance. At the same time, such a lack of transparency also feeds the category of so-called *snake-oil products*, i.e. products that intentionally carry no benefit, but instead intend to extract money from the pockets of an uninformed clientele exclusively through flowery advertising statements.

2.2.11 Energy Consumption of Digital Products – Example Blockchain

Due to advancing digitalisation, the energy consumption of digital products and infrastructure is also steadily increasing – both in absolute and relative terms, in spite that energy efficiency is also improving. One information technology method that seemed to have great potential in the recent past is blockchain (or distributed ledger) technology. It has also only become widely implementable due to advancing digitalisation. The goal is to avoid a central, reliable, and trustworthy server instance that stores certain facts and selectively offers them for access – an example of this would be any kind of account management. The blockchain lacks such a central instance, which is why the term *distributed ledger* was coined for the basic application model. All information is signed and chained in multiple copies at participants of the system to make forgeries nearly impossible. Deviating information can be easily recognised by the other system participants. By identifying the participants only by means of crypto keys, the pseudonymity of the transaction partners is guaranteed. It is important to note that only new entries are appended to the blockchain, but existing entries can never be changed.

The size of the Bitcoin cryptocurrency blockchain based on this principle was about 15 gigabytes (three feature films in HD resolution) in mid-2020. At that time, its size grew by about 1 megabyte per hour. However, Bitcoin is a fringe phenomenon in terms of its transaction volume – if the Visa credit card system were to be converted to blockchain technology, the growth of the blockchain would be assumed to be 1 gigabyte per second, thus exceeding the dimensions of today's IT technology within a day. Attempts have been made to reduce this growth through marginal losses in security, but the fundamental complexity could only be marginally influenced.

Moreover, the energy footprint of the blockchain is devastating. To compare: While 741 kWh were needed for a Bitcoin transaction in 2020, the same amount of energy would not have been needed to carry out even 100,000 credit card transactions on the VISA network – these would only have consumed 149 kWh of energy (Statista 2020).

2.3 Artificial Intelligence

Basically, the term "Artificial Intelligence" has been in use for 70 years and describes a melting pot of different techniques under this name. Therefore, it is comparatively unspecific. The following sections develop the term and subsequently narrow it down to machine learning and further to neural networks (see also Figure 2.10) – and thus to its current meaning.

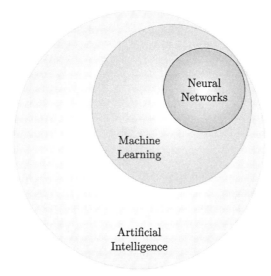

Fig. 2.10: Relationship of the terms Artificial Intelligence (AI), Machine Learning (ML), and Neural Networks (NN): NN is a true subset of ML and ML is a true subset of AI. (Source: The authors)

2.3.1 General Term: Artificial Intelligence (AI)

The term "Artificial Intelligence" is one of the typical buzz words of the digital age that has separated itself from the actual meaning of the word. It is driven by an intuitive understanding of the term "intelligence", which could roughly address

- multilevel, nontrivial insight, true to the etymological root *intellegere* (Latin, among other things "to show insight"),

- non-trivial, multi-stage reasoning involving hidden criteria and using formal-logical, but also heuristic methods,
- a surprising, initially mysterious planning move (analogous to the Queen's sacrifice in chess)

The latter was attributed to great explorers or generals such as Roald Amundsen or Napoléon Bonaparte.

The transition to artificial intelligence (AI) is already linked to the early days of electronic data processing, i.e. the "tube age", when researchers tended to exuberantly forecast the emergence of an electronic super being, the all-knowing, superior "electronic brain". In this enthusiasm, one of the founding fathers of computer science, Alan Turing, conceived the Turing test to determine the intelligence of a machine (Turing, 1950). In this test, machine intelligence is determined by a human interacting with hidden dialogue partners, some humans, some machines. If, after an appropriate time, the human cannot say with certainty which dialogue partner is a machine and which is a human, the machines (rather, the programmes driving the machines) have proven to display at least a basic measure of intelligence. At that time, such "chatbots" were still science fiction. Nowadays they are already reality, even if their abilities are not at the highest intellectual level.

With his ELIZA dialogue system (with an ironic twist) Joseph Weizenbaum demonstrated such a programme as early as 1966, which could pass the Turing test on a primitive level (Weizenbaum, 1966).

Since its conception, the use of the term "Artificial Intelligence" has undergone several fundamental modifications. Clearly, intelligence in the human, neurophysiological or cognitive sense is not to be expected, so the following quotations, albeit with a slightly cynical air, contribute to a better understanding of the term:

- "AI is whatever hasn't been done yet" (Larry Tesler according to Hofstadter (1999)).
- "Artificial intelligence has the same relation to intelligence as artificial flowers have to flowers. From the distance they may appear much alike, but when closely examined they are quite different. I don't think we can learn much about one by studying the other. AI offers no magic technology to solve our problem. Heuristic techniques do not yield systems that one can trust" (Parnas, 1985).

Thus, artificial intelligence in its conceptual vagueness was always a catch-all for the fringe areas of computer science still unexplored at any given time. Many methods that were still mysteriously attributed to AI 30 years ago have since passed into the canon of computer science or mathematics – for example, *objects* (originally frames (Minsky, 1974)), *genetic algorithms, linear and non-linear optimisation*, or even statistical methods such as cluster analysis.

Using AI, intelligence is simulated, and the underlying methods can be completely divergent having no connection to intelligence as understood by humans. In the history of computer science, artificial intelligence has appeared, vanished, and re-appeared again and again in waves. Between the hype waves were the so-called *AI winters*, i.e.

phases in which AI was almost completely disregarded. With each re-appearance, however, AI stimulated fantasies of superintelligence and, consequently, the end of humanity.

In the hype of the 1980s, the so-called "logic-based knowledge representation" was predominant: rule-based systems were based on the *modus ponens* as a method of inference: I know that the rooster crows (B) when the sun rises (A). So, if I see the sun rising, I can predict that the rooster is about to crow (A→B).

As an "execution model", rule-based systems dominated the market in the form of programming languages, such as *Prolog* with "executable" logic as its programming paradigm or as so-called "shells" with a motley assembly of logic-based methods. As end products, "expert systems" were offered that were supposed to perform in a manner comparable to human experts in a narrowly limited range of situations. They were also capable of justifying and explaining their assessments and decisions – this is very remarkable from today's perspective, since such an explanation capability is hardly found in currently prevailing AI systems (see Section 2.3.3).

2.3.2 Machine Learning (ML)

The AI bubble of the 1980s burst because it quickly became apparent that knowledge in factual or procedural form is subject to constant change over time, and that it is impossible to guarantee new knowledge is consistent with existing knowledge. The maintenance of an expert system, i.e. its regular adaptation to new conditions and requirements, very quickly turned out to be costly. A true simulation of a human expert also has to include a simulation of experience and further training, i.e. learning in the sense of adapting to changes in the environment, which could not be done by the AI of the 1980s.

By the sound of it, machine learning would provide a remedy to this deficiency, but of course this is only simulated learning. This kind of learning is merely a kind of indirect programming:

- Supervised learning: A system is presented with sample cases and sample solutions to "learn"; the "programmer" retains control over what a system is learning; the system is also learns the biases of the "programmer"- what it makes of them is initially unknown.
- Unsupervised learning: A system is presented with examples of cases; the machine develops its own conceptual system, and the people who operate the system then decide to what extent this might be useful. In case of doubt (and that will happen in many cases), they will discard unusable systems (*trial-and-error*).

Inductive methods have emerged based on symbols and logic which, however, often lead to the exponential growth of a rule base, since the consistency of the rule base is constantly in danger. At the same time, the ability of such a system to generalise tends to weaken. Mathematical-statistical methods are successful in finding regularities within mass data through regression or clustering. Therefore, they can simulate learning when executed on different or updated data sets.

The metaphor of the electron brain, as already emerged in the 1950s and constantly present in the background, led to the transition from the world of symbols to so-called "sub-symbolic" information processing.

2.3.3 Neural Networks

The metaphor underlying the term neural network refers to the neurophysiological elements and structures of the cerebral cortex and the hope that, instead of a few powerful processors as used in conventional information technology, a much higher number of primitive and relatively slow information processing units ("cells") could achieve a crude simulation of higher-order cognitive processes. These complexes of primitively simulated "brain cells" influence each other by weighted transmission of their internal state (see Figure 2.11 (Kratzer, 1994)).

External input is mapped into certain areas of such a structured network ("coded"), and the reactions of the network are tapped from other areas ("decoded"). Effective coding and decoding are very important for good overall performance of such a structure. "Learning" is essentially done by adjustment of the weights describing the influence of cells upon each other. Using methods of numerical mathematics, "learning" boils down to a search for extreme values in high-dimensional, nonlinear systems of equations.

Thus, a neural network is kind of a weight tensor, i.e. a highly complex constellation of weights which, in conjunction with the coding and decoding procedures, contain the "knowledge" of a network. The knowledge is represented in a way distant from human reasoning and language. The creator of such a network does not explicitly write down the weights; rather he/she does teaching sessions with case samples and sample solutions ("supervised learning") or allows the structure to watch its surroundings ("unsupervised learning"). We should always keep in mind that the use of the term "learning" in this context is rather pretentious – nothing more than a numerical extreme value search in complex, nonlinear systems of equations, far away from intelligence as defined by humans.

For example, the (admittedly simple) configuration shown in Figure 2.11 could be used to classify letters in a 6x6 black and white raster: For each connection between the cells, a weight is stored as a measure of the influence of one cell upon another. The black-and-white raster is mapped onto a so-called input layer with 36 cells. This, in turn, influences the eight cells in the intermediate layer, and these, again, influence the eight cells at the bottom where the classification results can be retrieved. The "knowledge" stored in a neural network depends on the:

- topology of the network (number and connection of processing units): How many input/output layers and how many intermediate layers are there? How many neurons are assigned to each?
- initialisation of the structure: What are the initial values for the weighted connections in the network?
- number, sequence, and quality of training samples;

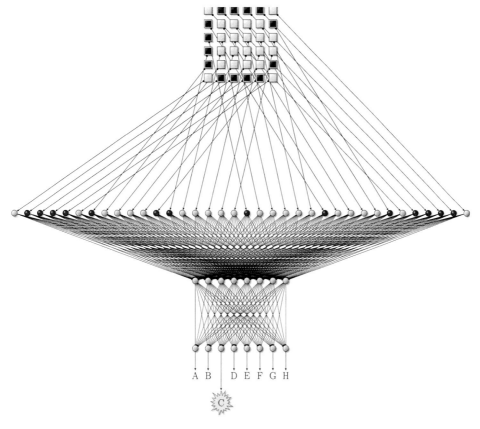

Fig. 2.11: Example of a simple neural network recognising letters based on pixels in an image. (Source: The authors)

- setting of parameters, for example: How will deviations from the ideal behaviour affect the weights between neurons? Abrupt weight changes may cause oscillations in the structure that cannot be controlled;
- method of coding input and decoding output.

From an external perspective, a neural network is a black box that, in tests and field trials, is either performing satisfactorily or unsatisfactorily. In most cases, a full-scale validation is not possible due to the underlying problem complexity. Many of the networks created by training are simply discarded because they do not fulfil expectations. The proponents of neural networks claim the inevitability of arriving at a solution by "training". Such a claim, for all practical purposes, is highly unrealistic. Since as a result of the high dimensionality of the weight tensor, a test is only possible selectively, large parts of the behaviour of a trained network remain in the dark. Thus, the reliability of such a network in practical use is narrowly limited, since sudden and inexplicable deviations from the intended behaviour might happen at any moment.

When designing and training a neural network, two dualisms usually emerge, for which a workable compromise has to be found by experimentation:

- *Reproduction vs. Generalisation*: Do I want my training samples to be reproduced cleanly and without deviations, with possible limitations in interpolative performance? Or is the focus on interpolative performance, with a possibly less than perfect reproduction of the training patterns and thus a possible reduction in reliability? These contradictory requirements are reflected in different network topologies, different parameter settings (depending on the network model), as well as different training strategies. The following example illustrates the problem. Let us assume that the following identification numbers are classified as a certain category:

  462 294 **193** 306 986
  348 202 206 776 872

  A reproducing network would react to exactly these identifiers and ignore any visible regularities, while a generalising network would recognise that these identifiers are mostly even numbers and ignore the exceptional case *193* as "noise". This unusual number might however be significant. From an application point of view, reproducing networks are not worthwhile because there would be cheaper ways of reproduction, but if the focus is on generalisation (ultimately a kind of interpolation), there is a danger that crucial facts will be "ironed out". Ultimately, this leads to the assessment of how many false-positive or false-negative classifications are acceptable.
- *Stability vs. Plasticity*: To what extent does the network follow new external influences that may contradict previously trained patterns, with the danger of an "oscillation" of the network behaviour? How stably does it behave in the face of such contradictory input – this is particularly important for neural systems that adapt continuously without human interaction.

No matter what decisions are made when creating a neural system, the actual performance is reduced to pure input/output behaviour. The "learning", or rather the adaptation, is based on the subsymbolic approach, which then leads to a neural structure developing its own terminology, which is not comprehensible to a human monitor. From this follows that there is no comprehensible explanation or justification of individual evaluations or decisions. Neural systems require blind trust. Depending on the operational scenario, this can have potentially significant, if not devastating, consequences for those affected.

2.3.3.1 Deep Learning

Deep Learning (Bengio, LeCun, and Hinton, 2015) sounds mysterious as a term at first, but it is not a conceptually novel approach. In principle, it addresses neural networks in which the number of neuron layers between input and output has been massively increased. This is not conceptual progress, but it allows training of certain parts of the network in advance. It is no coincidence that the indisputably great successes of such networks are in the field of image processing, in which so-called features, i.e. characteristic sub-areas, in combination lead to the classification of an overall image, such as "There is a kitten in the picture". The fact that neural networks and "Deep Learning" are currently being propagated as tried and tested methods for artificial intelligence cannot be attributed to an actual paradigm shift, but only to the immensely increased computing power available today compared to the 1980s, the much wider availability of data, even from the most intimate areas of life, as well as the willingness of broad sections of the population to share, assess, and react to digital stimuli.

2.3.4 Big Data

The term Big Data combines a wealth of different methods from computer science, but also from applied statistics, aiming at

- analysing very large amounts of data from different sources and revealing previously unknown regularities and correlations,
- deriving from such regularities predictions of the future behaviour of the society or of individuals, but also
- classifying individuals and phenomena.

Such visions have been developed very early – just think of Isaac Asimov's vision of a fictitious science, which he called *psychohistory*, set up to allow forecasts of social, economic, even military events and processes based on data from the past (Asimov, 1991). We are now, some 60 years after the advent of mass data processing, able to carry out such analyses in almost all areas of life, based on many criteria and a speed of processing much higher than would have been possible using human analytical work. This includes not only a quantitative dimension but also a qualitative dimension. Digital machines sometimes correlate data in ways that, at least at this point, no human has come up with, and which appear to better reflect reality. Of course, machine learning, especially neural information processing, has lately been increasingly used for this purpose, with its known pitfalls and especially the lack of justification for decisions. Applications are, for example, the "suggestion lists" familiar from Internet commerce or streaming services, but also credit ratings and classification of applications in recruitment procedures – right up to official procedures (e.g., entry into the Schengen area for immigrants without EU citizenship (Gallagher and Jona, 2019)).

It is important to note that information from many data sources, including contact and movement data, medical data from fitness bracelets, and – in case of physical contact – criteria such as eye movements, muscle twitches in facial expressions, or even the charging status of a mobile phone may be relevant.

2.3.5 AI and Us

The use of AI technology within application programmes targeting our personal sphere of life harbours a still drastically underestimated danger to our freedom and our interests, e.g., when touching on our health, our financial affairs, or our freedom. The danger is that we subject ourselves semi-consciously or unconsciously to an interest-driven, obscure catalogue of criteria to which we submit voluntarily or by law. This is especially true in the case of neural systems, since they refuse to explain a judgment or decision. Of course, free people cannot be prevented from submitting to such a regime by their own volition except when this is necessary for their own protection, or for the protection of the state and the legal system. Nevertheless, even in Germany there are areas, in which people have no real choice, for example, for a credit rating. In such cases, there must as a matter of principle always be an option for a complaint to a court. This has also been recognised by data protection agencies: Here, the first demand usually is for a disclosure of the methods used. However, this leads to the question as to what should be disclosed. The method itself, i.e. the algorithm, is usually known from the scientific literature and is therefore openly available – at least in its most basic form. The next step would be to reveal the underlying network structure and coding/decoding algorithms. However, this is of little value if the data used to "train" the network are not disclosed at the same time. Such training data are either deemed to be company secrets, or it is claimed that revealing the data would violate privacy rights of persons. Even if all that has been revealed, the soundness of an AI system is hard to judge even for experts – if at all.

Another approach to the absolutely necessary regulation of this new technology would be a requirement to explain and justify every single decision. However, this would practically exclude the use of neural technology, since such explanations and justifications are simply not available.

Proponents of AI technology often claim that such systems are neutral and objective. However, this depends on the data available for training, and it is quite possible to discriminate without using explicit criteria such as gender, age, or skin colour.

From text analysis, for example, it is statistically known that

- women use the word 'I' more often,
- men, on the other hand, definite articles,
- younger people more often use auxiliary verbs and
- older people use prepositions more often,

which opens the door to indirect discrimination, simply by analysis of a written text (Newman et al., 2008).

Neural systems are known to have developed sexism and racism simply by analysing social interactions. Artificial intelligence is thus not able to neutralise prejudices deeply rooted in society, or to observe universal values – instead, it represents a mirror of the recorded, possibly biased input data (UN News, 2020). However, there might also be deliberate manipulation of such systems: a well-known case was the chatbot Tay launched by Microsoft (Hunt, 2016), which was deliberately "re-educated" to racism by malicious users.

What is the relationship between such intelligent systems and their operators? It is often emphasised that these systems are "only supportive" and that "the human operator retains full control and the final decision". This goodwill attitude can be fairly assumed for the majority of engineers working in the field of AI. Looking closer, however, we quickly find out where this intellectual backdoor falls short: it is already known from the early days of decision-support systems that this "support" can rapidly give rise to pressure to justify why a machine assessment or recommendation was overridden, which would de facto turn the mere proposal of an assessment and recommendation into an order (Langer, König, and Busch, 2020, cf. also the following third chapter).

It is important to always keep in mind that any parallels between human intelligence and neural artificial intelligence are purely metaphorical. The determining factor in human action is not only plain rationality but also emotion, common values, often individual or socially shaped moral concepts, as well as the preservation of one's own existence or that of the group or species. A neural network, on the other hand, is a machine and as such implements a highly complex interpolation algorithm. The previously mentioned universal or personal values may or may not be reflected in the training examples – what then happens in non-trained situations, however, remains open. Artificial intelligence is a tool without life, self-reflection, bodily experience, the concept of lifetime and thus, without concern for one's own existence.

Commercial artificial intelligence systems predominantly deal with application fields whose complexity cannot be grasped by conventional programmes. Therefore, they cannot be tested and validated following the rules of conventional software development. In case of failures, the manufacturers usually promise improved performance in the indefinite future – but nothing more. This is not serious if the application field is harmless and no or only minor damage can occur (for example, in games or for harmless forms of marketing, such as suggestion lists). In critical fields of application such as autonomous driving, legal processes or complex surgical procedures, however, the already existing and tested approval and licencing procedures with the usual documentation and validation rules should ideally be applied. However, it must be clear this will make the use of machine learning considerably more difficult, if not impossible.

Finally, it should be mentioned that there are already substantial efforts in this area both nationally and internationally. These attempt to stem and limit the unregulated flood of classifications and analyses with negative consequences for those affected, and the marginalisation of people that is particularly evident in the labour market. Examples include:

- the European Union's "Ethics Guidelines for Trustworthy AI",
- the publications of the Data Ethics Commission of the Federal Government,
- the Ethics Guidelines of the German Informatics Society (GI).

All these reflect responsibility for our societies and the willingness to regulate AI to benefit all stakeholders. To what extent such principles will survive the confrontation with economic interests remains to be seen.

2.4 Summary and Outlook

In contrast to the widespread technocratic utopias, or dystopias, of a digitalised economic and social structure, the well-founded discussion of the underlying technology is rather sobering. However, the sometimes naïve or techno-optimistic overreliance on digitalisation and AI in the economy, society, politics, and the private sphere is based on such positive assumptions which, in most cases, are untenable on closer examination. In summary, the following points should be emphasised:

- The capacity of the underlying computing technology is limited. Capacity limits are constantly being expanded, but this expansion is struggling to keep pace with the demands of increasingly complex tasks.
- Nevertheless, it is possible to process and condense information on individuals and thus to characterise, advise, and in the worst case, scrutinise individuals to an ever-increasing degree.
- Digitalisation – similar to economisation since the 1980s – is increasingly penetrating weakly structured areas of the economy, society, politics, and the private sphere, which leads to decision-making spaces with a multitude of degrees of freedom. The underlying technology cannot keep pace with the complexity of human decision-making, but nonetheless might dominate it.
- Currently, so-called artificial intelligence is primarily driven by increased computer performance and only to a small extent by conceptual progress. We are still a long way from a superintelligence in the sense of the trans-/posthumanism presented in chapter five.
- The use of digitalisation is possible, regardless of the area of application. However, human individuals will perceive the behaviour of a digital system as increasingly erratic, surprising, or even threatening in application fields of high importance.
- The cohesion of our society is shaped by discourse: at an appropriate level, discourse is probably possible with the developers of an application system, but not with "intelligent" application systems themselves.

The penetration of digitalisation into a large part of our professional and private spheres has already taken place – a "rollback" to the analogue world would be unrealistic. An open democracy, however, needs a discussion of values that places both freedom and the dignity of every human individual at the centre of this discourse and which, in its implementation, leads to comprehensive regulation at all levels. This is exactly why we have written this book!

Chapter 3

Path Dependence and Lock-in

Jasmin S. A. Link

In the public debate, many suggest we are currently in the midst of a fourth industrial revolution. A main part of this revolution is the digital transformation of society. In order to stay internationally competitive, there appears to be no other way than a substantial digital transformation in Europe. The industrial sector is already digitised to a great extent. Politics and education would need to speed up to keep track. Therefore, schools should utilise the granted digitalisation funds even without a pedagogical concept to achieve a quicker digitalisation of the education system. But why? And where does this development lead?

Path dependence theory can be used as analytical basis to understand the speed of the development and the perceived inevitability of follow-up decisions. Path-dependent processes show exponential dynamics, historical manifestations and interdependencies, as well as cognitive and economical dependencies. There are, for instance, marketing agencies that try to use path dependence to retain customers through mutually dependent product packages. Others try to suggest one would already be in a locked-in situation and there would only be the choice to comply (but not to deny). Generally spoken, path-dependent processes generate dependencies and induce social dynamics. Furthermore, the analysis of path-dependent processes can help to identify future developmental dynamics reliably. In case of unintended direct or indirect effects of path dependence that cannot be anticipated, deliberate measures to counter such effects should be taken as early as possible.

Various definitions of path dependence can be important when focusing on digitalisation, IT-networks, and artificial intelligence (AI). Notations may vary depending on the object of concern that is analysed or discussed, the background of the scientist, and potentially also the observational level or the dynamics the object is embedded in.

When analysing path dependence in the context of digitalisation, IT-networks, and AI, the following diverse perspectives are used:

Digitalisation is coined by routines and supported by institutionalisation and standards. Some procedures do change – but in a coherent way – towards increasing digitalisation and digital analytics. Reactive sequences of historical events can have profound impacts, similarly organisational fields, individual experts[9], or social networks. Learning effects pay off and procedures are optimised incrementally.

---

[9]   Personal or individual experts do not necessarily count objectively as experts. However, these people function as experts for the individual decision-maker. In this sense, also a best friend that always has a clue or a mentor could be used as expert, even if that person does not have specific expertise to answer the given question.

In this context, networks are very important. In interconnected procedures, the digitalisation of one process necessitates the digitalisation of another one. Therefore, a well-functioning network needs a certain standard underlying equipment. Hardware and software need to be compatible, too. Then the network allows for a smooth and quick data exchange, data analysis, and data use, potentially even to a greater extent than is feasible in an analogue manner.

So-called AI can be applied in data preparation and data analysis. Increasing digitalisation and increasing interconnectedness are raising the availability of linked data, so that IT analysis tools are necessary to actually utilise these big data. Prior to decision-making, these big data need to be visualised, analysed by predefined criteria, and controlled for missing data or presumably erroneous data. Different hypotheses on these big data require different methods for testing. However, there is the substantial caveat that no one can personally recalculate or even check the iterative steps of a neuronal network. With increasing availability of electronic data, growing interconnectedness, and reinforced digitalisation, more and more human decision-making processes will depend on the results of the applied AI systems.

3.1 Definitions of Path Dependence and Lock-in

The general understanding of the term "path" does not necessarily coincide with the definition of path dependence and not every definition of path dependence includes the terminal state of a lock-in. After explaining the different perspectives of path dependence and several possible definitions of a path, the remainder of this chapter will use a process-oriented definition of path dependence: A path-dependent process is a self-reinforcing process with a tendency to reach the state of a lock-in. A process has reached the lock-in (i.e. it is locked in) if changes of the given state do not occur any longer or occur only incrementally.

3.1.1 Digression to Mathematics

Many scientific disciplines use mathematical definitions. Regarding path dependence, different understandings can be generated from different branches of mathematics. Therefore, the following small digression outlines these understandings in this otherwise sociological chapter.

For example, in graph theory in network analysis the notation "path" describes a chain of multiple edges in a given network (cf. Figure 2.1 in chapter 2): A path is a sequence of edges that connects nodes, which might not share an edge. The number of edges in this sequence is the length of the path[10]. Networks can be used to analyse many structures or bottlenecks, e.g. of a power grid, a logistic network for industrial products,

---

[10]   E.g., the longest path in the connected graph in the middle of Figure 2.1 has the length four. It connects two nodes that are not directly connected by an edge.

or a consumption network consisting of numerous households. Social networks can be analysed as well, e.g. in epidemic contact analyses or when analysing the diffusion of information. The topics of possible applications are diverse. However, this notation of a path is merely static and therefore, there is no understanding of an associated lock-in.

In a different subfield of mathematics, in stochastics, a path is also defined as a sequence of edges in a network, in which the nodes are events that occur with a certain probability. The structure of such networks is a tree, which means from a first event a certain number of edges branch as there are alternative subsequent events, from which again branches lead to further events. In this tree of connected events, a path denotes one potential sequence of events. Knowing the probability of each single event along the path allows for calculating the expected value of the final event of a path. If there is more than one path leading to the same final event, the expected value can vary depending on the path. The total expected value of a certain event depending on a specific starting event is the sum of all expected values of that particular event over all paths leading from the starting event to the given final event. Such expected values are for instance used in decision theory in business economics. Again, this understanding of path dependence does not show any dynamics. Therefore, no understanding of a manifestation such as a lock-in is included in the expected value theory.

Instead, if you consider a dynamic system, there is potentially no longer a countable set of alternatives. It is possible that different dynamics can be described, depending on the initial value. Some of these dynamics may have an attractor, which they approach and from which the dynamic process does not depart anymore after it has been reached. The plot of such dynamic can be interpreted as a path towards the attractor, the lock-in. If the process is accelerating the closer it gets to the attractor or is increasingly spatially limited close to the attractor, the process can be characterised as a self-reinforcing process with the tendency towards a lock-in. A physical example for such a process can be a pendulum that swings around a magnet, by which it is more and more attracted the closer it gets to that magnet.

In Sociology, these perspectives described above can be combined. The basic definition of path dependence includes the process perspective of a self-reinforcing process with the tendency towards a lock-in. Such processes are practically self-reinforcing dynamic processes on a weighted directed graph, which approach an attractor.

In sociological terminology, this matches the self-reinforcing processes with the tendency towards a lock-in, whose diffusive effects can be analysed in social networks, or weighted directed social networks, in which the weight of the edge correlates with the probability of a specific decision to act.

3.1.2 Path Dependence Explained by the Example of the QWERTY-Keyboard

As in Mathematics, sociologists use different notations of "paths" matching different subfields of Sociology or the perspective that is taken based on the neighbouring discipline as well. That way, examples are often analysed and discussed from multiple perspectives. Two classics in the sociological path dependence debate are the analysis of the development and diffusion of the QWERTY-keyboard (David, 1985) and the applicability of a Pólya urn model, in which the balls drawn are replaced and another ball of the colour just drawn is added as well (Arthur, 1994).

The historian Paul A. David assessed the development of the QWERTY-keyboard historically (David, 1985, 1997, 2001, 2007). He recognised that even on the early keyboards the placement of keys is the same as on the keyboards of later computers. But why?

David discovered that in the development of typewriters the typewriters were optimised in terms of writing processes, especially regarding the writing speed. When two bars were hit right one after the other, they had a higher probability of clashing or jamming if they were adjacent to each other – in contrast to bars further apart. Clashing bars always interfered with the writing dynamics because one needed to stop completely to release the bars before continuing to write. Thus, it was important to minimise the likelihood of bars clashing or jamming in order to optimise the overall writing speed. Accordingly, the English language has been analysed to calculate how often particular words occur and especially how often specific letter combinations occur in these words. Afterwards, the keys were placed on the bars to make it most unlikely that letters, which are often found adjacent in words, are placed next to each other on the typewriter as well. Additionally, one could remark that all letters of the word "TYPEWRITER" have been placed in the topmost row of the English keyboard. According to written records, this goes back to helping early salesmen to find the letters of their product name more quickly, even though they might not have been trained typists at that time.

But why did the placement of keys remain the same on the electronic typewriters, the computer keyboards, or even modern tablets or smartphones? There are no bars anymore that could clash or jam, nor does the brand name TYPEWRITER exist any longer. On tablets or smartphones there are not even keys anymore and it is likely that letters are not even typed with two hands, rather with one hand or even single fingers or even only one or two thumbs. Also, it can even be assumed that modern English has changed so much from the English at the time of the invention of the typewriter that even the likelihood of adjacent letters has changed in comparison to 100 years ago.

This phenomenon is called path dependence. "History matters" describes the distinctive influence of early decisions or events on later developments in a non-reversible chronology. And David (1985) has pointed out that "[...] while they [the agents] are, as we now say, perfectly 'free to choose', their behaviour, nevertheless, is held fast in the grip of events long forgotten[...]".[11]

---

[11] Some economists deny the mere existence of path dependence since their theories do not allow for any inefficiency (cf. Liebowitz & Margolis, 2014). Other economists come up with arguments how decisions in the historic context of the development of keyboards can be considered to be locally efficient nonetheless. There are learning effects, effects on the workforce, marketing effects, effects of scale, the theory of sunk costs, or a market efficiency at a higher level.

It can be assumed that once a person has learned to type on a QWERTY-keyboard, finding the correct keys on that particular keyboard layout is more efficient compared to an alternative layout such as the DVORAK-keyboard[12]. Thus, once early secretaries learned to write on the new typewriters using QWERTY-keyboards, retraining them would cost money that an employer could save when offering them a typewriter with the QWERTY-keyboard in the office. Furthermore, if an employee needs to be replaced, the replacement is likely chosen to be able to type on the available typewriter as well. Over time, the increase in demand of secretaries with typing abilities led to the official education of future secretaries in typing on typewriters. Accordingly, even more employers had to buy typewriters as basic equipment for their offices. Public awareness increased as well and at some point offices could no longer be imagined without typewriters. Exchanging an established typewriter by a model with a different keyboard layout would imply that previous investments in hardware and the education of the secretaries had to be accounted for as sunk costs.

Even with regard to the typewriter production it is likely that maintaining the established keyboard layout led to increasing returns to scale and reduced unit prices. Similar reasons, particularly learning effects and labour market dynamics, can be discovered in the subsequent substitution of mechanic typewriters by electronic ones as well as in later product developments, in which the mere "writing" was no longer an essential feature. Presenting the product in an advanced but still comforting traditional way might have helped marketing efforts and increasing consumer acceptance (Beyer, 2005).

The following examples highlight further perspectives to discover and analyse path dependence:

Historical sociologists describe reactive sequences, which are causal chains of events in which early ones trigger and increase the effects of subsequent events (Mahoney, 2000). Institutionalists stress that established institutions only change incrementally (North, 1990), which can be interpreted as an optimisation of the institution or as a lock-in effect. Political scientists apply the concept of path dependence to explain political dynamics (Pierson, 2000) and emphasise the unpredictability of the future path at the very beginning (Collier & Collier, 1991). However, innovation, management, and organisation researchers consider how any involved agents can learn to actively create paths, which implies initiating and shaping them (Garud & Karnøe, 2001). Sydow, Schreyögg and Koch (2005, 2009) have designed a three-phase-model (Figure 3.1) to visualise contingency at the beginning of a path as well as the self-reinforcing formation of the path in the second phase until only incremental changes are still possible in the

---

12   The so-called "DVORAK"-keyboard was designed to be particularly efficient with regard to ergonomic considerations. However, it could not overcome the market power of the established QWERTY-keyboard due to path-dependent processes. Until today, it occupies just a niche of the keyboard market and is used to only a limited extent. On mobile devices, there is nowadays also the possibility of an alphabetic keyboard layout. However, if you have learned to type quickly on the QWERTY-keyboard, your fingers usually find the correct keys almost by themselves. In that case, an alphabetic keyboard layout would be counterproductive, as you would have to think actively while typing where the keys are actually placed.

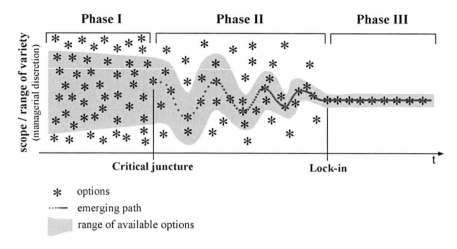

Fig. 3.1: Three-phase-model of the development of an organisational path (Sydow et al. 2009).

lock-in. However, taking a consulting perspective, the determination of a lock-in is dissatisfying because the three-phase-model does not consider any exit from the lock-in.

Organisational sociologists combine multiple perspectives by describing path dependence as a process, which is self-reinforced by accompanying mechanisms and manifested to the point of lock-in. The assessment of the logic of continuity of each mechanism also allows for the identification of chances for destabilisation (Beyer 2010, Table 1, cf. also Table 3.1).

3.1.3 Increasing Returns – Growth Dynamics That Can Result In a Lock-in

At about the same time as Paul A. David, W. Brian Arthur (1989, 1994) developed his theory of increasing returns. His fundamental question was how one could predict which technology would win in a market?

Even posing this question invokes economists' critique because according to Adam Smith the invisible hand as a free market mechanism selects the best technology. Thus, if there are clear preferences, complete information, and universal efficiency, the best technology succeeds in this perfect market. Arthur's question connotes that it is possible to think of markets, in which perhaps a suboptimal technology could win. That way, he contradicts the universal applicability of standard economic theory (Arthur, 2013).

Arthur observed increasing returns, i.e. growth trends intensified. Unit costs and thus unit prices could be reduced by increasing returns to scale in production processes, which can lead to further increases in sales. He formulated the hypothesis that a technology can win in a market even though it is not the best technology by objective criteria if increasing returns are present (Arthur, 1994).

Arthur uses a Pólya urn model to describe and demonstrate the positive feedback mechanisms (Arthur, 1994), a concept that is well known in stochastics.

Pólya urn model: At the beginning, there is an urn with two balls, one blue, and one red ball. In each step, one ball is drawn from the urn. After noting the colour of the ball that was drawn, the ball is returned. Furthermore, an additional ball of the colour that was just drawn is placed in the urn. Consequently, the probability of drawing a ball with a particular colour changes after each step. At the beginning, the ratio between colours is 1:1, at the next step it is 1:2. Then it is possible that the colour ratio is 2:2 but it is more likely that it is 1:3. The probability that a ball of the majority colour is drawn increases more and more so that existing discrepancies are manifested. However, balls drawn at the beginning have a more pronounced influence on the final ratio of colours. If there already are numerous balls in the urn, an additional ball only has a marginal effect. At a certain level, the colour ratio in the urn tends to stabilise. However, it is impossible to predict this final ratio at the beginning of the experiment.

3.1.4 Definition of Path Dependence and Lock-in in This Chapter

In this chapter, the definition of path dependence is: A path-dependent process is a self-reinforcing process with the tendency towards a lock-in (Link, 2018, p. 3).

This definition can be used as basis for the assessment of path dependence from various perspectives. A self-reinforcing process has the effect that early events play a more crucial role than later ones (history matters). The self-reinforcing process of the QWERTY-keyboard (David, 2001) can be determined in historical retrospective as having reached the lock-in very quickly, and manifesting mechanisms can be identified that stabilise the lock-in. "Increasing returns" (Arthur, 1994) are one of the mechanisms that can describe the self-reinforcing dynamics of a path-dependent process. Additionally, increasing returns can lead to a manifestation of a process towards the lock-in (cf. urn model, Arthur, 1994). Reactive sequences can describe the self-reinforcing dynamics by chains of subsequently triggered events and simultaneously highlight the non-reversal chronology by the tendency towards the lock-in.

This definition does not include any specifics at the beginning of the process. But by defining the self-reinforcing dynamics of the path-dependent process, even tiny differences in initial values can generate large impacts on the resulting lock-in very quickly (Arthur, 1994; Collier & Collier, 1991; Vergne & Durand, 2010).

In the following, this definition will be used to combine the perspective of agency with a network perspective to apply it to the digitalisation, IT-networks, and AI.

3.1.5 Path Dependence as Following Behaviour – Path-Dependent Networks

In this part, the micro level of a path-dependent process is observed and the question is what happens to agents who are involved in a path-dependent process? Subsequently, these insights from the micro level can be aggregated back to a meso level to generate a network perspective based on the definition of path dependence.

In the historical analysis of the development and the manifestation of the QWERTY-keyboard usage, Paul A. David already noted that even though agents are completely free in their decisions, their decisions depend strongly on former events. Remarkably, in this process even events can play a role, which none of the decision-making agents can directly remember. How can that be? These events are not part of legal texts or written on posters pinned to the wall with the note to always act accordingly no matter what happens. Instead, path dependency shapes the involved agents' actions so that they do not even think about alternatives. Therefore, they underestimate their own ability to act differently or are unaware of the necessity to leave the prescribed path-dependent process.

There are different mechanisms that are described in the previous parts in more detail, which lead to self-reinforcing dynamics in the path-dependent process. How does the path-dependent process affect a particular and any future decision-making processes? This can be analysed as follows (Kominek, 2012; Link, 2018, pp. 32ff.):

Looking more closely at the decision-making process of a single agent reveals that the agent has diverse alternatives to design the action decision-making process: for instance, spontaneously, value-rationally, purpose-rationally, or preference optimised according to an individual usability function. However, if this agent and the related decision are involved in a path-dependent process, this process will generate similar decision-making situations for the agent. Of course, the agent could again apply the same detailed decision-making process as before. But if several situations, in which decision-making is required, resemble each other, maybe the same action that has resulted from the first decision fits the second situation as well. Then simply applying the same action is quicker than starting an intensive decision-making process all over again.

The "least-effort-principle" (social psychology) states that whenever the brain has multiple ways of pursuing a goal it takes the one that requires the least effort. Applied to the decision-making situation of the considered agent, facing comparable decisions the agent increasingly merely applies the actions from former decisions again. Thereby, the agent develops a routine. Accordingly, also the agent's approach to new decision-making situations changes: Instead of acting value-rationally, purpose-rationally, or preference optimised like before, the decision-making process is shortened to a routine application.

But what happens, if this imprinted agent faces a new situation? At first, he might try to apply the old routine again. Either the old routine still works or it is no longer applicable. In the latter case, the agent needs a new routine that matches the new situation. However, developing routines is time-consuming. Instead of developing new routines, it is quicker to adopt an action draft from an external source that matches the new situation. This could be another agent, an institution, or the masses. Due to the least-effort-principle, the agent would choose to adopt a draft for his action. Consequently, even if the agent optimised his preferences in first place, he increasingly only applies available action drafts as his decision-making process is shaped by path dependence. The result is a following behaviour: Influenced by path dependence, the agent develops the tendency to follow other agents or the masses with his actions (Link 2018, p. 33).

Under the influence of path dependence, the tendency of one agent to follow other agents or the masses can turn into another routine for decision-making. Consequently, the agent can also apply this following behaviour in decisions to act, which are not directly related to the path-dependent process (Kominek, 2012; Link, 2018, Chapter 4). Such following behaviour can become a routine again, which further reduces the probability of even looking for alternative ways of decision-making.

Focusing on one agent who shows following behaviour, an egocentric network can be created that maps the sources of the action drafts such as other networks or indicators of aggregated behaviour of the masses. This path-dependent social network (PDSN) (Kominek & Scheffran, 2012) can be understood as an egocentric social network, in which e.g. the agent consults his computer expert, rests his consumer decisions on Amazon rankings, or uses the Google score as basis for spelling decisions. Likewise, at the airport, an agent A accidently recognises that a traveller B takes the same flight, follows B to get to the correct check-in and to the right departure terminal later-on. Traveller B does not necessarily need to recognise that person A is following him. At the same time, traveller B, who is followed by A, follows another agent C. Then it is possible to describe a PDSN for the second agent in the row, B, mapping his following behaviour, which agent C is part of. And by chance, agent A follows agent C indirectly, unknowingly, and without knowing C.

This has numerous consequences, for example:

- Via cascades of such coupled PDSNs, the German phase-out of nuclear power can be explained as a result of the effect of the Japanese earthquake that occurred close to Fukushima in 2011 (Kominek & Scheffran, 2012).
- Knowing such PDSNs and the probability with which the egocentric agent performs a following behaviour and applies this network accordingly, the agent's decision can be locally temporally approximated (Link, 2018, pp. 143ff.).

3.2 Path Dependence in the Digital Transformation

Based on the definition of path dependence as a path-dependent process, there are several levels available for the application to the digital transformation: a macro level perspective of dynamic processes, a micro level perspective of single agents who decide on actions, and a meso level perspective of interconnected PDSNs of agents, which essentially map following behaviour.

If you consider digitalisation as a process that includes the introduction of hardware and software, many mechanisms can be identified that stabilise the development path and lead to the self-reinforcing characteristics of the process. For example, learning effects occur as soon as many employees have the skills to work with digital methods, data, and documents. Then more and more digital processes can be included in the overall workload. At the same time, working processes that include digital processes such as digital documentation are going to ask for employees with the relevant skills once the positions are vacant again.

Educating new employees is also possible, but of course incurs costs. Furthermore, whenever digitalisation leads to better informed decisions or accelerates the proceedings, increasing returns occur: The more decisions are made based on analyses of digital data and the more working procedures are digitised, the more qualified and quicker the relevant processes become.

Further mechanisms are for example sunk costs, which refer to the value of previous investments in the progress towards digitalisation: Even if a change back to analogue processes was possible and knowledge gained from digital processes could help optimising analogue procedures, all previous investments in the digital transformation would have to be considered as exchange costs. These sunk costs manifest the digitalisation. In parallel, interfaces to other organisations within the organisational field are generated, which can trigger them to digitalise their organisational procedures or at least the interfaces to link processes between the organisations. Thus, the other organisations need to invest in relevant infrastructure as well.

A higher degree of digitalisation implies that agents work in or with the process of digitalisation more and more, developing the process by making alike decisions, and developing routines related to the digital transformation. Thus, applying path dependence theory to individual agents' behaviour, the path dependence of the process of the digital transformation shapes the behaviour of the affected agents by increasing their tendency to show following behaviour and reducing their perceived degrees of freedom. For these agents, supporting the digital transformation seems increasingly without alternative. The more digitalisation is applied in more and more parts of life, the more the tendency to reach decisions to act by merely showing following behaviour is extended to more and more areas of life as well. Additionally, the more parts of life are affected by following behaviour the more routines develop to reach decisions by following behaviour. Accordingly, agents are increasingly likely to apply following behaviour in other parts of life as well, which do not even need to be directly affected by digitalisation in the first place.

In general, the digital transformation supports decisions by following behaviour because for many topics aggregated data of the masses in form of ratings, quotas, likes, or clicks are available. Furthermore, channeled information can be consumed via online networks reducing the necessity to review them separately and sort them individually within the egocentric PDSN. For example, software information can directly be forwarded by a computer expert of a given individual agent or evaluated positively in online media, so that there is neither need to raise a question individually nor to actively consult the computer expert. Newsletters can function in the same way. But information that is dispersed via the social networks can be shaped and received more individually because the agent who follows another one is interested in the other's information and evaluation.

The more an agent involved in the digital transformation is actively present in social networks, the more his user profiles map his PDSN. The reason is that when showing following behaviour, it is useful to have the critical information right at hand, if necessary. Accordingly, it is helpful to always maintain mobile contact to these individually important sources, if needed. For example, someone driving a car can use a navigation system or if the navigation system is not available, call a friend who knows her way in that respective area, or someone with current access to a navigation system. Thus, via a messenger service he can ask for an expert or for the directions. Other options such as using a street map, or buying one at a gasoline station, or looking at the regional map of a bus stop become less and less likely. Based on the individual agent's routines, somebody who is informed predominantly digitally is likely to prefer digital visual or digital audible information.

3.3 Path Dependence, the Digital Transformation, and Networks

Networks can have different meanings: linked computers (IT-networks), connected social agents, data collection networks, analyses of interconnected data, interconnected analyses of interconnected data etc. Having these numerous applications of networks in mind, the relevance of networks in combination with path dependence and the digital transformation can be analysed from the macro level, the micro level, and the meso level perspectives.

In the path-dependent process of the digital transformation, computers or similar digital devices are increasingly connected to allow electronic exchange of data to work on projects together or to jointly use specific data. With increasing digitalisation, data collecting sensors can be linked to each other as well, so that interlinked data can be collected synchronically in different places at the same time. Combined data packages can then represent a specific point in time making separately collected time series from different sources obsolete. Such connected collections of data allow for prompt comparative assessments that make it possible to immediately detect and respond to variance in data points that can point to special events or measurement errors. If necessary, survey points can be checked and repaired or readjusted at an early stage. If there is a large amount of interlinked data, a team can work together on the same database from different locations, utilising interlinked assessments. That way, the inherent advantages of different locations can be used at the same time on the same dataset with regard to computing power, personal expertise, or sectorial insights. A core feature is that networking increases the work speed: The better the network, the quicker information or data can be exchanged.

For an individual agent who is involved in the path-dependent process of the digital transformation, electronic networks imply a quick, prompt, or temporally independent option to download information or exchange data. Accordingly, an agent with the tendency to show following behaviour can potentially use diverse options to find blueprints for his decisions to act. Nevertheless, at the same time, path dependence in its self-reinforcing process increases the likelihood of using practicable (available and not completely mismatching) blueprints again and again with the tendency to lock them in. Consequently, path dependence manifests or even reduces the number of used sources over time (Kominek, 2012; Link, 2018, pp. 38ff., 122ff.). In this sense, the number of sources of such blueprints tends to increase only indirectly again such as when one of the used sources expands its number of sources or via special recommendations of the sources used to actively add the recommendation to the draft portfolio of possible actions of the considered agent. Thus, additions to the portfolio of action blueprints can be referred to as closing "structural holes". These new links are rather an increase in the density than a real expansion of a network.

Considering the coupled PDSNs of agents involved in the digital transformation, IT networks allow for higher frequencies of reaction due to the quicker information and data exchange. That way, IT networks enable and accelerate cascades of information or action dynamics via these coupled PDSNs. One example of such a cascade is the effect of the Fukushima earthquake on the promptly following phase-out of nuclear power in Germany[13] followed by the political manifestation and the effect of the victory of the Green party in a regional election (Kominek & Scheffran, 2012; Link, 2018). Via coupled PDSNs, information of real events can diffuse and trigger other real events in totally disconnected regions and with different actors involved who are likely to not even know the agents originally affected by the first event. As long as data and information can be received and processed regionally independently via networks, the mere information about events can trigger real-life social dynamics, channelled via PDSNs.

3.4 Path Dependence, the Digital Transformation, Networks, and AI-tools

The stronger a given network, and the more intensive the analysis of linked data, and the quicker these data are analysed right after their collection, the more computer power is used in comparison to manpower. To achieve this speed, precoded procedures are developed such as neural networks that allow a quicker coding of input for a large computing capacity. Software routines are trained by machine learning to process more and more data scales or data types quickly to make it possible to promptly assess new data right after their collection, e.g. in analyses of X-rays or CT in medical science or a translation software in linguistics. Other software is specialised in dealing with very big data, even though such handling might be more time-intensive. Examples are meteorological or climate models. Also, there are deep learning algorithms that can use new data to automatically re-optimise the existing software or the application tools generated from machine learning.

The self-reinforcing process of the digital transformation more and more necessitates the utilisation of so-called AI-tools to enable the handling and use of digital big data in real time or with a certain data depth. The potential availability of such tools for analyses and the potential transfer to other applications of these tools can also expand the volume and intensity of digital data collection and therefore further promotes the

---

13 Path dependence reiterates the existing structures. In contrast to Germany, the reaction in France was to rather extend the use of nuclear power and to increase funding for research to make the handling and operation of nuclear energy even safer.

digital transformation. Accordingly, the tendency towards a lock-in in the digital trans-formation is connected to the tendency towards a lock-in in the use of these AI-tools for analysis because otherwise digital big data would be too big to handle. Analysing con-nected digital data is also quicker if AI-tools are used than if they were merely assessed by humans. Additionally, AI-tools are applied to discover particular clusters in data and thus, raise the connectedness of the data that are analysed. Therefore, AI-tool usage is a self-reinforcing process with the tendency towards a lock-in, thus it is path-dependent. At the same time, the path-dependent process of AI-tool usage is closely coupled with the path-dependent process of the digital transformation itself because each of the two enhances the other.

Agents who are substantially influenced by the digital transformation also heavily rely on the output of AI-tools if they receive them via their sources of actions (PDSN). Moreover, for each agent a dataset can be created consisting of all his digital activities and contacts as well as his real-life activities and contacts that are documented digitally. Based on all those data, there are AI-tools that can suggest further digital contacts, activities, or even text parts e.g. for E-mail messages. Further digital activities by these agents or direct reactions to the suggestions of the respective AI-tools can help to train and improve them even further. Suggestions for supplementary activities like advertise-ments, videos, news headlines, other web links, or even the writing of another message are not chosen like literature resources in the appendix of a text. In order to make agents spend more time online, AI-tools rather extrapolate from the developed patterns based on profile interests, search history, consumer behaviour, previous journeys, or likes and dislikes shared within social networks or social media. Agents who are used to show a particular following behaviour have the tendency to follow the suggestions of AI-tools. In this process, it can even happen that an AI-tool suggests blueprints for actions that the agent would not otherwise have considered at all such as watching more videos even though the agent initially wanted to watch only one very specific video.

Whenever PDSNs are digitally available, they offer AI-tools the chance to link any available profile data and the associated interests they determined – such as anxieties, likes, relevant topics, or wishes of an agent – with his sources of action blueprints and to analyse the resulting network. This way, changes and dynamics of these changes with regard to the respective topic or focus can be analysed and potentially it is even possi-ble to determine which source in the egocentric network of the agent triggered these changes. Diffusion dynamics in PDSNs can be monitored via topic related key words, likes, or clicks. Thus, cascades in connected PDSNs can be recognised and analysed, as long as AI-tools have access to the specific data for their analyses. For instance, an AI-tool reported the observation of the diffusion of a specific illness in China at the end of 2019 as part of an early warning system of the WHO (Bluedot, 2020; Niiler, 2020). A committee of experts carefully examined the data and initiated additional observations to decide which measures should be taken, whether further data would be needed, if action should be deferred to until the dynamics of the outbreak provide further insights, or if other nations or organisations needed to be warned.

In addition to monitoring the altered dynamics within a PDSN or observing real-life dynamics by monitoring information cascades in connected PDSNs, it is also possible to monitor cascades that spread through real-life events and digital information via connected PDSNs. One example of digital reports about real-life events that in turn can be analysed as the result of an informational cascade includes the platform TikTok. An information cascade triggered and enhanced via TikTok substantially impacted an event of the US campaign trail in 2020, which again was subject of many other digital reports and news spread via social media (Tagesschau, 2020).

3.5 Analysis, Options for Intervention, and Final Remarks

Based on the definition of path dependence as a self-reinforcing process with the tendency towards a lock-in, it can be deduced by using insights from Social Psychology that agents affected by path dependence develop the tendency to show some kind of following behaviour. So far, it has never been measured how strong this tendency actually is, depending on the kind of influence of path dependence. Sociological studies show that routines make up the largest part of our daily activities (Giddens, 1984; Link, 2018, p. 186). Furthermore, there are statistics that document the increase in average time spent online with the ongoing proceeding of the digital transformation. In general, suppliers of search engines, social media, or social networks are now able to store big data that can be analysed in a short period of time with an ever increasing quality of AI-tools. Influencers, followers, likes, and tweets are only examples of digital tools that incorporate and document following behaviour.

Even though it is possible in theory that path dependence can diffuse cascades via PDSNs, it can be observed that IT networks have even increased the likelihood of future cascades. In the general process of the digital transformation, PDSNs and cascades can be increasingly observed and documented digitally. Additionally, the path-dependent characteristic of the digitalisation increases the tendency of the involved agents to use PDSNs as basis for their actions. The path-dependent process of the development of AI tools is entwined with the path-dependent process of digitalisation. AI-tools are increasingly used and their necessity even grows concurrently the larger the interconnectedness and the amount of digitally available data become, which again increases within the self-reinforcing dynamics of the entwined path-dependent processes.

Generally, path dependence describes a reinforcement and manifestation of processes and behaviour. The manifestation implies that path dependence reduces the ability to adapt to new situations[14], i.e. to situations, in which current behaviour does not work anymore or even has counterproductive effects.

---

[14] In contrast, adaptation to small deviations can occur much faster than without path dependence, since the existing path-dependent processes and the hierarchies that result from following behaviour allow for a quick diffusion of the small changes.

One example is anthropogenic climate change. The particular characteristic of this process is that global effects of human behaviour are not automatically compensated by regional differences. Instead, all over the world consumption behaviour is based on industrial products, which generally produce greenhouse gas emissions during the processes of production, logistics, and even still do so during use of the product. All these accumulate in a total global effect. Global warming is the result and causes the necessity to change human behaviour: First, to avoid further global warming, second, to reduce the potential effects of global warming, and third, to adapt to the changing or future climate.

If human behaviour needs to be altered at a large scale – either by mitigation or adaptation – it might be most efficient to start to change action schemes of unintended effects as early as possible to efficiently reach a quick and substantial effect. Unfortunately, path dependence often restrains us from quick adaptation to new situations because in situations of lock-ins changes happen only incrementally (if at all). Real change is not impossible but it can be that alternatives to the customary are not recognised. Also, the stabilising mechanisms of the path-dependent processes rather block any tiny attempts of deviation. Thus, in order to achieve a big change of a path-dependent process it is important to analyse, which mechanisms self-reinforce the path-dependent process, which underlying logic is responsible for the continuity (Tab. 3.1), and what kind of interventions could destabilise, dampen, or change the dynamic towards a sustainable change of the process.

Recall that path dependence also encompasses the tendency to following behaviour. Some rather like to follow individual experts, i.e. their "neighbours", which resembles swarming behaviour, while others rather follow the masses, e.g. via information on ratings, and show herding behaviour. Once action schemes are set in a path-dependent process, it is not always easy to classify them as herding or swarming behaviour. But independent from specific processes, social dynamics can be recognised and can become societally relevant such as in flashmobs, hypes of speculation, e.g. on crypto-currencies like Bitcoin or high-risk stocks like GameStop, or in the US real estate bubble triggering the global financial crisis in 2008/2009 (Brunnermeier, 2009; Stein, 2011). These can start out by swarming behaviour initiating an exponential increase in participants triggering further dynamics. Once a critical mass is reached, people with an affinity to herding behaviour join in to cause an additional jump in the number of participants. Because of the large number of overall participants, even socially more distanced people practicing swarming behaviour can be reached, which further intensifies the cascade through the PDSNs and the associated dynamics (see the simulations in Link, 2018).

Apart from computer simulations, these dynamics of following behaviour resulting from path dependence have not been analysed empirically yet. Thus, no verified mechanism of intervention has been identified yet. Currently, many governments try to suppress unwished demonstrations or spontaneous large group events by blocking social communication media. In some situations, missing communication means cause diversification and therefore reduce the intensity of the dynamics (such was the case when access to Twitter or Facebook accounts were blocked during the siege of the US Capitol Building on January 06, 2021. However, in other situations, exactly that kind of shut-

Tab. 3.1: Overview of mechanisms capable of producing path-dependent continuity (cited from Beyer 2010, Table 1)

Mechanism	Continuity-ensuring logic	Destabilisation options
Increasing Returns	Self-reinforcing effect	Formation of adaptive expectations against established institutions; transaction costs of change are low and/or assessable; transgression of quantitative thresholds in combination with substantial efficiency gaps; transition to decreasing returns owing to change in the environment
Sequences	Irreversibility / Quasi-Irreversibility of event sequences	Overlay of effects; countersequences with annulling effect; termination of "reactive" sequences as soon as alternative options for action emerge
Functionality	Purposes, systemic requirements	Change of functional requirements caused by external circumstances; dysfunction as a result of functional compliance; emergence of significant "secondary effects", replacement by functional equivalents
Complementarity	Interaction effect	Domino effect following partial change; end of complementarity because of intervening factors; loss in the complementarity effect's relevance
Power	Power saving, power of veto	Formation of countervailing power; infiltration or "conversion"; influence or "layering" that suggest supplementation
Legitimacy	Belief in legitimacy, sanctions	Diverging interpretations and traditions; delegitimisation because of contradictions, e.g. inexpedience
Conformity	Exoneration from decision, mimetic isomorphism	Assertion of a new key conception e.g. because an innovation or crisis calls the old key conception into question

down of communication access led to an escalation of the situation because the reduced ability to communicate via technical devices caused an increased desire for meetings in person. This rather reinforced the evolving dynamics with powerful results such as the unrests in the Arab Spring, when demonstrations on Tahrir Square in Cairo, Egypt, caused the resignation of the Egyptian president (Hänska Ahy, 2016; Tufekci, 2018).

Theoretically, it can be deduced that there should be a possibility to prevent the occurrence of intensive following behaviour that is induced by path dependence in advance. To achieve this goal, it is necessary to e.g.

- reduce the number of path-dependent processes, in which the respective agents are involved;
- persistently try to maintain and point to alternatives and actively push people to search, keep, and deliberately choose and use alternatives;
- sustain variety;
- be alert, when the phrase "there is no alternative" appears and try to identify the relevant path-dependent processes, analyse and, if necessary, reduce their power by interventions in the mechanisms of continuity;
- let people actively and as objectively as possible analyse their decision-making processes and question retroactive rationalisations, if necessary;

- develop, maintain, and push to apply individual skills to select;
- inform people who are affected by path dependence about stabilising mechanisms and the path-dependent process so that they do not generalise their process-related following behaviour.

Path dependence and the resulting following behaviour can be efficient because decisions can be made quickly and generally, at least with acceptable quality. Therefore, an outsourcing of decision-making processes can be explained by the assumption that the person whom an agent follows has access to superior information. Furthermore, industrial standards, legal texts, and the constitutional pillars of a society form the foundation of smooth co-operation and allow a focus on new developments, based on the knowledge and decisions of previous generations. Substantial social following behaviour can include the sentiments of people who are generally less politically involved in daily politics, which shapes the agenda, and increases political sensitivity for societal problems. Likewise, substantial following behaviour can be problematic in democratic decisions: If each voter just regurgitates the political tenor of the person he or she follows, political decisions merely count majorities of polarised dynamics, regardless of any variety in voter's livelihoods. Accordingly, Facebook has noted that in many nations the candidate with the largest group of followers on Facebook has won the election in the recent past (Zuckerberg, 2017).

With regard to the digital transformation, IT networks, use of AI-tools and path dependence, this chapter has clearly laid out that the digital transformation can be explained as a path-dependent process that is stabilised and reinforced by IT networks. Also, the use of AI-tools as well as IT networks increases in the self-reinforcing mechanisms of digitalisation. In this current exponential growth of digital processes, it appears to be mandatory to encourage dynamic participation. At the same time, the expected effects are often described as disruptive, a notion that is supported by various perspectives presented in this book. It can be expected that the following behaviour of individuals is increased by digitalisation and particularly by the intensive use of information and communication technology. Simultaneously, global networks can lead to a globalisation of potential polarisations and contextual dynamics. Thus, current political agendas are shaped by connected dynamics[15] instead of by local societal problems. Then democratic decisions only represent a snapshot of the local share of potentially interconnected polarised dynamics. In these dynamics, the contents as well as the underlying societal forces can be controlled by AI-tools if voters use social media intensively.

At the beginning of a path-dependent process it is not possible to predict, at which level the lock-in eventually sets in, according to Arthur (1994) and the urn model and others who emphasise the contingency characteristic (Vergne & Durand, 2010). How-

---

[15] Accordingly, there were assessments whether the British people had indeed opted for Brexit, had the voters optimised their own preferences and thus reached a rational decision. Instead, social dynamics prior to the Brexit vote might have actually caused the majority of the voters to favour the British exit from the European Union (BBC, 2016; Clarke, Goodwin, & Whiteley, 2017; Usherwood, 2017).

ever, a classical urn model describes a process without intervention. In contrast to a ball in an urn, a human being can theoretically decide and act differently from what mere following behaviour would suggest.

Comparing countries, current levels of digitalisation vary substantially and there are parts of everyday life that are digitalised in some countries but not in others. An intensive use of smartphones to stay in touch with social networks is already common, which creates strong dependencies (Torevell, 2020). In Germany, for instance, there is currently an increasing tendency to include the use of smartphones in numerous daily activities to an even greater degree. The use of smartphone apps e.g. representing the personal identity card are encouraged and standards are set that make credit card payments without smartphone use substantially more difficult.

In contrast to the exponentially increasing majorities in an urn model, polarisations can reinforce both oppositions in real life. Thus, if there was strong opposition to the general digital transformation that discusses and preserves alternatives, e.g. to maintain flexibility for future activities, it could be perceived that the dynamics of the digital transformation changed from self-reinforcing exponentially to consciously reflected and in some application areas linearly instead. This change in dynamics could help societies to have more time to actively manage changes, prevent collateral damage, and reduce the disruptive consequences of adverse side effects. It may be possible to influence which topics should be connected to or disconnected from the exponential growth dynamics of the digital transformation to benefit economically, maintain scientific excellence, and preserve geostrategic autonomy. In this strategic and more proactive approach there should be a public debate, which parts of social life should perhaps actively remain exempt from the digitalisation process in order to assure the recreation and preservation of individual decision-making competencies and the functionality of democratic processes.

Chapter 4

Questions in the Philosophy of Technology

Stefan Bauberger[16]

Value-free Technology?

A naïve understanding of technology considers it as a tool that is intrinsically value neutral, but which can be used for good or for bad. Therefore, it is the use alone that determines the value of technology. A hammer can be used to hammer in a nail, but also to smash someone's head.

Behind this image is the correct intuition that technology does not have an end in itself but that it should serve people. However, a differentiated view shows that this intuition is not a description, but that it defines an ideal, and thus a task, namely the task of preserving this service character of technology.

Some techniques already carry the tendency towards a certain application. A nuclear bomb will never be used directly for good, although even there, good applications are conceivable. One such good application would be the utopian scenario of deflecting a large meteorite on a collision course with Earth from its path by means of a nuclear bomb. Even if such a scenario is conceivable, one still cannot meaningfully speak of the development of the nuclear bomb as being value neutral.

A reverse consideration can be made for the development of effective and affordable medicine, although these can be abused, too, for example, by being embedded in an unjust health system. However, one can assume that such medicine represents an example of technology that is usually good in itself.

In most cases, digital technologies are indeed ambivalent in the sense that their value depends on their application. However, the boundaries between development and application are fluid. An example of this is the development of artificial intelligence (AI) for military technology. These developments are not only further developments of already existing technology, but entirely new and frightening dimensions of war technology can be expected if these developments are not stopped by international agreements. Automated aerial drones, for example (which paradoxically become more dangerous the smaller they get), can be used for targeted killing or for acts of sabotage against infrastructures, whereby the attacker does not run any risk of his own and may not even be identified. They thus delimit warfare and lower the threshold for acts of war. In the 10th chapter, Götz Neuneck makes this forcefully clear.

---

[16] Cf. with other chapters by the author: Bauberger 2020a and Bauberger 2020b.

Greater Impact Through Embedding in Social and Economic Contexts

Furthermore, it should be noted against a naïve application of the tool metaphor that the effects of technical developments are often – and this is to be expected in connection with digitalisation and AI – not limited to the concrete individual case of application, but that they must be considered in relation to their overall social and economic impact.

The development of the steam engine in the 18th century and about 100 years later the technical utilisation of electricity and the invention of mass production with the assembly line each resulted in huge social upheavals, with economic winners and losers. Even more powerful was the effect of one of the oldest technical inventions, that of agriculture about 10,000 years ago. Only then dense settlements were possible, only then cities could arise, only then individuals or social groups could accumulate great wealth and so on.

An example of the relevance of this embedding in larger contexts arises – as with the earlier industrial revolutions – from the effects of digitalisation on the economy and the labour market. It is no coincidence that new monopolies and forms of platform economies have emerged in a very short time with the large digital corporations. The very low marginal costs in the application of digital technologies, as well as "positive" feedback through network effects, favour a "the-winner-takes-it-all" economy.

4.1 Technology, Modernity, and the Image of Man

The modern image of man is in many respects shaped by the natural sciences – that is, the widespread view. Naturalism, materialism, physicalism, and reductionism go hand in hand with the natural sciences, or at least they become a constant challenge to every image of man.

These positions are philosophical views. They are not directly the result of scientific knowledge. However, they are usually seen as a consequence of the natural sciences in the history of thought. The proponents of naturalism understand their view as a consequence of the results of natural science insofar as, according to their understanding, this provides a complete explanation of the world, including the position of the human being. The human organism is understood mechanically. The philosopher La Mettrie coined the expression "L'Homme Machine" as early as the 18th century (De La Mettrie, 1748).

From a philosophical perspective, these naturalistic and materialistic positions can be understood as absolutisations of scientific knowledge. They derive a claim of sole representation from the success of the natural science programme. Natural science is not based on pure objectivity, but is the result of an objectification that includes many preconditions that cannot be derived from natural science itself. In concrete terms, scientific knowledge always rests on the background of a life-world knowledge that is much broader, and which can never be completely objectified, but which must be presupposed for natural science (cf. van Fraassen, 2002 and Bauberger, 2011).

Ernst Kapp, a pioneer of the philosophy of technology, understood technology as an interplay. On the one hand, technical products are an "organ projection", i.e. an extension of the natural abilities of humans, shaped according to the pattern of these abilities. On the other hand, man understands himself anew in this projection (Kapp, 1877)[17], according to the model of his own creation. Technology therefore also shapes the image of man. Man's self-projection detaches itself from him as an image that, as a technical creation, is more perfect than he actually is in the respective area. And then, man understands himself from the image of his perfection in technology. However, this perfection is always only a partial perfection that picks out and enhances certain abilities of the human being. In this way, man's self-image is accentuated anew in the direction of this ideal that technology presents.

In this view, the materialistic and mechanistic world view can be understood as a consequence of mechanisation. In particular, the development of the steam engine and other prime movers (motors) clearly showed that movements can be caused purely materially. Through digitalisation and especially through AI, a new image is being formed: Man as an information processing system. Floridi, one of the leading digitalisation philosophers, speaks of the "infosphere" as a new model of human reality: "The next step is to rethink in terms of reality and to rethink more and more of its aspects in terms of information." (Floridi, 2015, p. 75) He speaks of a "dematerialisation of objects and processes." (Floridi, 2015, p. 101)[18] Weizenbaum and Janich pointed out some time ago that this is philosophically based on a hypostatisation and naturalisation of the concept of information that is factually incorrect (cf. Janich, 1997 and Weizenbaum, 1977, pp. 207–241).

Parallel to this, the penetration of the living world by computer technology leads to the cyber world being seen as the realisation and transcendence of what the human being actually is. Material reality is surpassed by an apparent spiritual reality. Wiegerling warns in this context of a "loss of resistance" (Wiegerling, 2011, pp. 26ff), due to the "loss of reality" (Wiegerling, 2011, p. 13).

The consistent exaggeration of this ideal leads to radical transhumanist fantasies. Ray Kurzweil, head of technical development at Google, developed the vision that robots and computers will eventually take control of the world, as they will overtake humans

---

[17] A still current example of this retrograde projection is the understanding of the heart as a pump (cf. Kapp, 1877, 98), i.e. modelled on the basis of mechanical pumps. I owe the hint of the significance of this retrograde projection on the self-image of the human being precisely connected to computer technology to Eugen Wissner, in the context of a master's thesis he wrote at the Munich School of Philosophy.

[18] Further quotes from Floridi to clarify: "According to the 'it-from-bit' hypothesis, even our bodies are, deep down, (composed) of information, not of some ultimate material substance distinct from what is immaterial." (Floridi, 2015, 101)
On privacy: "... so too privacy calls for a correspondingly fundamental reinterpretation that takes account of the fact that our selves and our interactions as inforgs are informational by their very nature." (Floridi, 2015, 159)

due to the progressive development of AI (Kurzweil, 2015). In his sense, this is not a loss, but it is in the good logic of evolution. Behind this is the idealisation of intelligence in the sense of data processing, and subsequently the reduction of humans to the same function.

The effects of digitalisation on the image of humans are particularly relevant, for example, in the field of medical technology and care for the elderly. In these areas, there is a high cost pressure (due to the politically set framework conditions), which elevates efficiency increase to the ideal. In addition, there is also a tendency towards mechanisation independent of digitalisation, which competes with the ideals of medicine and care centred on people. However, especially in care, people must be accepted in their bodily dimension and with their limitations, even with mental limitations such as dementia. If being human is defined according to the ideal image of an information-processing system, people in need of care can only be understood and treated as annoying marginal phenomena of society.

4.2 Mortification, Alienation, and Fascination

Günther Anders has described the human and social consequences of the development of technology as a sequence of ever-increasing mortifications and alienations (Anders, 1957, and Anders, 1980). Measured against the perfection of technology, for example, the power that machines unleash, humans experience themselves as imperfect and powerless. The nuclear bomb in particular demonstrated this powerlessness in a special way. Accordingly, in his philosophy of technology, Heidegger also criticises an alienation of man that arises from the power of technology detached from him (Heidegger, 1954).

The antithesis to this mortification is the fascination that comes from highly developed technology. The good thing about this fascination is that researchers drive the development of technology with great dedication. On the other hand, the fascination with technology carries the danger that technology becomes an end in itself.

4.3 Technical Solutions for Human Problems

"Instrumental reason has turned words into a fetish surrounded by black magic. And only the magicians have the rights of the initiators. And they play with words and deceive us" (Weizenbaum, 1977, p. 334). This is how Joseph Weizenbaum, one of the great pioneers of AI research, criticised the exaggerated hopes placed on computers as early as 1976. He pleads for a "reason that recalls its human dignity, authenticity, self-respect, and individual autonomy" (ibid.).

Weizenbaum criticises a paradigm of technical reason that combines with and absolutises scientific rationality. In this paradigm, problems are there to be solved, and they are ideally solved technically. This paradigm goes beyond the area of the application of technology; it has become independent in this respect. Weizenbaum cites as an example that in the American student riots, it was often suggested that the problems should be solved through better communication between the different parts of the universities. "But this view of the 'problem' (...) effectively hides and buries the existence of real conflicts" (Weizenbaum, 1977, p. 328). This "imperialism of instrumental reason" (Weizenbaum, 1977, p. 337) misses reality, or rather reduces this reality to everything that can be technically described and solved. Humanity is thus lost. In the human realm, many things are complex; even problems cannot be solved in many cases but must be endured. Human support is important far beyond the solution of technical problems.

Under the premise that technology should serve the well-being of humanity, the conclusion of the considerations in the philosophy of technology is that this tool character of technology is by no means a given. Instead, it is a permanent challenge to make technology serviceable. As a rule, but not always, technology is ambivalent. However, the use of technology must be considered in the social and economic context (with the associated interests), the danger of technology becoming an end in itself must be taken into account, as well as the effect on the image of man.

Again, the use of digital technology in medicine can be used as an example, especially if AI can take over many diagnostic tasks in the future. In the ideal case of good use of AI, the technical tasks of the medical art are transferred to technology, freeing doctors for the tasks associated with the human aspects of their work: as caring, compassionate people who, for example, discuss therapy options with patients, whereby decisions are made not only according to purely technical aspects, but also take into account the values and attitudes to life of the patients, as well as their environment. In the worst case, digitalisation reinforces the technical character of medicine. Economic constraints and lobbying interests within the health system will promote this bad case. In addition, there is the strongly technical image of man within this system (man as a biochemical machine) and the associated paradigm of technical reason.

Only with knowledge of these mechanisms can we succeed in turning the application of these ambivalent AI techniques, which also open up great new possibilities for medicine, to the good. For the good, these techniques promote the humanisation of medicine, so that the human and interpersonal aspects of healing gain value as more technical aspects of theories are transferred to machines.

Chapter 5

Digital Extensions, Transhumanism and Posthumanism

Frank Schmiedchen

The chapter follows up on the previous one and deals with our self-perception and the image of humanity with regard to the topic of the book and how human beings 'should be in the future'. It asks to what extent digital technology implanted into the human body can lead to changes of the person in essence, so that this technologically "upgraded" human being[19] becomes "a different being" (e.g. a cyborg) and what motivates such self-improvements. This is followed by a look at the essential statements of transhumanist and (technological) posthumanist ideas, also with regard to underlying images of the world and the human being. Subsequently, reflections on the intangible dignity of every human being are presented as counter-concept, and the various approaches are discussed.

5.1 Forms of Digital Self-Improvement

In the context of this chapter, digital enhancement is understood as digital technology permanently connected to the body to improve, support, or restore knowledge, skills, or abilities. Not all forms of digital self-improvement are relevant to the question of to what extent they trigger changes in human identity and personality. In particular, all neuroimprovements may have substantial effects on the consciousness and personality of the improved people and bear a relevant risk of being hacked and used to manipulate or control the human (cf. Clausen, 2006, p. 28).

Such improvements can be caused by assistive devices[20], prostheses[21], and neuroimplants as well as genetic and nanotechnological changes in humans.[22] Above all, neuroimplants to support cross-task skills, knowledge acquisition, and processing must be examined in principle for potentially triggering psychological, cognitive, and mental changes and having strong ethical implications (cf. Fenner, 2019, pp. 167–288). Such

---

[19] In the rest of the text, the inverted commas are omitted from words such as improvement, extension, etc. for the sake of text flow, even though the author is skeptical if they are improvements and who measures such improvement and how.

[20] They are often classic and analogue products replacing, supporting, or relieving human performance to prevent or compensate disabilities or to treat illnesses (e.g. visual and hearing aids, body replacements/ prostheses, orthopaedic aids, incontinence and stoma articles, and technical products that serve to introduce medicine or other therapeutic agents into the human body.

[21] While exoprostheses are body replacements that are attached but outside to the body (e.g. replacing limbs), endoprostheses are artificial materials implanted to the body.

[22] Non-digital genetic or nanotechnological enhancements raise analogous questions, but these are not discussed here.

neuroimplants are human-machine interfaces (brain computer interfaces, neurolinks) that directly connect human nerves with electronic technology and serve to exchange electrical signals between the brain and the respective technical devices (cf. Clausen, 2011, p. 3; Clausen, 2015, pp. 697–839; Jansen, 2015, pp. 226–234; Blumentritt/Milde, 2008, pp. 1753–1805). For example, performance-enhancing brain implants are used to improve memory performance. One or more interfaces implanted in the brain are connected to a chip in the brain or outside and are supposed to potentiate cognitive abilities. A distinction is made between (passive) deriving and (active) stimulating systems, whereby modern research assumes that we will soon arrive at integrated systems that serve both aspects (cf. Müller/Clausen/Maio, 2009).

In view of the scope of these interventions, the question must be asked whether the numerous and far-reaching upgrades, especially in the neurological field, are the historically contemporary form of human striving for self-improvement and self-conquest (cf. Fuller, 2020, pp. 44–55; Hansmann, 2016, pp. 28–30), or whether the digital enhancements available today (or soon) are generating a new quality finally putting into question humans as such? This new quality would always be the case when interventions lead to a fusion of technology and humans creating 'something new' by changing consciousness and identity leading to transformed creatures (transhumans, cyborgs). (cf. Rössl, 2014, pp. 21–28; Jansen, 2015, pp. 219–226; Sorgner, 2018, pp. 90–92.; Loh, 2018, pp. 58f).

But also noninvasive neurotechnologies capable to "read the brain and write to it" are essential (Schwab, 2018, p. 243). "A more precise influence on the brain could manipulate our self-perception, completely redefine what experiences mean, and fundamentally change what reality represents" (ibid., p. 244).

Enhancements of biological life through digital implants could potentially cause a convergence of man and machine that may ultimately lead to an all-encompassing network of life and the digital, raising the question of the extent to which such a result definitively transcends the boundaries of what it means to be human. This, in turn raises the question of whether and at what level there is an insurmountable difference between man and machine, and what this difference is based on.

However, the pursuit of self-improvement can also be very pragmatically motivated, free of such profound philosophical reflections, and explicitly or implicitly based on utilitarianist ethics. According to this, everything that is useful would also have to be ethically compelling.

5.2 Utilitarian Motivation

For utilitarianists, digital enhancements are a possible and useful, and thus mandatory improvement of the human being in order to reduce or eliminate observable deficits of the individual or of humanity as a whole. Only those improvements that benefit the individual but harm others or the public are to be rejected (cf. Birnbacher, 2013, pp. 153–158). Humans perceive the world, reflect, or abstract these sensory impressions, and then invent ways to better master the nature, in which they live or compensate for their deficits in terms of mastery. Humans must therefore complement themselves with things they create (cf. Rössl, 2014, p. 15). Humans develop their abilities by inventing things that improve their performance. Thus, the weak hand was first strengthened or replaced by simple tools (hand axe, sickle) and later by (automated) machines. The same applies to visual aids improving the spectrum of individual vision or the human ability to see per se. Without a telescope or microscope, humans would not be able to see anything far away or very small. Analogue aids, therefore, already fundamentally expand the spectrum of human possibilities.

Since the Renaissance and even more so since the end of the nineteenth century, the idea of competition has played a central role in these enhancements. People consider themselves to be in a permanent competition to see who is more successful, cleverer, more beautiful, etc. Today, this basic motive of individualised competition has become the source of a permanent sense of deficit due to potentially omnipresent digital-social networking. It is a subjectively felt deficit vis-à-vis believed or actual competitors, but also vis-à-vis the theoretically maximum possible (cf. Spreen/Flessner in Spreen et al., 2018, pp. 8–10).

As already explained in the last chapter, Günther Anders therefore speaks of a resulting fundamental mortification of man in the face of the machines he has created. This humiliation is the result of a growing gradient between one's own subjectively perceived imperfection and the ever-increasing perfection of the machines (Promethean gradient) (cf. Anders, 1956/1980, p. 16). In this respect, digitalised technology is only a further step, but as the book shows, also a qualitatively new step on the path people followed since they became self-aware.

Consequently, the optimisation of humans through the implantation of digital technology can be interpreted as a product of human invention for humans. From a utilitarian perspective, an omission of possible, digital improvements is unnatural and unethical, since an omission runs counter to man's continued, historical decisions or, more broadly, his natural law compulsion to constantly improve (cf. Nida-Rümelin/Weidenfeld, 2018, pp. 64–70). In this understanding, exceptions are only permissible where dangers for others or humanity as a whole are deemed too great to continue using a technology once it has been invented/discovered. This corresponds more or less to the experiences of the 20th century, in which only those technological paths were at least partially abandoned that have proven that they can cause mass destruction (e.g., toxic gases), or where alternative technologies show fewer side effects (e.g., CFCs, nuclear fission).

The majority of pragmatically oriented self-improvementists will affirm that human life should be improved in its natural life span in every possible way. They will probably also agree that at the same time, whenever possible, people's life spans should be extended. Thus, pragmatic utilitarians also dream of immortality and do not ask whether such a development is desirable, but affirm this assumption without reflection (cf. also Fenner, 2019, pp. 148–159).

5.3 Transhumanist Motivation

The most important transhumanists[23] go beyond such purely pragmatic understanding of a "better-is-always-useful" utilitarianism. For them, it is a matter of supplementing or replacing biological evolution on earth (and in the cosmos in general) by cultural-civilisational-technical means to accelerate evolution and improve the result (cf. extropianism: Erdmann, 2005).

The idea of transhumanism ("beyond-human"), which essentially only[24] emerged in the 20th century, aims to evolve *homo sapiens sapiens* as a species (cf. More, 2013; Kurzweil, 2014; Sorgner, 2018). Numerous original voices and manifestos profess this common fundamental belief.[25] Transhumanist concepts share a view in which humanity no longer feels bound to its approximately 300,000-year existing form (cf. Gunz, 2017). Rather, possible sociotechnical evolutionary steps, including those that lead beyond humans, are regarded as naturally human. Thus, the biologically constituted human being is supposed to improve himself in every technical form possible to him. This explicitly affirms that the result of such a synthesis is in fact another creature (transhuman, cyborg) or that the contents of human consciousness only exist in digitalised form; a dream shared by trans- and technological posthumanists. Such a view then applies to all technical interventions that humans undertake in order to (co-)shape evolution. The nature of these interventions (e.g., genetical, nanotechnical, or digital) is irrelevant. The changes stem from human creativity and are therefore natural in this view (cf. Sorgner, 2018).

Transhumanist concepts are not solidified doctrines of ideas, ideologies, or theories. Instead, they are diverse and partly contradictory ideas that are based on different basic philosophical convictions. However, the spectrum is dominated by strictly atheistic technologically oriented thinkers but can also be found in Protestant-Calvinist or Jewish-derived beliefs. Their common basic belief is that humans are not fully developed but in a permanent state of further development, in which they themselves actively carry evolution beyond their previous humanity through scientific knowledge and technical progress. In this context, for some of them, transhumanism represents a kind of intermediate stage to posthumanism, so that the boundaries are blurred here.

Transhumanist concepts are based on a paradoxical image of man, which describes human as imperfect and emphasises the necessity for his self-improvement and self-conquest, but also sees humankind as the "awakener" of the cosmos who

---

[23] e.g. Nick Bostrom, Fereidoun Esfandiary (FM-2030), Eric Drexler, Ben Goertzel, Aubrey de Grey, Yuval Harari, James Hughes, Julian Huxley, Zoltan Istvan, Saul Kent, Timothy Leary, Ralph Merkle, Max More, Elon Musk, Gerard O'Neill, Larry Page, Martine Rothblatt, Anders Sandberg, Pierre Teilhard de Chardin, Peter Thiel, Natasha Vita-More, Ken Warwick

[24] Some transhumanists refer to ancient sources. However, since there is not even now a stringent transhumanist theory, only the period in which most transhumansitic ideas were developed is relevant.

[25] Examples: https://humanityplus.org/; https://humanityplus.org/philosophy/transhumanist-declaration/; http://www.transhumanismus.demokratietheorie.de/docs/transhumanismus.pdf; https://transhumane-partei.de/was-ist-transhumanismus/; (each 03 January 2021)

provides it with a meaning. In this respect, transhumanism is also transcendentally oriented and aims at that which lies beyond the human being, but can be reached by him (cf. Coeckelbergh, 2018). Man, as a weak and imperfect living being, needs constant improvement to be whole. Humans can and should gradually improve their genetic makeup, physical and mental health, perception, emotional and cognitive abilities, and skills, and significantly extend their lifespan (cf. Boström, 2008, pp. 107–136). This is achieved through the permanent incorporation of technology (genetic, digital, nanotech, etc.) (cf. Hayes, 2014).

The core element of this conviction emphasises that it is human intelligence, his cognitive capabilities, that calls man to higher things. In contrast, biological corporeality, including emotions, is only an expression of previous inadequacy of "blind" evolution up to our present. Nevertheless, all transhumanist concepts are ultimately entangled in a paradox of quasi-religious-transcendental and materialist thinking (cf. Coeckelbergh, 2018). Ultimately, it is about a physical vision of human expansion, even if this is reduced to the brain content, which as the sum of its physical manifestation fully constitutes consciousness – the mind. In this sense, uploadable consciousness is the sum of available and processed information. Only for this reason can the mind be uploaded onto a computer, because for most transhumanists it is materialised information and nothing more than that.

Despite this reductionist, materialist understanding, there is an inherently quasireligious interpretation of man overcoming himself and only thereby becoming himself, but also numerous philosophical and historical points of contact, for example with Protestantism, especially in its Calvinist and evangelical forms (e.g., currently strong among the Mormons) or in Judaism (cf. Samuelson/Tirosh-Samuelson, 2012, pp. 105–132; Krüger, 2019, pp. 110–115). Since the Enlightenment, human reason, intelligence, and linguistic ability were identified as expressions and signs of man's calling and capacity for higher things. These spiritual qualities indicate that man is the image of God.

Transhumanists tend to make references to the thinking of Herder, who describes humans as deficient beings. For they are the "most miserable of the animals". This results in the possibility and necessity for him to constantly improve and elevate himself by means of his intellect and language (cf. Herder, 1769, Part Two, First Law of Nature). In contrast, it continued to be true, and since the Victorian 19th century even more so, that man's natural physicality and the associated drives (especially sexuality) and feelings were the expression of eternal sinfulness and inferiority that needed to be overcome or redeemed[26] (cf. Lüthy, 2013, pp. 11–25; Mulder, 2013, pp. 30–43; Samuelson, Tirosh-Samuelson, 2012, pp. 105–132).

Then, at the end of the nineteenth century, in the context of the growth of capitalist convictions, the Protestant postulate was further developed, especially in the USA, into "the-successful-is-successful-because-God-loves-him". Finally, we also find ample refer-

---

[26] In contrast, Luther emphasises in his *Diputatio de Homine* the impossibility of determining reason as a defining characteristic of man, since it could just as well be an expression of his "sinfulness".

ences to Nietzsche and his need for the creation of the divine superman. The creation of the Nietzschean superman advertises a tantalising transhumanist proposition: If the first cognitive revolution made the "insignificant African ape" master of the world, the second cognitive revolution will elevate man to "master of the galaxy", quite peacefully with the help of genetic engineering, nanotechnology, and interfaces between brain and computer (cf. More 2010; Harari, 2017, p. 476f.). In its quintessence, this leads to a posthumanist view that evolution as a whole should be regarded as imperfect and therefore be supplemented or replaced by scientific-technological evolution (cf. Sorgner, 2018; Jansen, 2015, pp. 219–234).

5.4 Posthumanism

Posthumanism, which is also extremely heterogeneous, is divided into a technologically oriented majority camp[27] and a critically oriented, small minority camp[28]. Apart from the relatively uninfluential small group of critical posthumanists, all authors understand socio-cultural progress as a consequence of technological developments.[29]

Technological posthumanists argue that the (final) human creation (*our final invention*) in the form of the singular superintelligence and its then "descendants" created without human intervention will awaken the cosmos, but not humans themselves (cf. Barrat, 2013). Their approach thus leads them to the normative demand of the final and complete overcoming of man by his technical inventions, since these are superior to him. Accordingly, humanity should make room for a (digital) superintelligence (singularity) created by it, which, superior to all humanity as a de facto God, creates paradise on earth and in the cosmos – at its discretion, with or without human beings.

Although "real" transhumanists only aim at the improvement of human life on a higher evolutionary level and do not want to leave the Anthropocene but extend it, many authors who also express themselves as being transhuman are ultimately technological posthumanists[30]and see humanity only as a necessary intermediate step towards higher, nonhuman but human-induced intelligence or life forms that will ultimately awaken the cosmos. Without simplifying too much, technological posthumanism can be described as a current of thought/quasi-religion that is hostile to human beings. The way of dealing with people who actively strive for technological posthumanism should therefore be discussed socially as soon as possible.

In this context, the following is also significant. In the confrontation between "real" transhumanists and technological posthumanists, a conflict of epic proportions is perhaps brewing, which will not only be fought on paper. In this already emerging, pos-

---

[27] e.g. Raymond Kurzweil, Marvin Minsky, Hans Moravec, Vernor Vinge, Eliezer Yudkowsky
[28] e.g. Rosi Braidotti, Janina Loh, Stefan Sorgner
[29] https://whatistranshumanism.org/ (02 March 2021)
[30] The boundaries between the two groups can only be determined philosophically-analytically anyway. See on this: Loh, 2018

sible (hot) war, the transhumanist side fights for the continued existence and further development of a technically improved humanity in the sense of an expanded concept of philosophical humanism, while the technological posthumanists want to overcome humanity in an antihumanist way through a new "technology species". Elon Musk explicitly justified the founding of his company Neuralink by saying that humanity would have to arm itself for the inevitable war against an artificial intelligent singularity through human-machine interfaces to survive[31].

Besides the technological posthumanists, there is a small group that calls itself critical posthumanists with a completely different vision. This group does not strive for the technological overcoming of humans, but wants a fundamental ethical human renewal by overcoming humanistic, anthropocentric ways of thinking and behaving, with which humans have illegitimately elevated themselves to being the murderous master of the world and all its life forms (cf. Braidotti, 2014; Loh, 2018). However, technological tools are not rejected along the way. What the critical posthumanists have in common with the technological posthumanists is merely the demand to overcome the Anthropocene. But critical posthumanists only want to push humans down from the throne of the crown of creation so that they can consciously integrate themselves into the overall organic structure of the planet, in which they are always and completely embedded anyhow. This way, socially and ecologically reprehensible actions committed in the past and present should be avoided in the future (cf. Braidotti, 2014, p. 197f.). Thus, critical posthumanism is a current of the pseudoleft (il)liberalism that we know from the identity politics, antihumanist debates. Critical posthumanism is hardly noticed in mainstream trans/posthumanist discussions, especially not in the US and the P. R. China. Since critical humanism is also not a (primarily) technology-oriented current of thought, it is only mentioned in this book for the sake of completeness.

5.5 Man as a Self-Purpose – Machine as a Tool

If one wanted to make it (too) simple, the human striving for complete control over one's own life, the environment, and one's own death could be psychologically described as infantile, narcissistic, and characterised by hybrids (cf. Nida-Rümelin/Weidenfeld, 2018, pp. 188–197). A more serious engagement with the radical views put forward on human self-improvement and on trans- and posthumanism[32] requires a clear vision regarding one's own self-image as well as the underlying image of humanity.

We contrast the trans-/posthumanist view of human beings, which is only roughly sketched here, with a view that values the "imperfection" of human beings and biological processes as something natural, intentional, and meaningful. In this view, illness and

---

[31] https://neuralink.com/; https://www.sciencemediacenter.de/alle-angebote/press-briefing/details/news/brain-computer-interfaces-hintergruende-zu-forschungsstand-und-praxis/; https://industrie-magazin.at/a/neuralink-warum-elon-musk-computerchips-direkt-ins-hirn-pflanzen-will

[32] In the following sections, posthumanism always only refers to technological posthumanism

death are natural phenomena that are meaningful (cf. Bauberger/Schmiedchen, 2019; Coekelbergh, 2013, pp. 21–23).

Such an acceptance of the consequences arising from the complete and indissoluble interconnection of humans with nature is considered suboptimal by trans/posthumanists. For them, it is morally reprehensible to resign unnecessarily. For them, a view, in which the human being is an integral, indivisible manifestation of the ecosystem earth (and the cosmos) in its physical constitution that goes from birth through growth, reproduction, and decay to death, is an inexcusable capitulation in the face of the assumed creative powers of man to change his destiny.

The religious scholar Oliver Krüger speaks of posthumanism as the attempt to "heal" the four grievances of the human being. In doing so, he refers to the work of the philosopher Johannes Rohbeck who in turn based his ideas on thoughts of Günther Anders, Hannah Ahrendt, and Sigmund Freud (cf. Krüger, 2019, p. 423; also: Flessner, 2018). He continues to say that the alternative to these grievances "lies in the uncomfortable acceptance of human imperfection, death, and ageing". In this sense, it can be argued that transhumanist thinking leads to a loss of the specific value of humans as biological life form, because in transhumanism "human nature finds a sensitive devaluation in its vulnerability, its transience, and above all its replaceability" (Hansmann, 2016, p. 109). We resume here and focus on the self-purposefulness of life (cf. Bauberger/Schmiedchen, 2019). In our view, technology always remains instrumental (see the following chapter). The usefulness of technology for humans is therefore an essential basic condition for the justification of technology.

Those who do not want to follow our reasoning can help themselves as follows: How do transhumanists want to prevent an increasing "inertia" of necessary adaptation processes in case of a massive extension of lifespan and the (presumably unrealistic) use of cryonics (cf. Fuller, 2020, pp. 184ff.). Sexual reproduction contains a "built-in" mechanism for better adaptation in an apparently permanently changing universe. This ensures the preservation of the species. Even more mundane is that even the greatest minds of humanity get tired over the decades and tend to no longer produce qualitatively new ideas. Only a new generation can do this relatively effortlessly and also produce great minds again. But even if there were no qualitative advances in knowledge in a generation, the natural process of generational succession would be mathematically more reliable than an insistence on massive life extension of Nobel laureates.

Those following our reflections can resume with the self-purposefulness of the human being. Ultimately, all human beings (like all beings) are interconnected and related to each other. This includes the fact that every human being has an intrinsic value that cannot be relativised, which can be politically described as human dignity. This individual dignity is neither negotiable nor weighable but absolute (cf. Nida-Rümelin/Weidenfeld, 2018, p. 64ff). "The guarantee of dignity (…) is thus a central ethical guideline for the design of digitalisation" (cf. WBGU, 2019, p. 44[translated]). In humans, this is combined with subjective ethical insight that cannot be traced back to scientific laws or facts. Being a person, with ethical responsibility founded therein and the human interconnectedness with all living beings always includes the emotionality and physicality of humans. People show respect for each other and appreciate each other in their materiality. (cf. Bauberger/Schmiedchen, 2019).

Machines, on the other hand, cannot act autonomously, make decisions, or bear responsibility. Their "autonomy" is always only derived. Therefore, the goals of all technical products must be relativised to those values that are assigned to them by humans (cf. Bauberger/Schmiedchen, 2019). Even when considering compassion or consideration, we humans as sentient beings are fundamentally closer to all other living beings than to any machine we have created, no matter how intelligent it may be (ibid.).

The end of human life in itself does not allow any external pressure for self-perfection, as it contradicts human dignity. It is even more forbidden to give people future life value only in an improved form. Likewise, it must be categorically rejected if humanity is to be overcome in its biological evolution and replaced by an artificial life form created by him.

5.6 Where Should the Journey Go and Why?

In the last 250 years, humanity has succeeded in achieving tremendous improvements, some of them incredible, in almost all socio-economic areas: An increasing number of humans have not died of hunger, childhood diseases, appendicitis, or childbirth. All this has been achieved through human progress and social struggle. And it is important to make clear that scientific positivism, economic capitalism, and constant warfare since the mid 19th century have been the framework in which this has happened.

However, it is equally important to point out that only the successful empowerment of the poorer and weaker (e.g. with the help of trade unions, critical media, free and equal elections) have led to a socially accepted distribution of these technological achievements, at least in tendency on a national/regional level. In order to be able to discuss ethical questions of neurological self-improvements and trans/posthumanism even between different cultures, common starting points and questions are needed, on which people in all regions of the world can agree. To be able to answer these questions sensibly and in a way that is relevant to action, we in turn need answers that are not only suitable for Sunday speeches, but also survive in the cold realpolitik of interest representation.

Possible starting questions could be:

- Is life, as it is, good in principle, and is it sustained and justified by itself without any further intervention (e.g. merits, achievements, efforts)?
- What do we humans want to be like in 2050, 2075, or 2121?

A possible minimal consensus could be that the primary meaning and purpose of human existence is to constantly renew life without causing too much damage; primarily through reproduction, secondarily through cultural achievements (carrying the torch forward).

An internationally more controversial, but probably necessary aspect, is that scientific findings since the beginning of the 20th century support the view of reality, in which everything that exists, at every point in space-time, is in constant connection and exchange with everything else that is. Consequently, the perfection of the individual is only a meaningful process if at least no harm is done to others as a result. However, this can only be claimed if possibly required scarce resources for this perfection process "are not lacking elsewhere" or would not make a "higher" contribution there to the "improvement" of humanity or the common good. However, the advocates of limitless human optimisation logically fail to provide this proof. So far, everything points to the fact that the said improvements are directly and indirectly at the expense of those people who (for whatever reason) are not able or willing to have these improvements made to themselves. Since this is not problematised by enhancement/trans/posthumanism advocates, there is a well-founded initial suspicion that the possible potential for creating a worldwide two-class society is either not seen or condoned. While the rich are acquiring every opportunity to continuously upgrade themselves technically and prolong their lives in an unprecedented manner, the non-improved are becoming increasingly useless and disruptive.

In economics, we know the principle of marginal utility and the principle of opportunity cost. When using these two economic principles, we must consider what the desired improvements of and for the people will bring about and whether the trans-/posthumanist path means more for humanity as a whole. A few simple considerations already point to serious ethical problems: improvements and life extensions lead c.p. to the individual being able to be active longer. This can be accompanied by a reduction in the birth rate or by a constant or even an increasing birth rate. In the first case, humanity would become older through enhancement and life extension. In the other two cases, more people would live at the same time than would be the case with normal mortality. As shown above, there is always a serious risk that humanity will slowly freeze socioculturally as the proportion of younger people who keep giving birth to new ideas decreases relatively. In the latter case, the Earth would reach the objective resource limits even faster.

Technological posthumanism triggers much more serious problems:

- An artificial intelligence that reaches the stage of a singularity can, following its basic programming, be "benevolent" to humans and "behave" in the interest of humanity. In this (logically not very probable) case, we would have created a paternalistic "God" who protects and develops human life.
- But the singularity could also be equally "benevolent" towards humanity and the well-being of the planet, and "realise" that humans usually do not behave for their own good. Then this paternalistic "God" would take us by the hand and lead us into a future, which he deems as correct.
- Or, the singularity feels more "obliged" to the well-being of planetary life as a whole. In that case, people would have to accept numerous restrictions, would

possibly be decimated, or, in extreme cases, wiped out because they have (so far) behaved in a harmful manner without understanding.

Each of the just-outlined alternatives shows that the benefit-danger ratio cannot really be considered to be beneficial for humanity. Accordingly, the final question remains to be asked: Who would benefit from a trans-/posthumanist future?

Part II

Legal Frameworks and Standards of Digitalisation

Introduction

Jasmin S. A. Link

Some years ago, I have asked a peace researcher and member of the Federation of German Scientists (VDW): If you had known that nuclear bombs could be developed, would you have sought to discover nuclear power nonetheless? His answer was: Yes. Because that way global peace could be guaranteed for a long time via international treaties. It could be expected that everything that can be done, will be done. Thus, if he had not been the one to make a certain discovery, somebody else would have been the one. But the peace researchers' achievements have demonstrated that it is possible to use a new technology for the international good such as international stability and peace, which would be assured by bilateral or multilateral treaties.[33]

Just as important is the VDW activity with regard to the technological impact assessment of the digital transformation. Early requests to define the necessary legal framework of digitalisation, digital IT networks, and research and development and implementation of so-called artificial intelligence systems (AI) can have crucial impacts on society. Part I of this compendium has shown that digitalisation, digital networks, and AI development can cause disruptive changes and can pose fundamental challenges for technical, democratic, functional, economic, human health, and social development security. The digital developments can even trigger initiatives that question the right of human existence. Potentially, within many societies the ongoing digital transformation already seems "without alternatives".

In part II, the legal aspects of the digital transformation are assessed. Who is to be made responsible for potential "collateral damage", e.g. of autonomous cars? Can machines be made responsible? If so, do they have "rights" in exchange, too? Accordingly, would human rights need to be applied to machines as well? Who is right in contradictory testimonies: humans or machines?

Which social standards are already developed to use existing political structures for the design of the process of digitalisation? Which additional legal measures are planned? Can and should there still be a copyright for intellectual property? Which and how?

And how can lethal autonomous weapon systems be designed to fulfill existing martial law acts to avoid civil victims? How does the development or existence of autonomous weapons change not only warfare but also international diplomacy?

The following chapters of part II will discuss these and further questions in detail.

In chapter 6, Stefan Bauberger contrasts the transhumanistic perspective, which was presented in chapter 5, with a machine-ethical perspective. He scrutinises the ethical

---

[33] This does not mean that all peace researchers at VDW share this opinion. It is ethically problematic to just do everything that is possible because otherwise someone else would do it instead. This requires further scientific debate with regard to the responsibilities of science in general.

aspects of so-called autonomous actions of a machine that is controlled by AI. Particularly, when checking whether the concept of human dignity can be applied to machines he discovers ethical problems. Even the question whether a machine controlled by AI can develop consciousness is discussed briefly.

In chapter 7, Christoph Spennemann assesses from a legal perspective who can be made accountable for potential malfunctions of AI controlled systems, robots, or autonomous cars. Existing laws are evaluated with regard to their applicability in the AI context and it is analysed whether they can be extended to include the digital transformation or whether the EU needs to design new legal frameworks to assure the insurability of AI products on markets.

The subject of standards in the digital world will be addressed in chapter 8 in two parts. In part A, Eberhard K. Seifert describes classical standardisation processes and organisations. It is explained how national and international mechanisms interact. Applied to the example of digitalisation, it is assessed which goals can be addressed and how some mechanisms of action can potentially create undesirable side effects. For example, the bottom-up self-organisation can result in an economic politically motivated top-down effect on EU member states.

Additionally, in chapter 8 B, Michael Barth analyses the international standardisation structures focusing on digitalisation, and he points out how powerful the politico-economic effect of standardisation processes can be, especially in an international dynamic. He describes which organisations and nations are currently taking which international positions in the context of the digital transformation, digital networks, and AI. He emphasises why this positioning is of increasing fundamental concern for Germany and the EU.

In chapter 9, Christoph Spennemann analyses the legal dimensions of which kind and how intellectual property is protectable in the context of the digital transformation, digital networks, and especially the application of AI. He is aware of internationally diverse legal frameworks shaping the competition in innovation in the development of AI systems, and also in connection with the potential development of strong AI. The trade-off between innovation capacity and property rights may become necessary, just as the founding of a committee of external experts, as well as the analysis of potential indirect international future effects of regional legislation.

In chapter 10, Götz Neuneck critically assesses the current development of drones and autonomous arms systems from a technological and peace research perspective, particularly with regard to possible consequences for warfare, military operations, and impacts on civilians. Even though already existing arms treaties might be interpreted in a particular way or may be extended to include autonomous weapons, humans still have a long way to go to internationally regulate such arms systems. Therefore, it is necessary to first create awareness of the real relevance of potential scenarios of armed conflict that may result from the possible development or malfunction of AI.

When reading these chapters, please note that the perspectives presented originate from different scientific disciplines. Consequently, the exact meanings of the used terms such as robots, autonomous machines, or artificial intelligence are based on the scientific literature of the particular discipline. Therefore, the specific use of the terminology may vary from the technical definitions that were presented in chapter 2 by a computer scientist.

Chapter 6

Machine Rights

Stefan Bauberger[34]

There are various calls in the literature for a personality status for robots or corresponding rights. The European Parliament's call for this in 2017 is particularly significant:

> "The European Parliament (…) 59. Calls on the Commission, when carrying out an impact assessment of its future legislative instrument, to explore, analyse and consider the implications of all possible legal solutions, such as: (…) f) creating a specific legal status for robots in the long run, so that at least the most sophisticated autonomous robots could be established as having the status of electronic persons responsible for making good any damage they may cause, and possibly applying electronic personality to cases where robots make autonomous decisions or otherwise interact with third parties independently" (European Parliament, 2017).

This demand is related to other proposals for regulating claims for damages in the event of damage caused by robots.[35] The demanded status of the "electronic person" is thus seen as analogous to the status of the legal person of partnerships or to the Anglo-Saxon trust concept. However, the justification ("where robots make autonomous decisions") also hints at transhumanist arguments and thus goes beyond the context of claims for damages. The only demand was to examine whether such a regulation makes sense.

As shown below in chapter nine by Christoph Spennemann on intellectual property rights, there is another argument to consider regarding the issue of robot rights, namely the unresolved attribution of copyright of AI products.

An open letter from more than 150 "political leaders, AI/robotics researchers and industry leaders, physical and mental health specialists, law and ethics experts" articulated widespread opposition to this demand from the European Parliament (Nevejans et. al., 2018). The European Commission did not follow the call.

The open letter rejects the claim to give robots a legal status corresponding to a natural person, but also the claim to recognise robots as legal persons. Both demands can be discussed separately.

6.1 Artificial Intelligence as a Legal Entity?

Shawn Bayern as well as several European authors have pointed out that there are already loopholes in US and possibly also in European laws, through which an AI (which does not necessarily have to be connected to a robot) can obtain the status of a legal entity. This makes clarification of this issue all the more important (see Bayern, 2015

---

[34] Cf. with another chapter by the author: Bauberger, Stefan: Welche KI? Munich: Hanser-Verlag 2020. And a submitted contribution to the planned Handbuch KI.

[35] Cf. Chapter 9 of this book, by Christoph Spennemann.

and Bayern et.al., 2017 and Burri, 2018). Kaplan also conceives of the scenario of an AI (possibly a robot) that buys itself, in analogy to how, at the end of slavery in the USA, slaves ransomed themselves (cf. Kaplan, 2017, p. 123).

The construction of the "legal person" was created for business purposes so that not all actions of a company or association or other entities have to be attributed to the respective natural persons behind them. A legal person ensures a stability of contractual relationships beyond the respective natural persons and prevents the need for complicated division of legal obligations (e.g. in the case of a public corporation among the many shareholders). Furthermore, it enables a limitation of the liability of the respective natural persons. The goals of the legal person, generally economic goals, are given to it by the natural persons, but are then assigned to the respective legal person. Every legal person needs – according to the previous concept – natural persons as decision-making bodies.

This last point is the starting point for the idea that AIs could act as electronic persons along the lines of the already established legal persons, if they can make "autonomous decisions" (formulation of the European Parliament, cf. above), i.e. if they no longer need human decision-makers. The main purpose of the electronic person would then be an appropriate regulation of liability claims in case of damages whose causation cannot be attributed to a natural person because the decisions in question do not originate from a natural person, or the attribution of copyright or patent rights to AI products.

In order to assess this concept, it must be clarified, on the one hand, what "autonomous action" by machines means and, on the other hand, whether such a regulation brings societal benefits. For this purpose, the following argument is limited to the area of liability claims.

A certain form of autonomy can be attributed to all self-moving machines. The significant difference of AI (AI in the sense of machine learning) is that it acts inherently unpredictably. The purpose of machine learning is to deploy the corresponding systems in high-dimensional state spaces that cannot, in principle, be fully anticipated. Machine learning then means either that the systems behave analogously but in a generalised manner to given training data (pattern recognition), or that they adapt their behaviour through "self-learning", i.e. on the basis of given rules and optimisation parameters.

As shown in chapter nine, when AI is used in areas with a high potential for damage, such as motor vehicle liability insurance, an (at least largely) unlimited claim for compensation by the injured parties is socially desirable, without the injured parties having to prove intentional misconduct on the part of the manufacturers or the users.

In addition to the question of liability insurance, it must also be clarified whether and to what extent product liability is possible for AI products (cf. Chapter 9). There is the argument that product liability for AI is anti-innovation, especially because the risks are not always foreseeable. In this regard, however, it should be noted that although manufacturers cannot assess these risks completely, they are in the best position of all actors to do that. They are the only ones with access to the design principles and training data for an AI. Both will generally be trade secrets. Therefore, from a macroeconomic perspective a good balance between the risks and opportunities of AI can only be enabled by extensive product liability.

It follows from these considerations that a separate legal status for the legal entity could bring advantages to manufacturers, but that these advantages would be associated with considerable disadvantages for society in terms of liability regulations. This status has the potential of socialising the risks, which no one can assess, in favour of the producers. It should therefore be ethically rejected from this perspective.

6.2 AI as a Natural Person?

It has already been mentioned above that the text of the European Parliament can also be read as being based on transhumanist arguments, as presented in chapter five. This means that because machines are capable of "autonomous decisions" and "independent interaction with third parties", no difference in principle can be found between humans and machines. In addition, there is the concept in the corresponding literature that advanced AIs can be programmed to behave ethically so that they take responsibility for their actions.

The rejection of such concepts is sometimes criticised as anthropocentric. The point of comparison invoked is the often-repeated struggle for equality throughout history, e.g. in the course of the abolition of slavery and the civil rights movement in the USA, through which personal rights and civil rights were extended further and further. In retrospect, the previous restriction of personal rights is clearly identifiable as unjust. A correspondingly backward and unjust position is assumed by advocates of personal rights for robots by those who criticise these rights. The robot Sophia was granted citizenship in Saudi Arabia in 2017, but given Sophia's real capabilities, this can only be understood as a symbolic marketing action.

The radical transhumanist view of AI (cf. e.g. Moravic, 1999; Kurzweil, 2015) even assumes that robots will overtake humans in development and thus replace them as the highest stage of evolution. When development has "progressed" to this point, it is ethically imperative that humanity relinquishes its claim to leadership and hands it over to AI (cf. Chapter 5).

Such considerations build on a reductionist view of the world and human beings. Within this view, the concept of a "higher stage of development" cannot be defined as an ethical category, but only on the basis of the factual fact of who prevails in development against other forms of life or existence. Ethically motivated demands of robot rights are therefore self-contradictory if they are based on the postulated higher development.

Two arguments still need to be considered: The autonomy of AIs and the question of whether AIs can develop consciousness.

"Autonomy" of AI means – as analysed above – practical unpredictability of action due to great complexity. If autonomy is understood in this way and put forward as an argument for continuity between human action and the "action" of robots, then this presupposes that the autonomy of action of humans is reduced to the unpredictability of this action (due to the complexity of the organism, especially the brain). This reduction contradicts the foundations of ethical action, which Kant defines as follows: "Autonomy of the will is the constitution of the will, by which the same is a law to itself (independent of all constitution of the objects of the will)" (Kant, 2012). This conception of autonomy also underlies all current legal systems, which understand the free action of man in such a way that new causes are set in it. Only then can these actions be assigned to the respective autonomous subject who bears responsibility for them.

In this respect, a justification of personal rights for AIs based on this empirical concept of autonomy dissolves the foundations of ethics and jurisprudence in their present form and must be rejected from an ethical perspective. To base an ethical claim to personal rights for robots on this is again self-contradictory.

A stronger argument for a person status of AIs is not based on this empirical notion of autonomy, but on a claimed ontological continuity between humans and AIs. In particular, it is pointed out that AIs could develop consciousness. However, this argument suffers from the problem that the existence of consciousness cannot be operationalised. Consciousness is only accessible in the perspective of the 1st person, the ego, not in an objective external view (Nagel, 1974). The concrete argument that AI can develop consciousness is based on the reductionist thesis that consciousness, including human consciousness, is a result or concomitant of complex information processing. This paradigm is widespread in the philosophy of mind. However, the concept of information cannot be justified in naturalistic terms (Janich, 2006), which means that this paradigm is based on an inversion of the logical reasoning. Furthermore, it is highly questionable to link consciousness to information processing. Rather, consciousness precedes discursive thinking and is thus not dependent on it.

6.3 Human Dignity

Ultimately, a demand for rights for robots or for AIs in general (corresponding to a person status) must be based on granting them a dignity corresponding to human dignity.

However, as Frank Schmiedchen already shows above, it is highly questionable and does not correspond to the established legal view to attach human dignity to intelligence. Mentally handicapped people would otherwise have less dignity and, in general, the dignity of people would be graded according to their intelligence. But that would be the logical consequence if robot rights were claimed on the basis of intelligent behaviour of robots.

Human dignity is based on the end in itself of human beings. In Kant's formulation, this reads as follows: "Act in such a way that you use humanity, both in your person and in the person of everyone else, at all times as an end, never merely as a means" (Kant, 2004, p. 79). This self-purposefulness also applies to other living beings in a graduated manner (or, according to certain bioethical views, in the same manner). It is based on the fact that the goals of their actions are inherent in living beings. In view of the development of robots and AIs that imitate human action, it is important, in contrast to the self-purposefulness of living beings, to strengthen the techno-ethical principle that machines do not have a purpose in themselves, that rather their meaningfulness depends on the goals that are given to them by humans, which I have already explained above. The autonomy of AIs in this context means – as shown above – only a pragmatic unpredictability of the behaviour of these machines, and it has as an ethical consequence to demand high safety standards for their development and application.

6.4 Outlook

It is to be expected that the discussion on a legal status for AI, which is just beginning, will gain in importance with the spread of applications of AI.

Firstly, the economic interests outlined above will grow to escape product liability with such a construct, especially for devices whose behaviour is not completely pre-dictable. The companies involved will claim a public interest in not falling behind their competitors in international competition in this innovative field.

Secondly, psychological research shows that people tend to project that everyday objects in their everyday world, and especially robots, are alive (cf. Darling, 2016 and other articles by this author). This natural human tendency can combine with the concept of machines as natural persons with philosophical speculations of a completely different content, which question any special position of humans from a naturalistic position.

Against these tendencies, two principles should be strengthened at an early stage. The first of these principles is that technology should not be an end in itself. Technology is created by humans to make life easier and must always be directed towards this end (cf. Chapter 4).

The second principle is human dignity and in particular the aspect that this dignity does not depend on the intelligence of the human being. This principle arises from the rootedness of this human dignity in the autonomy of the living. It leads, properly understood, in the debate on the status of AI, to humans rediscovering their kinship with other living beings, especially sentient living beings, as opposed to their supposed kinship with machines, however intelligent they may be.

Chapter 7

Liability Issues

Christoph Spennemann

A variety of liability issues can become relevant in the context of digitalisation. The chapter therefore builds on the philosophical foundations of the previous one. It focuses on liability issues in the specific context of AI and is based on German law. AI can violate tangible (life, health, property) as well as intangible legal assets (intellectual property rights).

7.1 AI and Liability for Material Damage

It is much-discussed case studies, which have already been mentioned in several places in the book or will still appear, that are relevant here: A self-driving car controlled by AI injures a passer-by against the will of the driver. A surgery performed with the help of a robot leads to an injury to the patient against the will of the doctor performing the surgery. In both cases, the AI users followed the instructions given to them. The AI was faulty, which was not recognisable to the user. Here the question arises as to who is liable for the damage that occurred. The manufacturer of the product containing the defective AI? The programmer of the software on which the AI is based? The AI user? In any case, AI is not liable itself due to the lack of its own legal personality.[36]

7.1.1 Liability of the Manufacturer or the Programmer

The basis for a liability claim by the injured party is usually non-contractual, as there is no contractual relationship whatsoever between the manufacturer and the victim of the defective AI, as in the aforementioned examples of the pedestrian injured by an intelligent vehicle and the patient injured by a surgical robot. Therefore, contract-independent claims under the Product Liability Act (ProdHaftG) and the German Civil Code (BGB) come into consideration. The ProdHaftG has introduced a so-called "strict liability" of the manufacturer, which arises even without fault only due to placing a danger on the market.[37] It is true that the prevailing opinion in research and practice assumes that AI is not to be regarded as a product within the meaning of the ProdHaftG, as this presupposes a movable object and it is doubtful whether software and algorithms in their non-physical nature fulfil this prerequisite (Federal Ministry for Economic Affairs and Energy [BMWi],

---

36　See below, Chapter 9 on Intellectual Property Rights. For more details on the problem of a separate legal personality for AI, see Chapter 6, Machine Rights.

37　§ 1 (1) ProdHaftG. Liability extends to damage to life, physical integrity, health and, with certain restrictions, property.

2019, p. 16). This means that the programmer is not liable under the ProdHaftG. In combination with a physical object such as a motor vehicle or a surgical robot, however, such a product exists in any case and the manufacturer of the motor vehicle or the robot that installed the AI can be considered liable. However, product liability presupposes the existence of a defect in the product, and this defect must be proven by the injured party.[38] This is likely to prove very difficult for an outsider in the field of AI, where not even the manufacturer may have a comprehensive understanding of the inner workings of the AI.

If the injured party succeeds in establishing the difficult proof, the product manufacturer can in turn try to exclude his liability by invoking various grounds for exculpation.[39] In the field of AI, the ground mentioned in § 1 (2) No. 2 ProdHaftG could be particularly relevant, according to which liability is excluded if it can be assumed that the product was not yet defective when the manufacturer placed it on the market. AI can change in unforeseen ways in the course of its operating time and achieve unplanned results or cause damage. Although it has been pointed out that in such cases the fault does not lie in the knowledge trained into the AI, but in the programming that was faulty from the outset, which made undesirable developments possible in the first place (BMWi, 2019, p. 17). However, cases are conceivable, in which the original programming cannot be considered faulty. Accordingly, a manufacturer can be exculpated if he can prove that the defect could not have been detected according to the state of the art in science and technology at the time the manufacturer placed the product on the market.[40]

An alternative to product liability is recourse to "producer liability" under the German Civil Code.[41] It is true that the producer is only liable here in the case of fault (intent or negligence). However, in the case of producer's liability in particular, such fault is presumed due to a reversal of the burden of proof. If the producer is unable to rebut this presumption, he is liable for the defective AI. However, this presumption only comes into effect after the injured party has first proven the defect. In this respect, producer liability presents the injured party with the same difficulty as product liability, namely, proving an error in an algorithm, the exact functioning of which may also be unclear to the manufacturer.

7.1.2 User Liability

Here, too, the basis for a claim is the tort law of the BGB. However, in contrast to the special producer liability (see above), the conventional rules of the burden of proof apply to the AI user. The injured party must therefore prove that the user acted intentionally or negligently in handling the AI.[42] If the user has adhered to the operating instructions

---

[38] § 1 (4) ProdHaftG.

[39] Cf. § 1 (2) and (3) ProdHaftG.

[40] § 1 (2) No. 5 ProdHaftG.

[41] This is an instrument developed by case law on the basis of § 823 (1) BGB (law of torts), the exact form of which is not expressly formulated in law.

[42] The driver of the vehicle is only liable to pay damages in case of fault (intent or negligence), cf. § 18 (1) Road Traffic Act (StVG).

of the AI, such proof of fault will not be possible. User liability can therefore be virtually ruled out under German law.

7.1.3 Special Case: Liability of the Vehicle Owner

In the Road Traffic Act (StVG), the legislator has established a strict liability of the vehicle owner (n.b. not the vehicle driver[43]), which applies in the case of self-propelled vehicles. Liability is linked to the existing liability insurance of all vehicle owners. As with the ProdHaftG, a motor vehicle owner is liable solely for the operation of a source of danger, regardless of his or her own fault. Unlike under the ProdHaftG, however, the injured party does not need to prove a fault in the AI. The keeper is liable solely because of his capacity as keeper. In the case of self-driving vehicles, injured parties therefore have the best prospects for compensation.

7.1.4 Contractual Recourse Claims

In the case of liability of the manufacturer or the vehicle owner, the liable party (or the insurance company that pays for the damage) will have an interest in being reimbursed for the damage incurred. Contractual claims for damages against the seller of the product containing the defective AI (e.g. a car or a surgical robot) may be considered. The latter will in turn address its own contractual partners upstream in the manufacturing chain, e.g. a car supplier, in order to obtain compensation for damages due to defective performance. At the end of the chain of recourse is most likely the AI programmer. However, a problem arises here that is comparable to the enforcement of claims from product liability: each injured party must prove the existence of a defect to the respective contractual partner ahead of him in the production chain. If the AI itself was programmed incorrectly, such proof could be very difficult. And even if the error in the AI can be conclusively shown and the claims for damages finally reach the software programmer, another difficulty arises here: programmers in many cases use so-called *open source software* (OSS), the various elements of which (and thus also incorrect programming) cannot be clearly assigned to a specific programmer. Accordingly, contractual liability of individual developers is excluded in the terms of use of OSS (BMWi, 2019, p. 17). The innovation advantages of the division of labour in OSS thus prove to be a disadvantage for contractual claims for damages.

Summary of Legal *Status Quo* and Existing Proposals for Reform

The prospects of a person injured by AI to obtain compensation are not always equally favourable. Very good prospects exist in the special area of road traffic, as the vehicle owner is subject to strict liability. Less promising are claims for damages against the

---

43 § 7 StVG.

manufacturer of AI-operated devices outside road traffic. Both the producer liability of the BGB and the product liability of the ProdHaftG require the injured party to prove a defect. Such proof can certainly succeed in the case of analogue products whose mode of operation is clearly transparent. In the case of misguided algorithms, on the other hand, this proof seems unreasonably difficult. The possibilities for exculpation of the manufacturer provided for in the ProdHaftG (in particular the proof that the product is free of defects before it is placed on the market) create additional legal uncertainty for injured parties.

Claims against users of the AI, such as a motor vehicle driver or a surgeon, appear even more difficult, as it will hardly be possible to prove fault on their part. This also applies to possible tort claims against the programmer of the AI (§ 823 BGB). Claims for damages against the programmer based on product liability are also ruled out, since software is not a product in the sense of the ProdHaftG according to today's predominant understanding.

Both the European Commission and the Federal Ministry of Economics and Technology (BMWi) have commented on questions of liability for defective AI. The BMWi considers such cases to be covered in principle by the ProdHaftG and producer liability and therefore sees no need for action by the legislator as long as there is no independent AI (BMWi, 2019, p. 19)[44]. However, in its analysis, the BMWi does not address the practical difficulties with the burden of proof that affect the injured party. This problem is discussed by the EU Commission. In a report from February 2020, it provides comprehensive considerations on the applicability of existing liability regimes to AI-powered technologies (European Commission, 2020). In this context, the Commission does not yet make any recommendations but indicates, which approaches it would like to pursue further and secure by seeking opinions. In doing so, it draws on an expert report from 2019 (European Commission, 2019). The following approaches of the Commission will be briefly discussed here:

- In order to ease the burden of proof problems described above for the injured parties of AI-operated technologies, the Commission is considering strict liability for operators of AI technologies, comparable in result to motor vehicle owner liability. Both liability of the direct operator and of persons who exercise permanent control over the function of the technology, for example through ongoing updates of the operator's software, are possible. The person who exercises the strongest control over the technology, either directly or indirectly, should be liable (European Commission, 2019). However, the Commission advocates strict liability only for those products and services that are used in public areas and may expose the public to significant risks to life, health, and property. An

---

[44] However, the BMWi advocates an expansion of the definition of error in the ProdHaftG. So far, this only covers defective programming of algorithms, but not the training of an algorithm with erroneous learning data (ibid., p. 18).

extension of strict liability to all technologies operated by AI is rejected. In this context, the Commission expresses the fear that too far-reaching strict liability could delay the introduction of new AI technologies (European Commission, 2020)[45]. The 2019 expert report also doubts whether insurance companies would be prepared to comprehensively cover the risks caused by AI technologies if the quantification of a loss and thus the sum insured should prove too complex due to a lack of experience with new technologies (European Commission, 2019, p. 61).

- For the operation of all other AI applications, which in the Commission's view "are likely to constitute the vast majority [...]", the Commission is considering adapting the rules of evidence of existing liability regimes to the specificities of new technologies:

  o In the case of product liability, the manufacturer of AI technology could, in contrast to the currently valid ProdHaftG, also be liable for such defects of the product that only materialised after the product was placed on the market, for example through subsequent updating software of the AI over which the manufacturer exercised control, or in the case of subsequent changes to the AI originally placed on the market due to the self-learning properties programmed before the product was placed on the market. Once the injured party has proven damage caused to him by the digital technology, also in derogation from the current ProdHaftG, the fault is presumed to be at the manufacturer's expense if it was unreasonable for the injured party to prove that the manufacturer failed to comply with certain safety standards due to associated costs or disproportionate practical difficulties (European Commission, 2019, p. 42). The same applies if the manufacturer fails to comply with an obligation to provide documentation that can reveal errors in the AI software (ibid., p. 47). Conversely, the Commission considers contributory negligence on the part of the injured party if the latter has failed to update the software that has damaged him in a reasonable manner (European Commission, 2020, p. 18).

  o Under fault-based liability, operators should be obliged to select the appropriate AI system, monitor it and maintain it. Manufacturers should be required to design, describe and market AI systems in such a way that an operator can comply with the above obligations. In addition, manufacturers should adequately monitor AI technology after it has been placed on the market. (European Commission, 2020, p. 44). The concrete definition of these obligations would determine the extent of the burden of proof on the injured party. If he meets this burden, fault on the part of the operator or manufacturer would be presumed. From the point of view of German liability law, this would not be an innovation with regard to the manufacturer. As described

---

45 With reference to European Commission, 2019, and the analysis found there on p. 61.

above, the ProdHaftG waives fault. The producer liability drafted by case law has already established a presumption of fault. With regard to user liability, it is also questionable whether the Commission's proposal would sustainably facilitate the evidence situation of the injured party. After all, the injured party would have to prove the AI operator's defective selection, monitoring, or maintenance of an AI system. Here, the above-mentioned approach of strict strict liability for technologies used in the public sector might prove to be much more advantageous for the injured party.

- In addition to easing the burden of proof, the Commission is considering other important changes, such as in particular an expansion of the concept of product in the EU Product Liability Directive. As already stated, the prevailing opinion in Germany is that software is not covered by the ProdHaftG. Only if the concept of product is explicitly extended to software, software programmers could also become liable to recourse, for example against a liable manufacturer who has incorporated defective software into a product that has injured the injured party.[46]

7.1.5 Comment / Recommendations for Action

Assuming that in the future AI will not only enable autonomous driving but also influence other areas of technology, it seems desirable to ensure uniform liability for material damage. Why should obtaining compensation for a failed surgery be more complicated than in the case of a car accident, even though the legal interests of the injured are the same? It is therefore obvious to introduce strict liability construed as the liability of motor vehicle owners also in areas outside road traffic. The injured party should always be able to claim this against the person who directly injured him. However, this would require the introduction of a general insurance obligation for AI-related damage, without which doctors or hospitals, for example, would hardly engage in AI-based treatment.

As there are hardly any liability cases due to faulty AI at this point in time, it cannot be clearly assessed whether the Commission's concerns regarding strict liability affecting all AI applications are justified. The example of motor vehicle owner liability shows that strict liability with simultaneous insurance obligation need not be an obstacle to the introduction of new technologies. Even the operation of a motor vehicle can result in very high losses for the insurer, without this being detrimental to the development of the car industry. However, we have to cede to the Commission that at this stage it is also unclear how well insurers will be able to determine the risk caused by largely unknown AI technologies. It would probably make sense to introduce strict liability in some areas first, as proposed by the Commission.

In the interest of effective consumer protection, these areas should cover cases where the user is potentially exposed to a high degree of harm to life, health or property, such

---

46 Cf. § 5 ProdHaftG in conjunction with § 426 (2) BGB.

as in the examples of self-driving cars mentioned earlier, but also in the use of robots in surgery and in nursing and care of the elderly. The liability requirement mentioned by the Commission for the use of AI-powered products and services in public areas must not lead to an exclusion of liability for use by private carriers or private individuals (e.g. in the use of care robots in private households). Rather, strict liability should kick in when someone operates or controls a technology that is available for use by the public.

With regard to other AI applications, i.e. if in particular there is no serious risk to life, health, or property, European product liability law should be amended in accordance with the Commission's proposals. A reversal of the burden of proof regarding the existence of a product defect in favour of the injured party seems particularly important here (see above). This would make it much easier for the injured party to sue for damages. However, it also seems fair for the manufacturer to prove contributory negligence on the part of the user if the consumer has failed to update the software in a reasonable manner.

However, one should be open to further adjustments of liability law in the sense of extending strict liability to all areas of technology if it becomes apparent that compulsory insurance for the operation of selected technologies works, i.e. does not slow down technological innovation and is also well accepted by the insurance industry.

The EU-wide introduction of consumer-friendly liability rules could, because of the size of the European market, lead to insurance companies also offering insurance for the operation of AI-based technologies outside the EU.

7.2 AI and Liability for Immaterial Damage

If intellectual property rights are infringed by AI – e.g. by unauthorised copying of copyright-protected works in the context of an AI-driven mass data analysis – the stakes are somewhat different than in the case of material damage. Unlike in the case of failed operations or a misguided motor vehicle, the AI functions perfectly, but for that very reason infringes certain intellectual property rights. The rights holder can therefore only address the person who uses the error-free AI in a way that infringes rights, e.g. a researcher who analyses masses of data for commercial purposes and infringes copyrights in the process.

In this context, there are civil law claims to cease and desist from further infringements and, in the case of culpable (i.e. intentional or negligent) infringement, claims for compensation for the damage caused by the infringement.[47] Culpable infringements of copyright and trademark rights are also criminal offences and can be punished with imprisonment.[48]

---

[47] For the field of patent law, see Section 139 (1) and (2), Patent Act. For the area of copyright law, see Section 97 (1) and (2), Copyright Act.

[48] See Sections 106 et seq. of the Copyright Act; Sections 143 et seq. of the Trade Marks and Other Distinctive Signs Act.

Depending on the individual case, the degree of independence of the AI can also cause ambiguity here: the user of the AI could object that he or she can no longer understand the internal processes of the AI and therefore acted without fault. However, this would have no effect on claims for injunctive relief that do not require fault. Moreover, the credibility of such an objection is doubtful in many cases, as the user is usually likely to be aware of the tasks the AI can take over. This usually justifies the assumption of at least negligent infringement. Thus, the applicable liability regime for immaterial damages does not currently require any special adaptation to new AI technologies.

7.3 Concluding Observations

Well-functioning liability rules are an essential component of future digitalisation. They provide the legal certainty without which manufacturers and users would hesitate to invest in or use AI-powered products. Since internal technical processes and functionalities of algorithms and software are generally more difficult to understand than in the analogue sector, existing rules must be changed to enable AI users to be compensated for the damages they have incurred. The EU Commission's proposals for an AI-oriented revision of EU product liability law are a step in the right direction. The EU should move forward swiftly to adapt its legal framework. Coordination with other global actors in international forums should be sought, but should not influence the pace of EU legal changes. Rather, as in the case of the General Data Protection Regulation, EU standards can be expected to influence corporate behaviour far beyond the EU thanks to the size of the Single Market. For example, the successful introduction of liability insurance for producers and users of AI-based products and services could encourage insurance companies to offer similar business models in other parts of the world. By moving forward decisively, the EU can set new global standards here.

Chapter 8

Norms and Standards

A. Norms and Standards for Digitalisation

Eberhard K. Seifert

Introduction

Regulations based on consensus standards are of crucial importance for the further development of digitalisation, networking and artificial intelligence (AI): because whoever sets the standards determines the market! The economic benefit of standardisation is estimated at around 17 billion Euros annually in an updated study for DIN on the first decade of this millennium (cf. Blind, K. et.al., o. J.).

At the beginning of 2018, the German Institute for Standardisation (DIN) proclaimed in a four-page position paper that digitalisation only succeeds with norms and standards – successful digital transformation through active standardisation:

> "1. norms and standards are the first means of choice to achieve technology transfer and combine it with global market penetration.
> 2. cross-industry and inter-disciplinary cooperation platforms, e.g. the German 'Industrie 4.0' platform, are particularly suitable for expanding Germany's competitive positions if sufficient attention is paid to standardisation. Policymakers should therefore support the development and use of voluntary standards.
> 3. Start-ups and Small and Medium Enterprises (SME) should be encouraged to utilise and support standardisation via government funding initiatives. This will make market access easier and also raise the existing innovation potential in terms of technology transfer" (DIN, 2018, Preamble).

Contrary to their actual enormous economic and, in some cases, societal significance, standardisation and the processes by which it comes about are hardly known and discussed – not even in research and teaching.[49] Responsible institutions and actors, governance structures, and the processes of national and international negotiations for generating, editing, or further developing standards are largely unknown.[50]

Section 8.1 of Part A of this chapter therefore first provides a brief insight into the function, structure and institutional landscape of the usual privately organised standardisation processes at national, European and international level. Based on this, Section 8.2 provides insights into (inter)national standardisation activities on Industry 4.0 and artificial intelligence. Due to the complexity and technical nature of the subject matter, only the essentials can be addressed here, but this overview is extremely important for

---

49 An exception is the European network EURAS initiated/registered in Germany (https://www.euras.org/).
50 See, for example, the instructive anthologies by Hawkins, et.al. (Eds.) (1995) and by Hesser, et. al. (Eds.) (2006).

the purpose of the book. Part A of the chapter concludes in 8.3 with references to possible restructurings of standardisation processes that have been common up to now. This is followed in Part B by an assessment of the growing geostrategic significance of norms and standards outside the standardisation organisations by Michael Barth.[51]

8.1 Insights Into the World of Norms and Standards

8.1.1 Setting Norms and Standards

Norms and standards are linguistically distinguished only in Germany, while internationally they are generally referred to as standards.[52] The scope of norms and standards set by recognised standards organisations, such as DIN in this country, lies between laws or directives on the one hand, and particular business standards or company-owned standards (such as apple plugs) on the other, and can be illustrated by the following pyramid (Fig. 8 A.1):

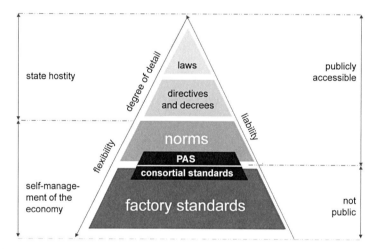

Fig. 8A.1: Pyramid of scope of standards and norms. PAS stands for publicly available specification. (Source: DIN / DKE)

A DIN standard is a document that specifies requirements for organisations, products, services, or processes. For standards to be accepted, broad participation of so-called interested parties, transparency and consensus are basic principles. In theory, all those

---

51   I would like to thank Michael Barth for his useful comments as well as Frank Schmiedchen, especially for his congenial editing and necessary but painful shortening of the chapter. More detailed explanations and additions, especially on the ongoing standardisation activities in the platform examples, are therefore reserved for a later publication.

52   All of the following descriptions of DIN and DIN/DKE/VDE and international activities are taken from the public announcements of these organisations (especially freely accessible websites) and are therefore provided without more specific references.

interested in a topic are given the opportunity to participate and to contribute their expertise to appropriate committees or to propose such committees. Before final adoption, draft standards are also made known to the public, which can formulate comments or even objections to be finally dealt with by the responsible committee. The experts involved in committees that draft standards should reach a consensus on the final content before they give their approval. Standards are reviewed at least every five years to meet the state of the art.

A pre-standard is the result of standardisation work, which is not published as a standard by DIN because of certain reservations regarding the content, because of a different drafting procedure compared to a classic standard, or because of the European framework conditions. Pre-standards offer the public the opportunity to use the results of standardisation projects that cannot be published as DIN standards, e.g. the revised DIN VDE V 0826-1 with detailed information on safety in smart home applications.

The faster way to prepare a standardisation and publish initial regulations is a DIN specification (SPEC) according to the so-called PAS (Publicly Available Specification) procedure, the content of which is developed by at least three parties (without needing consensus) and then published by DIN. It provides manufacturers with sufficient security for market testing.

DIN SPECs are listed, for example, on current topics (such as Building Information Modelling (BIM), blockchains, autonomous driving, artificial intelligence or digitalised parking). As the results of standardisation processes, DIN SPECs are strategic means of developing, establishing and disseminating innovative solutions more quickly. A SPEC should not conflict with existing standard(s). A SPEC can, however, be published as a supplement to an existing standard. A SPEC can be the basis for a new standard if it forms the basis for international standardisation projects proposed by national standards bodies (NSBs) as national preparatory work.[53]

Official standardisation, on the other hand, aims to formulate, issue and apply regulations, guidelines, or characteristics and should ideally be based on the established findings of science, technology, and experience (the so-called 'state of the art') and aim to promote benefits for society. In contrast to legally binding regulations, norms and standards are basically of a voluntary nature. Nevertheless, industrial practice shows that standards are very important and are applied for the sole purpose of securing market access for a consistent product.

Standards are only indirectly legally-binding, if laws or legal ordinances (e.g. EU directives) adopt them or if contracting parties make the application of standards binding under private law in agreements. Even in cases where DIN standards have not been made part of a contract, or not prescribed by law, they can serve as an aid in dispute settlements (e.g. in questions of liability under warranty law, tort law or product liability law), in order to assess whether a manufacturer has complied with

---

53   internationally then referred to as PAS

the generally recognised rules of technology and has applied reasonable care (see also Chapter 7 above). Compliance with standards thus offers a certain degree of legal protection against possible liability. For this reason, voluntary regulatory procedures have proven their worth. Technical norms and standards are increasingly being developed and applied worldwide. Historically, norms and standards functioned as pure technical agreements. In the last decades, an increasing number of standards have included socio-political aspects (such as on the topics of quality, environmental protection, and sustainability).

Founded in 1917, DIN is the most important national standards organisation in Germany. In addition, the German Commission for Electrical, Electronic & Information Technologies (DKE), as an organ of DIN and the Association for Electrical, Electronic & Information Technologies (VDE), is the organisation responsible in Germany for the development of standards, norms and safety regulations, especially in the field of digitalisation.

As a platform for electrotechnical standardisation projects, DKE is the central competence centre for electrotechnical standardisation and is therefore responsible for representing German interests in the European and international standardisation organisations. These private institutions provide the administrative and logistical preconditions for standardisation activities, while technical content of norms and standards are developed by external interest-driven experts, who traditionally come from industry and bring their respective company or association interests to reach balance of consensus.

Depending on the topic, other stakeholders may also participate in such committees (e.g. government, trade unions, academia, standards users, consumers, and/or civil society). The national committees (usually a maximum of 21 members) also represent the so-called mirror committees for international standardisation projects and nominate the national delegates to the international bodies from their ranks. In this respect, it is a kind of bottom-up approach to sub-governmental regulation. Since the early 1990s, socially relevant interests (environmental protection, climate change, and other sustainability issues, especially adressing the UN SDGs (Sustainable Development Goals)) have increasingly being considered in standardisation activities, in addition to technical and economic aspects.

8.1.2 Institutional Interlocking at European and International Levels

At the European level, there are three major European Standards Organisations (EOS): CEN (European Committee for Standardisation), CENELEC (European Committee for Electrotechnical Standardisation) and ETSI (European Telecommunications Standards Institute). As a politically important European speciality, it should be emphasised that CEN can also implement legally binding, so-called harmonised standards at the instigation of the EU Commission (see below). An overview shows the mirror structure of national responsibilities on the national, the European and international levels:

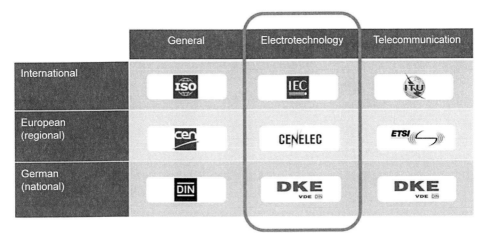

Fig. 8A.2: Overview of national and international standards organisations with incorporated-framed responsibility of DKE / CENELEC / IEC, technical focal points relevant to digitalisation. (Source: DIN/DKE)

- Founded in 1946, the International Organisation for Standardisation (ISO) is the central global association of standards organisations.[54]
- However, it is not responsible for the areas of digitalisation and electronics, which falls within the purview of the International Electrotechnical Commission (IEC).
- The development of telecommunication standards falls within the purview of a UN organisation, the International Telecommunication Union (ITU).

Together, these three organisations, ISO, IEC, and ITU form the WSC (World Standards Cooperation).

In accordance with the Vienna and Frankfurt Agreements on Technical Cooperation between ISO and CEN, or IEC and CENELEC, duplication of work is to be avoided and simultaneous recognition as an International Standard and as a European Standard is to be achieved. Priority is given to ISO and IEC. In general, the principle of country representation applies to ISO participation.

8.1.3 European Standards System: CE Marking and Harmonised Standards

The EU is a globally unique special case with its own standardisation system, which is currently also highly relevant for the world's first legal regulation of AI applications (see section 8.3). The former "New Approach" (NA) of the European Community (1985), which was updated in 2008 by the New Legislative Framework (NLF), pursues in the EU the basic approach of a state relief[55] for the development of a uniform economic

---

54 DIN has been a member of ISO since 1951, being founded in the former Federal Republic of Germany.
55 Basic idea of the New Approach, i.e. clear separation between sovereign legislation and private standardisation.

and goods traffic area as well as for the marketing of products on the basis of (EEC, EC and today EU) so-called directives and the associated CE marking. Not all technical details should and can be regulated top-down in detail for all member states: Only the framework conditions can be decided at the EU level, while the designs are the responsibility of the private EOS.

Today, there are approximately 30 EU directives covering more or less large product groups (e.g. electrical equipment, medical devices, hazardous substances in electrical appliances, energy-related products). They are limited to so-called basic safety requirements, while the technical specifications of the contents for the respective products are to be defined by technical rules (standards). The above-mentioned European standardisation organisations CEN, CENELEC, and ETSI are responsible for the development of these standards. The CE marking has been defined with regard to the elements and the individual modules of the conformity assessment: All actors acting in supply and distribution chains must ensure that only products that comply with the applicable legislation enter the EU market (this is the placing on the market). The CE marking thus serves to control the permissible marketing of products and thus represents a kind of passport for products in the European internal market. However, it is not a seal of quality and is not aimed at end consumers.[56] It is not maintained by DIN.

Here, the otherwise unusual concept of presumption of conformity plays a central role. The manufacturer or the person placing the product on the market (e.g., in retail) marks the product clearly visible with a CE label. In doing so, it states that the product fully complies with all corresponding requirements. The presumption of correctness of this statement is now automatically assumed (keyword: presumption of correctness). If it subsequently turns out that this is not the case or that damage has occurred, this has immediate legal consequences.

European standards (EN) are regulations that have been ratified by one of the three European committees for standardisation mentioned above. All ENs are the result of a public standardisation process. Here, too, the process begins with a standardisation proposal submitted by a national member of the European standardisation organisations, such as DIN, or by the European Commission or by European or international organisations. The national standards organisations decide on adoption as a European standard in a final vote lasting two months. For adoption, at least 71 % of the weighted votes of the CEN/CENELEC members are required (unlike in ISO, where the rule is: one country, one vote). Ratification of a European standard takes place automatically one month after a positive voting result. Thereafter, a European standard must be adopted unchanged as a national standard by the national standards organisations. Conflicting national standards have to be withdrawn to avoid double standardisation.

---

56 The CE marking is not a mark of conformity with standards, but an EU directive conformity mark with a function as a supervisory mark, which is intended, for example, to make it easier for the trade inspectors in the EU countries to check whether the products are being marketed (placed on the market) in a permissible manner.

In Germany, for example, each adopted European standard is published as a DIN EN standard with a national foreword. The national foreword serves as an additional source of information on the respective standard for the user of the standard and is prepared by the responsible German mirror committee.[57]

Harmonised and Mandated Standards

The term (European) harmonised standard (eHN) has a definition established by the European Commission in the context of the New Approach (see above):

- the standard has a mandate or standardisation request from the European Commission and EFTA to CEN, CENELEC or ETSI, and
- the reference of the standard has been published by the European Commission in the EU Official Journal.

All standards prepared by CEN, CENELEC and ETSI are the result of European harmonisation and in this sense are harmonised throughout Europe. However, only those which fulfil the two above-mentioned requirements are to be regarded as eHN within the framework of an EU directive and the legal definition of the European Commission. Mandated standards result from a politically motivated and (co-)financed order (mandate) of the EU Commission to develop certain European standards. For more than 4600 standards, the EU and EFTA have issued so-called mandates or standardisation directives to CEN, CENELEC and ETSI, largely under directives issued under the former NA and NLF, respectively.[58]

8.2 Norms and Standards for Digitalisation

In Germany, the German government is promoting specialist discussions on the digital transformation, also with a view to setting corresponding standards. To this end, the German Federal Ministry for Economic Affairs (BMWi) has co-initiated ten platforms with respective focus groups, which are intended to develop the digital policy challenges as well as approaches to solutions from a technical and political perspective with a view to necessary standardisation activities for the digital transformation in business and society.[59]

---

[57] https://de.wikipedia.org/wiki/Europäische_Norm

[58] https://www.eu-richtlinien-online.de/de/informationen/harmonisierte-und-mandatierte-normen

[59] Platform 1 "Digital Networks and Mobility," Platform 2 "Innovative Digitalisation of the Economy," Platform 3 "Industry 4.0," Platform 4 "Learning Systems," Platform 5 "Digital World of Work," Platform 6 "Digital Administration and Public IT," Platform 7 "Digitalisation in Education and Science," Platform 8 "Culture and Media," Platform 9 "Security, Protection and Trust for Society and the Economy," Platform 10 "Consumer Policy in the Digital World." The 14th annual 'Digital Summit' Nov./Dec. 2020 was, due to Corona, the first purely virtual one – see also the extensive program of lectures and 3 forums in the brochure of the BMWi: Digital Gipfel 2020 – Digital nachhaltiger Leben (Berlin, Nov. 2020); the still physical 2019 Summit in Dortmund (end of Oct.) under the motto 'PlattFORM die Zukunft' in Zeiten der Plattformökonomie, had accordingly an annual focus topic 'Platforms', introduced since 2016, see program brochure of the BMWi (Berlin as of Oct. 25, 2019).

Platforms 3 (Industry 4.0) and 4 (Learning Systems), which are of particular relevance to this book, can only be presented exemplarily in some detail.[60]

8.2.1 Industry 4.0

The BMWi has committed to joint European projects, for example on the topics of microelectronics and communication technologies, which have various standardisation interrelationships. The first edition of a so-called "Industrie 4.0" standardisation roadmap was published in 2013, the fourth and so far last Industrie 4.0 in March 2020 (DIN/DKE, 2020).[61] This explains the cooperation with the German "Plattform Industrie 4.0", with regards to the fundamental innovation and transformation approach of industrial value creation and the 2030 mission statement for Industrie 4.0 (DIN/DKE, 2020, p. 12 f.). The Industrie 4.0 platform was founded by three German industry associations (BITKOM, VDMA, ZVEI) and is headed by two German ministries. In addition to standardisation, there are also the fields of action research and innovation, security of networked systems, legal framework conditions, and work and training. DIN is involved in contributing results at the international level.

The following diagram (Fig. 8A.3) shows the interconnectedness of DIN's activities of the key players shaping the digital transformation for Industrie 4.0: Plattform Industrie 4.0, Standardisation Council Industrie 4.0 (SCI 4.0), and Labs Network Industrie 4.0 (LNI 4.0).[62]

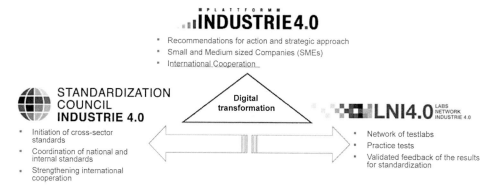

Fig. 8A.3: Interplay of central standardisation actors in Industrie 4.0. (Source: DIN[63])

The standardisation roadmap (NRM) addresses the implications of the fourth industrial revolution on the organisation and management of the entire value chain. The techno-

60 On the socially controversial topic of platform 1 'mobility', for example, Sabautzki (2020) provided a political critique regarding 'lobbying'.
61 Only exemplary outlines can be given here for this publication of around 140 pages.
62 https://www.sci40.com/sci-4-0/über-uns/
63 https://www.din.de/de/forschung-und-innovation/themen/industrie4-0/arbeitskreise

logical merging of IT (Information Technology) and OT (Operational Technology) is leading to the overlapping of previously separate areas of standardisation.[64]

It is assumed that issues, requirements, and working methods that were previously relevant to the information and communication technologies sector are now increasingly affecting all sectors. The central questions here are: What will a global digital value creation system look like? How can the normative framework conditions for this be identified and implemented? The 2030 mission statement of the Industrie 4.0 platform proposes a holistic approach to the design of so-called digital ecosystems. The goal of Industrie 4.0 is to replace rigid and firmly defined value chains with flexible, highly dynamic, and globally networked value networks with new types of cooperation. Based on the specific prerequisites and traditional strengths of Germany as an industrial location, the framework of a future data economy is to be reconciled with the requirements of a social market economy. This model primarily addresses Germany as an industrial and business location, but explicitly emphasises openness and an orientation toward cooperation with partners in Europe and the rest of the world.

Three strategic fields of action and their close interconnection are postulated as particularly central to the successful implementation of Industrie 4.0: sovereignty, interoperability, and sustainability. With the BMWi prestige project GAIA-X[65], the Industrie 4.0 platform has proposed a basis for a distributed, open data infrastructure for the whole of Europe based on European values. This project is being pursued in SCI 4.0 (above diagram) with the aim of promoting interoperability. This is intended to ensure networking across company and industry boundaries, for which standards and integration are necessary. In addition, a uniform regulatory framework on decentralised systems and artificial intelligence is required so that companies and business models from Europe can be globally competitive. This is important for digital sovereignty and, above all, also offers trustworthy securities for users of cloud services.

The former German Minister for Economy Peter Altmaier saw in this comparatively fast-moving initiative opportunities for an export hit for Europe, whose open ecosystem would also be joined by non-European users.[66]

Six working groups have been set up to implement the entire NRM Industry 4.0.[67]

Here, for example, WG 1 Reference Architecture, Standardisation formulates its approach as follows: Standardisation in industry is not a new phenomenon, but Industry 4.0 is bringing about a major change. Standards that regulate a small section of production would no longer be sufficient for networked production. In the standards, hardware and software, user and supplier industries, and product design to recycling

---

64 https://www.plattform-i40.de/PI40/Navigation/DE/Industrie40/WasIndustrie40/was-ist-industrie-40.html

65 (https://www.bmwi.de/Redaktion/DE/Dossier/gaia-x.html); (https://www.plattform-i40.de/PI40/Navigation/DE/Industrie40/ WasIndustrie40/was- ist-industrie-40.html)

66 Thus again in the panel of the BDI Industry Day on 22 June 2021 'How can Gaia-X and European data spaces promote innovation?' (live-stream: www.bdi.eu/tdi)

67 https://www.din.de/de/forschung-und-innovation/themen/industrie4-0/arbeitskreise

must be thought together. This is the only way that different components in digital ecosystems can work together smoothly (keyword: interoperability). There are two focal points for this work:

a) on the part of the "ISO/IEC Joint Working Group21" (ISO/IEC/JWG21) for the harmonisation of Industry 4.0 reference models with a Technical Report (TR);

b) regarding the adoption of the standardisation proposal for the administration shell by IEC/TC65 to set the course for this to become the central integration plug for so designated digital ecosystems.

on a) Industry 4.0 reference architecture model – RAMI 4.0

The interaction and communication between factories with their machines transcends company and enterprise boundaries. For this reason, production companies from different sectors are to be comprehensively networked with suppliers, logistics companies, and others in a value creation system. To achieve this, interfaces must be harmonised. This in turn requires internationally coordinated norms and standards for these interfaces.

A reference architecture model is intended to provide a uniform conceptual and methodological structure as a basis for the experts involved from the various disciplines to master complexity and speak a common language. It creates a common structure for the uniform description and specification of concrete system architectures. The Reference Architecture Model Industry 4.0 (RAMI 4.0) developed for this purpose in Germany represents such a model. This model has already been successfully introduced into the international standardisation landscape and published as IEC PAS 63088.

on b) Management shell: structure and its submodules

To ensure semantic interoperability[68] of hardware and software components in production (machines, stations and individual assemblies within machines), the concept of the management shell was developed in Germany. In order to help the management shell to achieve a breakthrough in international standardisation, the concept was pre-coordinated with partners from France, Italy, and China, among others, under the coordination of SCI 4.0. With the acceptance of the standardisation application for IEC 63278-1 ED1 Asset administration shell for industrial applications – Part 1: Administration shell structure at IEC/TC 65, a first step was taken to make the administration shell the central integration plug for the digital ecosystems described in this way.

The GoGlobal Industrie 4.0 funding project also supports the global harmonisation of national Industrie 4.0 concepts through the SCI 4.0. International collaborations that go beyond cooperation on the previous topics are intended to take account of the profound changes in organisational and value creation structures in the fourth industrial revolution (cf. Chapter 14).

---

68 Definition of suitable data structures for the exchange of data and their defined meaning.

In view of the high level of interest in AI, an Expert Council for AI in Industrial Applications[69] was established in 2020 by the SCI to serve as a hub for normative discussions and coordination in the designated area, both nationally and internationally.

8.2.2 Platform Learning Systems – Artificial Intelligence – Standardisation Roadmap AI

This 4th platform was initiated in 2017 by the Federal Ministry of Education and Research (BMBF) at the suggestion of the Autonomous Systems Expert Forum of the High-Tech Forum and the German National Academy for Science and Engineering (acatech). It is intended to serve as a venue for exchange and to promote the implementation of the German government's AI strategy.[70]

In seven thematic working groups, opportunities, challenges, and framework conditions for the development and responsible use of learning systems will be addressed. In addition to specific fields of application such as medicine and mobility, the focus is also on cross-cutting topics such as human-machine interaction or legal issues. At the beginning of 2018, the interdisciplinary Artificial Intelligence Working Committee was founded at DIN. In addition, DIN, together with DKE, prepared a white paper on "Ethics and Artificial Intelligence: What can technical standards and norms achieve?" on behalf of the BMWi. (DIN/DKE/VDE-without year). On Aug. 1, 2020, a steering group for a standardisation roadmap on AI was established under the leadership of the BMWi and DIN to pave the way for the expansion of Germany as an AI location.

The setting of norms and standards also plays a central role in the German government's artificial intelligence strategy[71] at all three levels (national, European, and international), although this is primarily seen as a task for industry (BuReg, 2020, p. 41). In its update "Fortschreibung 2020" (Dec. 2020), the government announces its mandate to DIN/VDE (DKE) to have a comprehensive standardisation roadmap for AI drawn up, which should be presented together with company representatives, trade associations and leading scientists at the Digital Summit 2020 (BuReg, 2020, p. 21).

This NRM AI is to be a central building block in the AI strategy in order to open up international markets for German companies and their innovations. To this end, the NRM AI is to provide an overview of existing norms and standards on AI aspects and, in particular, make recommendations with regard to future activities that are still necessary. It will be compiled by the respective interested parties from business, science, the public sector, and society. And DIN is to organise this cooperation in the sense of a neutral platform. In this respect, DIN is also to implement the German government's AI strategy. Standardisation should promote the rapid transfer of technologies from research to application.

---

[69] See https://www.sci40.com/themenfelder/ki-k%C3%BCnstliche-intelligenz/
[70] (https://www.plattform-i40.de/PI40/Redaktion/DE/Downloads/Publikation/Leitbild-2030-f%C3%BCr-Industry-4.0.html)
[71] www.ki-strategie-deutschland.de/2018 74 www.din.de/go/normungsroadmapki

The initial results of the BMWi-funded AI standardisation roadmap were presented at the Federal Government's Digital Summit in November 2020 and handed over to the Federal Government as an overview of the status quo, requirements and challenges, as well as standardisation needs for seven key topics: Foundations, Ethics/Responsible AI; Quality, Conformity Assessment, and Certification; IT Security in AI Systems; Industrial Automation; Mobility and Logistics; and AI in Medicine.[72]

To this end, five overarching and central recommendations for action were named (full recommendations and action requirements in the AI standardisation roadmap), the implementation of which is intended to build trust in AI and support the development of this future technology:

- implement data reference models for AI system interoperability;
- create horizontal AI basic safety standard;
- design practice-oriented initial criticality testing of AI systems;
- national implementation program "Trusted AI";
- analyse and evaluate use cases for standardisation needs.

For this purpose, identified needs for topics (such as fundamentals or ethics / Responsible AI) were discussed in each case in workshops and participants as well as other interests were invited (at: Kuenstliche.Intelligenz@din.de) to participate in their implementation, i.e., in national investigations in further processes for which no times or processes had yet been announced at the time of writing.

8.2.3 International Standardisation Approaches on Artificial Intelligence

With a 35-member DIN mirror committee, Germany is thus represented on various ISO committees on AI. The responsible ISO subcommittee ISO / IEC JTC1 / SC42 (established in 2017) has already developed and published eight standards on AI. Another 22 are under development. The secretariat is provided by the US-American National Standards Institute/ANSI and a US representative also serves as chair. It is worth mentioning that at the international level (by ISO and IEC), the first AI-specific document published by these bodies in a joint JTC 1/SC42 was a Technical Report (TR) on AI and trustworthiness (ISO/IEC TR 24028 Information Technology – Artificial Intelligence- Overview of trustworthiness in artificial intelligence), with the aim of supporting standardisation activities on gap identification.

It should also be emphasised that recently, for the first time in this field, a so-called management system standard (MSS) has been prepared for a future ISO 42001, whereby the end number "1" indicates the special feature that only these few MSS in the respective series of standards with requirements for external certification have been prepared to date and are therefore regarded as flagships.[73]

---

[72] www.din.de/go/normungsroadmapki
[73] The annual ISO surveys on generic MSS such as ISO 90001 on 'Quality', ISO 14001 on 'Environment', ISO 50001 on 'Energy' as well as on individual sector-specific MSS, e.g. in the automotive sector, show

World's First Proposal by the EU Commission on the Artificial Intelligence Act (AIA)

On April 21, 2021, the EU Commission presented a world-first regulatory proposal for AI in the form of a Proposal for a Regulation of the European Parliament and of the Council laying down harmonised regulation on artificial intelligence (EU COM, 2021). With this, the EU COM intends to prescribe basic requirements for AI systems, for example on risk management, transparency, robustness, IT security, and human supervision of AI.

The regulation would have immediate legal force and implies the corresponding mandates for harmonised European standards within the framework of the NLF mentioned above (see section 8.1) by the European (private) standardisation organisations. In this respect, this Commission initiative is of extraordinary, strategic importance both for the EU and, where appropriate, for international follow-up activities.

The AIA proposal expresses a risk-based approach that distinguishes four categories:

- risk-free AI systems should not be regulated;
- low-risk applications (e.g., chatbots) are to meet transparency requirements;
- dangerous AI applications (e.g., social scoring) are to be banned;
- the intermediate, high-risk AI applications are to be marketed according to the principal of the NLF are to be placed on the market.

For the technical concretisation of such requirements, reference is made to harmonised European standards (hEN), which are to be developed by the EOS (CEN/ENELEC and ETSI) on the basis of a corresponding standardisation mandate of the EU Commission. If these European standards are complied with by manufacturers, the following will apply, as stated in section 8.1: it is presumed that this would also meet the requirements of the AIA, i.e. the so-called presumption of conformity exists. With the required, visible affixing of the CE marking, the conformity of the manufacturer or the distributor with the applicable legal act and the corresponding eHN is declared and the product is placed on the European internal market with the CE marking. In this respect, therefore, the familiar procedure for the European system of harmonised standards with then considerable, worldwide first-time requirements for manufacturers of AI applications. DIN and DKE commented on this in a joint position paper on June 9, 2021 (DIN/DKE, 2021) under the heading: "Standards as a central building block of European AI regulation". The paper welcomes the fact that the Commission is following the NLF with this regulatory proposal, which has proven its worth for the European single market. As usual, all interested parties can participate in the standardisation processes to be set up for this purpose, which contributes to a high level of acceptance of standards on the market and at the same time prevents or helps to reduce non-tariff barriers to trade, since all national NSBs undertake to adopt these

---

considerable global certification figures, especially for the generic ones (ISO 2019), which at the same time indicates the considerable business field activities that are triggered and associated with such MSS for consultants, certifiers, accreditation bodies.

standards unchanged in their respective national body of standards and thus also to withdraw any conflicting standards.

In addition, however, demands are also made in the AIA proposal:

- continuation and further development of the cooperation between the standardisation bodies and the EU Commission;
- the timely development of standardisation requests for AI by the EU COM in cooperation with the European Standardisation Organisations;
- the deletion of Article 41 "common specifications", which is included for a legal framework on AI and is criticised as non-specified enacting legislation[74];
- the involvement of the European standards organisations in the work of the 'European Artificial Intelligence Board' to be established.

From DIN/DKE's point of view, the German Standardisation Roadmap AI could serve as a basis for forthcoming standardisation work. The overview of existing norms and standards contained therein, as well as the listing of further standardisation requirements, is described as a basis for forthcoming standardisation mandates. Practical work on the implementation of the AIA will have to show whether DIN/DKE will be successful with this. In any case, the AIA represents a major challenge for the interested expert groups to assume ethical and socio-political responsibility far beyond purely technical issues. This opens up an exciting field of observation and possibly also of activity for the Federation of German Scientists (VDW).

8.3 Quo Vadis Digitalisation Standardisation?

This chapter 8A has provided an inside view of the complex world of standardisation and has shown why standards and standard setting are also central to further digitalisation. This was highlighted by the comments on Industry 4.0 and artificial intelligence. Initial experience shows trends for standardisation and digitalisation:

At the national (DIN) level, the traditional rule-based standardisation bodies and organisational processes can continue to be used effectively, also to successfully introduce German considerations into international (ISO) processes in preparation, which is also supported by officials. However, the traditionally private standardisation activities with their non-political self-organisation on the part of industry are being supported and promoted by the state to an unprecedented extent in the case of digitalisation topics, and the original bottom-up self-organisation of individual economic interests is thus being transformed into top-down influence by politically organised and controlled governance structures.

The accompanying politicisation through the broad involvement of actors in the sense of multi-stakeholder processes is, on the one hand, democratically welcome, but on the other hand also poses problems in consensus-finding processes that should not be underestimated in the highly competitive environment, or raises suspicions of po-

---

74 21-06_DIN_KE_position paper_Artificial Intelligence Act.pdf

litical legitimisation or showcase events or even lobbying activities, as criticised in the example of the Mobility Platform (Sabautzki 2020).

There could be a tendency for a kind of culture to develop in ISO bodies, especially in the case of AI topics, to generate and already implement new work topics and projects themselves from within the existing committees, instead of developing them from proposals from individual NSBs for international coordination processes, as envisaged. This can be countered for the corresponding ISO working structures by sufficient national active participation and NWIP submissions/voting. Germany should implement the self-confidence it has expressed with regards to both Industry 4.0 and AI by consistently representing its interests, especially in the EU framework.[75]

So far, European standardisation by CEN/CENELEC has not yet played an important role for digitalisation standardisation (also due to avoidance of duplication of work according to the Vienna Agreement on ISO activities), but with the EU Commission, a political actor and standardisation mandator is present who can pursue the above-mentioned legal powers, as now shown by its new AIA initiative, which takes up the socio-political mega-topic AI. Strengthening the European standardisation system is as imperative for the EU as an economic superpower with the largest single market in the world, as it is for achieving the EU's goal of developing a digital and sustainable union.

In general – without closer insights into other European countries and their national standardisation practices – it can be asked whether standardisation processes on such cross-cutting issues affecting the entire economy and society as digitalisation in Germany and at EU level could tend to outgrow the traditional private organisational processes and structures in favour of a hybrid character of state-political use and influence? Or can control by a private quasi-sub-state and voluntary self-regulatory instrument like the traditional private organisational structures of standardisation be preserved?

In any case, the offensive efforts of international competitors, especially the USA and China, to assert their national standards on economic players in other countries require greater vigilance, precise analyses, and appropriate measures. The question is which structures and processes help to assert the interests of the EU more effectively within Europe by strengthening the European standardisation system. But how can European interests also be strengthened through broad participation and assertion in international standardisation processes? In this respect, it can be seen as a political exclamation mark that on June 7, 2021, a public hearing was held by the Foreign Affairs Committee of the German Bundestag on "Innovative Technologies and Standardisation in a Geopolitical Perspective."[76][77]

---

[75] Wahlster, head of the steering group for the AI standardisation roadmap, confidently states: "AI research in Germany is among the global leaders. Norms and standards pave the way for developing innovative products from the results, which can become export hits for our economy" (Wahlster: 'Künstliche Intelligenz: Ohne Normen und Standards geht es nicht', https://www.din.de/de/forschung-und-innovation/themen/kuenstliche-intelligenz).

[76] https://www.bundestag.de/auswaertiges#url=L2F1c3NjaHVlc3NlL2EwMy9BbmhvZXJ1bmdlbi84NDM2MjgtODQzNjI4&mod=mod538410.

[77] HBS 2020: Technical standardisation, China and the future international order – A European perspective

This forms the bridge to more geostrategic considerations with regard to norms and standards, especially outside the international rule-based standardisation processes, as will be presented in the following Part B of this Chapter 8.

Postscriptum to Englisch Edition

- Meanwhile, the DIN organised some meetings and communications on the Chinese ISO-strategies and practices, see: S. Gabler: China und die Normung. Ein Rückblick auf die DIN/DKE-China-Frühstückreihe, in: DIN-Mitteilungen 1/2022 (pp. 11–18)
- Further, referring to the new EU-Commission 'Standardisation Strategy', presented 2 February 2022 (https://ec.europa.eu/commission/presscorner/detail/en/ip_22_661), DIN published a first evaluation by S. Gabler: 'EU-Standardisierungsstrategie veröffentlicht: viel Licht und wenig Schatten', in: DIN-Mitteilungen 3/2022 (pp. 7–9)
- Concerning the German activities on AI standardisation, DIN/DKE is conducting a broad multi-stakeholder process in 2022 on the second AI roadmap in eight working groups (three of them new to the first one on: 'Socio-technical systems', 'Financial services' and 'Energy and Environment') and additional subgroups. This roadmap will develop guidance for practical standardisations and will be presented to the next 'Digital Summit' in November 2022.

B. Standardisation as a Geopolitical-Technological Instrument of Power

Michael Barth

The world of standardisation, which was originally remote from politics, has increasingly developed into an instrument of global influence, especially in connection with the increasing comprehensive networking through information and communication technologies. This is certainly due, on the one hand, to the already mentioned universal character of these technology areas, as they are cross-sectional technologies that affect not only individual sectors, but the entire economy, the state and society, and, on the other hand, certainly also to the underlying rules of information and communication technology concerning market and investment dynamics:

- high innovation promises and strong growth rates entail disproportionately higher investments than in classic industrial sectors;
- rapid growth leads to market dominance ("The winner takes it all");
- lock-in effect of the user as part of many business models.

In this context, the momentous differences between norms and standards in areas of information and communication technologies and standards in other technology areas must be carefully analysed. One major difference, for example, is that the effects of the extensive use of digital technologies that cannot be controlled nationally lead to dependencies much more quickly than would be the case, for example, with mechanical engineering components. The example of services available via cloud technologies illustrates this particularly clearly in comparison to mechanical engineering components that are important to industry: Once components have been installed in systems, they can easily run for several decades with the corresponding mechanical maintenance cycles. With added services using digital remote maintenance options, on the other hand, it is easy to exclude these components from predictive maintenance and control mechanisms for maintaining performance from one moment to the next. This becomes even clearer with purely digital services, which exist in B2C business just as much as in B2B. When payment ends, access to the service and thus also to the data used ends. This also occurs when systems have to be shut down due to an accident (or hacker attack). It is no coincidence that an important requirement of business customers for their service provider is data portability to other services, which global players in the digital industry often know how to prevent for lack of an alternative.

In terms of security policy, this situation is made transparent by the example of sanctions or export embargoes: machines can continue to run for some time after sanctions have been triggered, and with a lack of spare parts, their performance decreases over a longer period of time. In the case of digital products, access ends from one moment to the next with the decision of the providing company or even the state behind it, which makes sanctions much easier to enforce and also more effective.

The control of standardisation and thus the sustainable influence of standards in one's own interest is thus suitable as a promising and effective instrument of power politics, especially in the field of digital and information technologies (cf. Rühlig, 2021).

These connections manifest themselves in the struggle for digital sovereignty, which has now become a global phenomenon, especially in those regions of the world that are not home to global players in the digital industry. They are also evident where "cultural areas" of digitalisation meet, for example the United States of America and the European Union or the United States of America and the People's Republic of China. Here it can be seen that these states are trying to establish their own digital "cultural spaces", to demarcate them from one another and ultimately to expand them. This happens, for example, through regulation or social norms that attempt to control the use of digital technologies and services nationally, or to impose conditions that promote or force adaptation to the respective "digital cultural space". At the European level, for example, the General Data Protection Regulation (GDPR) should be mentioned here. In the People's Republic of China, this is done through cybersecurity legislation, which promises the state and its administration full control over the technologies used. In the USA, for example, this is reflected in the procurement power of the public sector, which relies on "Buy American" not only for digital products and categorically excludes suppliers from other legal systems. Variations of these tendencies can be found in all the systems discussed here.

This clash of systems is evident not only in trade policy, technological, and cultural approaches, but also in the area of standardisation considered in more detail here. Europe, as in many other subject areas, relies on a multi-stakeholder approach. As described in Section A by Eberhard Seifert, the European approach is strongly rule-based. Standards and norms are first discussed and drawn up at national level, then negotiated and adopted at European level, and only then taken to international bodies.

Basically, the approach in the United States of America corresponds to the European approach. Here, too, there are large standardisation organisations that carry American developments to the international level. The main difference, however, is that large bodies such as the American National Standards Institute (ANSI) certify other bodies as standardisation bodies, thus creating a multitude of different standardisation organisations, currently more than 600 at ANSI alone. Each of these bodies can submit standardisation proposals which, if they meet certain requirements, first become a national standard and are then brought to the international level. In this way, ANSI wants to create as much inclusivity as possible and also represent the diversity of the US economy. At the same time, the standardisation umbrella organisation ANSI also has the official task of representing US interests in important markets such as China or Europe, which means that an increasing mix between politics and standardisation can also be seen in the USA. This is also reflected in the historical function of standardisation organisations in the USA. For example,[78] the National Institute of Standards and Technology was founded at the time to counteract the economic and technological superiority of Great Britain, Germany, and other competing systems.[79]

However, the practice of numerous smaller certified standardisation organisations also makes it easier for large, market-dominant companies to control the activities of individual standardisation organisations and to influence their results more strongly in their own interests. Thus, increasingly strong individual companies or consortia in the ICT economy determine the standards to which economic participants must orient themselves, especially if they have an interest in interoperability with the market leader. The United States thus relies here on the innovative power of the IT companies based there, which are pioneers in many areas.

This means that digital technologies in many fields already elude the traditional and practised mechanisms and the committee culture that are usually found in the field of standardisation. Although the creation of norms and standards is still important here, the consensus within the community is much more strongly brought about by the respective market dominator or industry consortia. For example, it has a strong influence on the usability of certain certificate types in browsers when Alphabet stipulates which

---

[78] On the role and self-image of ANSI see: https://ansi.org/about/roles
[79] https://www.nist.gov/about-nist

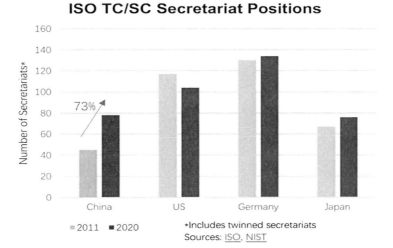

Fig. 8.B.1: Comparison of the number of staffed secretariats in international standardisation bodies by China, USA, Germany and Japan in 2011 and 2020.

certificates they accept for the Chrome browser. Meanwhile, this mechanism can also be seen in concrete national attempts to influence standardisation processes.

One example that illustrates this particularly well is the People's Republic of China. Formerly a pure production location for the world, it has outgrown this status both in its self-image and in its external perception. Its global power-political ambitions can be seen not only in its exchanges with other states, in the development and securing of raw materials worldwide, or in its dealings with influential technology providers (cf. Bartsch, 2016). They are also reflected in the world of standardisation. For example, the number of secretariats and chair positions occupied alone shows a strong increase in Chinese ambition (cf. Rühlig, 2020).

This is further underlined by the consistent application of Chinese representatives for vacant leadership positions in international standardisation bodies (cf. Steiger, 2020).

This finds an even stronger expression in the strategic documents derived from the determinations of the medium and long-term plans of the state leadership. In direct relation to our topic, for example, the national standardisation authority has derived a Standardisation Strategy 2035. "China Standards 2035" is directly related to the national industrial strategy "Made in China 2025" and is intended to support and secure it in the medium term. All in all, these development plans are embedded in the proclaimed goal of the People's Republic to be a world leader "politically, culturally, ethically, socially and ecologically" by 2050, the 100th year of its founding. China's head of state Xi Jinping formulated that the country must be the global leader in innovation by 2035 (cf. Lamade, 2020). The foreign and economic policy of the People's Republic is already causing concern not only among its direct neighbours, but is also forcing the European Union, for example, to rethink. For a long time now, China has not only been using direct investments to secure influence in developing and emerging countries, but also member states such as Hungary, Greece, or Italy are exposing themselves to greater Chinese influence.

While the standards for 5G, for example, were still developed in cross-industry consortia and providers such as Huawei used the existing standards in a large number of patents here, "China Standards 2035" is intended to establish the next step for the People's Republic of China here, namely the setting of standards in important future topics such as information technology (here above all the fields of artificial intelligence and cyber security) (cf. Arcesati, 2019), biotechnology, high-end production, but also completely different important future fields such as environmental protection, agriculture, or the standardisation of urbanisation.

With a total of 117 individual measures, China wants to achieve a top level of standardisation both nationally and internationally. The main focus is on five fields:

- strengthening the strategic positioning of standardisation;
- intensification of standardisation reform and development;
- strengthening the standardisation system and improving the capacity for leading high quality development;
- taking the lead in international standards and improving the internationalisation of Chinese standards;
- strengthening science management and improving the effectiveness of standardisation efforts.

All fields are aimed at strengthening the People's Republic of China's ability to create standards and to make them valid not only nationally but also internationally. While many individual measures are simply about enabling the Chinese economy and the national standardisation system to create good standardisation and to use it to improve production in a wide variety of fields, the measures in the last two fields in particular are strictly geared towards converting national improvements into international standardisation successes. In this context, great attention is also paid to state support for domestic research and international standardisation activities from the Chinese side. This is in line with the strategy of transforming China from the workbench of the world to a nation that creates so-called tier-one companies. According to Chinese diction, tier-one companies set standards, while second-tier companies develop technologies and third-tier companies merely build products for others. The Chinese leadership thus sees standard-setting as a way to control technologies and products. This in turn fits in as a technology-driven facet of the rule-based power projection of the current going-abroad strategy of the People's Republic of China (cf. de la Bruyère/Picarsic, 2020).

The important role that the People's Republic of China plays in high technologies is already evident in examples such as 5G, artificial intelligence[80], or the Internet of

---

[80] On standardisation in the context of AI and the US-China Balance Act: Ding, Jeffrey: Balancing Standards: U.S. and Chinese Strategies for Developing Technical Standards in AI, Oxford 2020, available

Things. This is also evidenced by the suspicion that Western governments have of China's technological pioneering role – most recently observable in the ban on Huawei components in the UK's 5G network or the debate about the IT Security Act 2.0 on the topic of critical components (cf. Schallbruch, 2020). The law does not mention the Middle Kingdom at any point, but it was clear to all participants in the discussion that took place beforehand that it was actually about how to keep "non-trustworthy" components out of German networks – for example, those of the market-dominant Chinese network suppliers. Thus, in the IT Security Act 2.0, it is not only the technical requirements in terms of trustworthiness that play a role, but also political assessments of the respective countries of origin of components. In the future, the Ministry of the Interior, with its IT security competence, as well as the Ministry of Economics and the Federal Foreign Office will be involved in the decision as to which components from private providers may be installed in economically important installations such as communication networks. This represents a paradigm shift in the economic and industrial policy of the Federal Republic of Germany, which is otherwise oriented towards market-based mechanisms.[81] Recent involvement of the Bundestag's Foreign Affairs Committee, which otherwise deals with rather non-technical topics, also fits in with this change (cf. Bundestag, 2021).

The extent to which the Chinese intention will be crowned with success is not yet foreseeable. In general, it is in the interest of the international community of states, but also of the economy, to have Chinese companies at the "standardisation table", as this promises a broader technological consensus and also a smoother procedure in global trade and technology transfer. The strong state interference, however, makes both the industry associations of the Western world and the decision-makers on both sides of the Atlantic uncomfortable, also because China has recently increasingly tried to enforce national standards in other nations that are beneficiaries of Chinese investments. In this respect, regardless of the prospects of success, another "battle of the systems" is to be expected here, in which norms and standards will be fought over.[82]

The newly enacted Chinese strategy on this topic has not yet had a direct impact on the reality of standardisation, because the "mills" of the standardisation bodies do not grind like the geopolitical situation. Nevertheless, with the described and self-documented ambitions, the People's Republic of China makes it clear in which direction it wants to go and

---

online at: https://www.nbr.org/publication/balancing-standards-u-s-and-chinese-strategies-for-developing-technical-standards-in-ai/

81 German industry also clearly expresses its concern about the increasing influence of the People's Republic of China: Bundesverband der deutschen Industrie (ed.): Grundsatzpapier China, Partner und systemischer Wettbewerber – Wie gehen wir mit Chinas staatlich gelenkter Volkswirtschaft um, Berlin 2019, available online at: https://www.politico.eu/wp-content/uploads/2019/01/BDI-Grundsatzpapier_China.pdf

82 On this conflict from the US perspective also extensive: Strategy of the United States towards the People's Republic of China, published in German on the homepage of the US Embassy in Germany: https://de.usembassy.gov/de/strategie-der-vereinigten-staaten-gegenueber-der-volksrepublik-china/

what the standardisation approaches of the United States and the European Union should be prepared for. Here, on the one hand, very rule-based approaches and, in part, highly dynamic standardisation intentions driven by individual companies stand in contrast to a strongly state-supported standardisation policy interwoven with geopolitical interests, which is equipped with extensive resources, both human and financial, and pursues a clearly formulated goal that is not necessarily that of the most interoperable technology. To this end, the US and EU power systems have to deal with many other global challenges (cf. Semerijan, 2016/2019). This will be all the more difficult in a system that relies on consolidated assessment and agreement on norms and standards by many industry participants. Europe is thus also under pressure in this field due to its consensual approach between two system approaches (that of the United States and that of the People's Republic of China). However, the strategic assessment of the European Commission with regard to standardisation already clearly shows the need for action, especially in the field of information and communication technologies. Here the crucial fields are named and it is also underlined how important the influence on technologies is in a changing geopolitical landscape (cf. EU COM, 2021). In terms of speed of development and the build-up of technological pressure, the US and Chinese approaches are probably superior to the European system[83], which unfortunately also means for Europe: "Being caught between a rock and a hard place. However, the link between the political agenda and the standardisation agenda at European level alone shows that the need for action has been recognised.

---

[83] In addition, NIST is already carrying out initial activities to assess the Chinese activities and derive possible responses: https://www.nextgov.com/emerging-tech/2021/05/nist-wants-help-assessing-chinas-influence-emerging-technology-standards/174052/

Chapter 9

Intellectual Property Rights

Christoph Spennemann

Introduction

Intellectual property rights (or IP rights or IP for intellectual property) play an important role in the creation and protection of technologies that make up the digital economy. Copyrights, patents, trade secrets, and designs protect, to varying degrees, computer software, digital platforms, information and communication technology (ICT), and their devices and applications. In addition, IP rights protect traded assets, e.g. digital music and literature. While digital technologies have developed rapidly, the legal framework for intellectual property lags somewhat behind. This means both opportunities and challenges for our societies. To ensure legal certainty, it is necessary to adequately adapt IP legal systems to the new technologies to ensure their use and development. At the same time, the protection of digital business models can marginalise traditional exploitation of ideas and endanger economic livelihoods. The adaptation of intellectual property rights to the digital world should therefore be done cautiously and allow nation states the necessary space to take into account their cultural and industrial specificities. This is especially true for developing countries, which usually lack specific policy guidelines on how to strike a fair balance between digital rights holders and users. This chapter first provides an overview of the problems that arise in applying intellectual property rights to digital technologies. It then looks specifically at the role that IPRs play in the protection and use of economic data.

9.1 Overview: Intellectual Property Rights and Digital Technologies

IP systems were originally developed for the analogue age. Their basic purpose is to strike a proper balance between the interests of creators and inventors on the one hand and users and consumers on the other. This has become much more difficult in the digital environment. On the one hand, it is technically possible to produce and distribute electronic copies of original works in unlimited quantities, which can threaten traditional business in publishing, printing, and book trade. Furthermore, digital copies can be distributed across national borders, while intellectual property rights are restricted by national jurisdictions. On the other hand, there is the question of how the rights of consumers and competitors can be transferred from the analogue to the digital context. For example, someone who buys a patented or copyrighted physical product is free to resell it to third parties. Can someone who lawfully acquires a digital copy of a piece of music or a film also sell it to others, within the technical possibilities of unlimited electronic reproduction?[84] Also, to what extent is the reconstruction (*reverse engineering*) of protected computer programs, which is essential for software developers, especially in developing countries, comparable to simply reading a copyrighted book in an analogue context? Some developed countries have addressed these issues in the digital environment by limiting the scope that copyright offers in the analogue context (Samuelson/Scotchmer, 2001).

The balance of interests between rights holders and users is not only important for digital technologies, but also for the data that can be generated from these technologies via online platforms. While data collected by a search engine cannot be protected by a patent, data can be protected as a trade secret under certain conditions, and copyright or certain *sui generis* database protection rights can be considered for the protection of data collections or databases (see the analysis later in this chapter). Policy makers face the task of striking a fair balance between incentives to create data and datasets (e.g. to promote AI applications) and the need to share data to promote big data analysis and the improvement of products and services. In addition, ways must be found to reconcile the potentially conflicting goals of innovation-enhancing data sharing on the one hand and data protection on the other as far as possible. Digital platforms can lead to difficulties in IP enforcement, but also in consumer protection. For example, are platforms liable for IP-infringing content by their users? The EU Directive on Copyright in the Digital Single Market, adopted in 2019, met with considerable resistance in the legislative process because of its alleged obligation for platforms to install "upload filters" to select and block infringing content as distinct from non-infringing content. The discussion contrasts arguments for effective IP enforcement with concerns about automated decisions that replace human case-by-case consideration and may inadvertently ban content not protected by IP.

Digital platforms and software producers have also raised concerns about abuse of dominance by some global companies. The competition proceedings launched by the EU Commission against *Google* and *Microsoft* are based on complaints about digital "lock-ins" that effectively tie consumers to a company's products. Leading competition authorities differ on the conditions, under which holders of intellectual property rights in digital technologies are liable for abuse of a dominant position.

In some cases where legislation has been deemed too slow, the private sector has tried to close the gap by introducing voluntary commitments in the digital context. Open source software (OSS) is based on copyright, but rights holders authorise third parties to modify and distribute the programme under certain conditions. This reflects the belief that in a sharing economy, the consumer can be a creator at the same time, contributing to the continuous improvement of the underlying technology.

---

[84] Rights holders have responded by increasingly licensing digital content rather than transferring ownership, thus reserving the right to control the further distribution of the content (Okediji, 2018, p. 30).

A cooperative approach is also necessary when the development of new products depends on the interoperability of digital technologies from different rights holders. Interoperability is ensured by technical standards developed by standardisation organisations such as the International Telecommunication Union or private organisations. For example, mobile phone standards (most recently 5G) include a large number of IP rights. Standards developers rely on patent right owners to disclose all IP rights claims and provide licenses on fair, reasonable, and non-discriminatory (FRAND) terms. If IP owners hide their claims or refuse FRAND, the use of the standard is at risk unless IP laws or competition rules fix the problem. For example, the US Federal Trade Commission sued *Qualcomm*, which owns essential standard patents related to 4G technology. The reason for the suit was *Qualcomm's* refusal to grant its customers, such as *Apple,* access to 4G on FRAND terms. While this case illustrates the complex intersection between IP and competition law, the worldwide and multi-year conflict between *Apple* and *Samsung* over IP-protected technologies and external design of smartphones and tablets shows how courts in different jurisdictions can draw different conclusions about the infringement of software and device patents by competing technologies due to the territorial nature of IP law.

In conclusion of this section, it should be mentioned that the questions addressed here are answered very differently – or not yet at all – in national legal systems. The multilateral legal framework on intellectual property, the TRIPS Agreement of the World Trade Organisation WTO, provides the necessary scope for this. However, the TRIPS Agreement offers little guidance on how to shape national frameworks for digital value creation. This poses considerable challenges for developing countries in particular and justifies the need for multilateral exchange of experience in the WTO (cf. Brazil and Argentina, Joint Statement, 2018).

Protection and Use of Data and Artificial Intelligence: What Role Do Intellectual Property Rights Play? What Is the Relationship to Private User Rights?

This section will now examine the regulation of the innovation potential of data and the artificial intelligence (AI) generated by it. Data is treated as a broad term that includes "isolated or isolable entities that can be processed and analysed by machines", such as statistics, financial data, measurement data, information available in lists, structured and unstructured texts, and multimedia productions (Swiss Federal Administration, 2019–2023).

Here, intellectual property rights as a classic instrument for promoting innovation move to the centre of considerations, since AI and data are not about tangible assets, but about the application of ideas through algorithms and their results. The decisive factor here will be how intellectual property rights are to be structured in order to guarantee legal certainty and investment protection on the one hand, but on the other hand to encourage data to be made available to other economic participants. The more willingly data is made available (in technical jargon, "shared"), the greater the innovation potential of this data and the AI based on it. This approach is also the basis of the European data strategy published by the EU Commission in February 2020 (European Commission, 2020). In it, the Commission states the following (ibid., pp. 7/8):

> "The real value of data lies in its use and re-use. There is currently not enough data available for innovative reuse of data, including for the development of artificial intelligence.
> [...]
> Despite its economic potential, data sharing between companies has not yet become sufficiently widespread. This is due to a lack of economic incentives (including the fear of losing competitive advantage), a lack of trust between economic operators regarding the actual contractual use of data, unequal bargaining power, fear of misuse of data by third parties, and a lack of legal certainty about who is allowed to do what with the data (e.g. in the case of jointly generated data such as that from the Internet of Things – IoT)."

The following questions are analysed in this section:

1. Are intellectual property rights suitable for protecting both AI and its underlying data?
2. Do conflicts arise between the protection of economic operators through intellectual property rights on the one hand and the protection of users under the General Data Protection Regulation (GDPR) on the other?

9.2 On the Applicability of Intellectual Property Rights

9.2.1 Applicability of Intellectual Property Rights to AI

- **Patents** protect certain inventions. Patent protection can be considered on the one hand for the AI itself; on the other hand for the products created by AI. Let us first consider the former. The crucial element of AI, the patentability of which is at issue here, is the algorithm. This is a procedure (digitally programmed in the context we are interested in) to solve a problem (Czernik, 2016). An algorithm thus falls under the category of programs for data processing systems, scientific theories, and mathematical methods, which are in principle not patentable.[85] The European Patent Office therefore denies the patentability of AI *per se*. However, it affirms patentability if the AI in question and the underlying algorithm serve a concrete technical application (cf. Free, 2019/2020, p. 32).[86] Algorithms alone, i.e. in abstract form, are thus not patentable. Similar to mathematical methods, the underlying idea here is that the general public should not be denied access to such building blocks of science and innovation through exclusive rights. This idea should also apply to the ideas and principles underlying a computer program, including the interfaces between different program elements. However, general

---

[85] For example, § 1 (3) of the German Patent Act. Comparable provisions can be found in the laws of other states.

[86] The US *Supreme Court* takes a similar approach (Levy/Fussell/Streff Bonner, 2019/2020, p. 31).

algorithms can be protected as **trade secrets.** Unlike a patent, this does not grant absolute copy protection, but merely protects against unlawful acquisition, use or disclosure of secret information by a competitor. However, the latter is not prevented from developing and using the algorithm itself through fair business practices (see further discussion below). Another peculiarity of patenting AI arises from the patent applicant's obligation to disclose the invention "sufficiently clearly and completely to enable a person skilled in the art to carry it out."[87] Depending on the individual case, it is conceivable that AI may become self-executing to a certain extent through data-driven learning and that the patent applicant does not have full insight into the exact processes of the invention. In such a case, full disclosure of the invention is not possible. It is still open how patent offices deal with this problem. In the author's view, the duty to disclose the invention concerns the AI at the development stage, at which the inventor has developed it and can still fully understand it. It is this development stage, and not possibly further, independent ones, that must serve as the assessment basis for patentability, i.e. in particular the examination of novelty, inventive step, and industrial applicability. In this context, the *International Association for the Protection of Intellectual Property* has rightly expressed the view that this problem only concerns those inventions that consist essentially of AI itself, but not those that are merely made possible by AI, but then exist and function independently of the AI (such as the aerodynamic shape of a car body made possible by AI) (*Association Internationale pour la Protection de la Propriété Intellectuelle*, 2020, paragraphs 7 and 8).

This leads to the second question of this section, namely the patentability of products created by AI. In case of such inventions, the question arises as to whether they can still be attributed to the human programmer of the AI or its user, or are already to be regarded as the result of a self-determined AI. According to general opinion, German patent law presupposes a natural human being as inventor. Autonomously acting machines that independently produce something new through AI therefore do not qualify as inventors, so that a regulatory gap opens up for this area. The Federal Ministry of Economics still assumes that cases of such autonomous AI are extremely rare (BMWi, 2019, p. 24). But who should be entitled to the patent in the cases of non-independent AI that frequently occur today and through which an invention is created? The programmer of the underlying software? With respect to this and the underlying AI, the programmer may already have a patent claim. It seems excessive to grant him an additional patent on the AI-generated products. Companies that are to use AI to innovate would probably be reluctant to invest in such AI if they were not entitled to patent claims for AI-generated products (BMWi, ibid.). Accordingly, it is obvious to regard the user of the AI as the inventor of the product under patent law. Unlike

---

[87] Section 34 (4), German Patent Act.

the programmer, the user has applied the AI to a specific individual case in order to create a specific innovative product or service.

- **Copyrights** expressly extend to computer programs to the extent that they are considered personal intellectual creations of the programmer.[88] By contrast, copyrights do not cover ideas and principles that underly an element of a computer program, including the ideas and principles that underly the interfaces.[89] The requirement of personal intellectual creation must also be met by AI. Copyright protection comes into consideration on the one hand for the AI itself – i.e. an algorithm for the digital solution of a problem – and on the other hand for works produced by this AI – for example photos, music, texts, further software. While the protection of the AI itself would benefit the software developer, the rights holders to the works produced by AI would be users of AI, such as musicians or developers of further software. The AI itself cannot be considered a personal intellectual creation (and thus protectable by copyright) in (rare) cases of independence, i.e. when its processes can no longer be understood by its programmer and are distinct from the basic AI. An exception only applies if the programmer has already provided for certain options of independence in the basic AI (cf. Schürmann/Rosenthal/Dreyer, 2019). According to current legal opinion, copyright protection requires a natural person as the rights holder – due to the requirement of personal intellectual creation (cf. Schönenberger, 2017). The legal situation is the same with regard to the protection of AI products. AI users who produce works such as music or texts with the help of AI can claim copyright protection for them. This no longer applies if the AI creates certain works independently and the AI user no longer has any influence on their design. (cf. Herfurth, 2019) As in the area of patent law, there is therefore a regulatory gap here. However, as already mentioned, this is irrelevant as long as the technical possibilities to enable independent AI remain limited.
- The **right to protect trade secrets** can extend to any AI. This is particularly interesting for AI that cannot claim copyright or patent protection for the reasons mentioned above. Of much more practical relevance than the hitherto rather rare cases of independent AI, these are in particular the ideas and principles of an algorithm, including the interfaces between different elements of a program. An EU directive from 2016 has harmonised the legal situation in the EU in this regard.[90] The programmer of the AI itself or a user of the AI can claim such protection. The right protects against unlawful acquisition, unlawful use, or dis-

---

[88] Section 2 (1) 1. and (2), Section 69a (3), German Copyright Act (UrhG), and Article 10.1, TRIPS Agreement.

[89] Section 69a (2) UrhG.

[90] *Directive (EU) 2016/943 of the European Parliament and of the Council of 8 June 2016 on the protection of confidential know-how and business information (trade secrets) against unlawful acquisition, use and disclosure.* (hereinafter: Trade Secrets Directive).

closure of the AI, in particular the underlying algorithm. However, the protection only exists as long as the AI is kept secret. If a valuable algorithm is directly part of a marketed product (for example, the software of a medical diagnostic device), a trade secret does not protect against attempts to fairly decipher the AI and in particular the underlying algorithm. This is a significant limitation of this legal institution. The situation is different in cases where the valuable algorithm does not become externally tangible, for example, it merely optimises an internal manufacturing process for a product. Here, a trade secret can offer valuable protection.

- In summary, it can be stated that the existing system of intellectual property rights is in principle applicable to AI. The decisive factor in both patent and copyright law is whether the respective algorithm was developed specifically as a solution to a particular problem. In many cases, this will be the case. Abstract theories, principles, and ideas that are detached from concrete applications, as well as programme interfaces, are neither patentable nor copyrightable, nor are the rare cases of independent AI. Here, trade secret law only offers valuable protection if the corresponding algorithms or underlying ideas are not publicly accessible, e.g. they concern parts of an internal manufacturing process.

Comment / Recommendations for Action

In terms of legal policy, it seems desirable to promote the further development of AI in order to enable our society to reap the benefits of digitalisation, create promising jobs and secure Germany's and Europe's competitiveness. However, a future development of independent ("strong") AI should be accompanied by ethical considerations as to which consequences of digitalisation are undesirable (cf. Schmiedchen et al. 2018). This would go beyond the purpose of this chapter. Intellectual property rights are crucial to determining the focus of investment incentives for AI. Exclusive rights play an important role here. On the other hand, especially in the ICT and software sector, exchange with other developers is also essential for innovation. Intellectual property law should therefore be appropriately balanced to promote innovation through both property rights and increased exchange. The exemption of the theories and ideas underlying algorithms from the scope of protection of the most important intellectual property rights enables their use by competitors and researchers. Traditional IP rights are widely applicable to AI, so important investment incentives are available in principle. However, there is a regulatory gap for independent ("strong") AI. Due to the current lack of technical possibilities to actually create and apply strong AI, there is no immediate need for action. However, consideration should already be given to whether and how strong AI should one day be protectable as an intellectual property right. Two questions should be addressed here, namely (1) How can intellectual property law be designed in such a way that the creation of strong AI itself proceeds along controllable lines? (2) Should intellectual property law be adapted to protect the products of strong AI?

(1) As already explained, there are significant concerns about the patentability or copyrightability of strong AI according to current legal opinion (among other things, due to the lack of a human inventor or author of an algorithm). Strong AI could, however, be protected as a trade secret. Would such protection encourage uncontrolled development of AI, which could lead to a scenario in which humans lose influence over AI? Unlike a patent, a trade secret cannot in principle prevent a competitor from independently developing the secret technology. A trade secret may be researched and deciphered as long as fair means are used and industrial espionage is not resorted to. This insight is important for the permissibility of reconstructing source codes that can provide information on the construction of computer software. This makes it possible to discover malfunctions and to control the further development of an algorithm.[91] However, under current EU law, the reconstruction of software is not permitted without limitation, but is subject to certain limits determined by the copyright on the programme.[92] Accordingly, the decompilation of a programme is only permitted for the purpose of establishing interoperability between the reconstructed programme and another independently created programme.[93] For the purpose of verifying the functionality of a programme, it may be observed, examined, and tested by loading, displaying, running, transmitting, or storing it.[94] However, if these actions require even temporary or partial reproduction of the programme, the consent of the rights holder must be obtained.[95] Since reconstruction of the source code usually requires copying of the programme (this can be inferred from: Samuelson/Scotchmer, 2001, p. 1609), this consent requirement may prevent a programme review or, if a paid license must first be acquired, act as a financial deterrent. Not every author may be interested in having their software programmes reviewed.

---

[91] It has been pointed out in the literature that reconstructing the source code alone is not sufficient to fully understand how a programme works. Accordingly, further steps are required, which will not be discussed here (Samuelson/Scotchmer, 2001, p. 1613).

[92] Unlike the source code underlying the programme, which contains an idea that cannot be protected by copyright, a software programme represents the creative expression of this idea and can thus be protected by copyright.

[93] Article 6(2)(a) Directive 24/09/EC of the European Parliament and of the Council of 23 April 2009 on the legal protection of computer programs. Available at https://eur-lex.europa.eu/legal-content/DE/TXT/PDF/?uri=CELEX:32009L0024&from=en

[94] Ibid, Article 5(3).

[95] Ibid, Article 4(1)(a). The exception to the reservation of consent formulated in Article 5(3) is so broad that it does not create an actual exception but a circular argument: "The person authorised to use a copy of a program may, **without having to obtain the authorisation of the rightholder**, observe, study, or test the functioning of that program in order to determine the ideas and principles underlying a program element, if he does so by performing acts of loading, displaying, running, transmitting, or storing the program which **he is authorised to do**." (emphasis by the author). It is self-evident that someone does not need the authorisation of the rights holder to do what they are entitled to do. What third parties acting without authorisation are entitled to do is clear from Article 4(1)(a), namely to load, display, run, transmit, or store the programme, but only so long as this does not require reproduction of it.

In view of future developments in strong AI, a revision of the EU Directive on the legal protection of computer programs seems necessary. What is needed is a clear exemption in favour of any activities that serve to verify the functionality of a program. This seems necessary to critically accompany the future development of source codes and to prevent their functioning from being increasingly removed from human understanding and influence.

Also from a legal policy perspective, the review of a programme should be able to take place independently of the author's consent. Copyright protection does not extend to the idea underlying the creation,[96] in this case the source code. Duplication of the program cannot be considered copyright infringement if the duplication is not the actual goal, but serves to understand the source code, which is not protected by copyright. A more restrictive application of copyright law also leads to an impermissible restriction of the freedom provided by a trade secret existing in a source code. Accordingly, the decryption of the protected secret by bona fide means is permitted. However, such decryption is made illegal if copyright law prohibits the reconstruction of the program for the purpose of its verification.

In the section on the applicability of intellectual property rights to AI, it has been pointed out that the permissible decryption of a trade secret is only possible if the relevant technology is publicly accessible, for example, if it is integrated into a marketed product. However, if the AI is part of a purely internal manufacturing process, for example, third parties have no access and the considerations just made on the revision of the EU Directive on the Legal Protection of Computer Programs offer little help. For such constellations, a proposal by the Federation of German Scientists (VDW) seems interesting. According to this proposal, AI-relevant design information and source codes should be stored in public institutions to make them widely accessible in the long term and thereby document malfunctions without gaps and optimise the chances of success for repairs (Schmiedchen et al., 2018, p. 17). If the owner of a trade secret consents, no problems under intellectual property law are apparent. However, if there is no consent, the protection of trade secrets provides protection against public access for the purpose of maintaining competitiveness. There is one possibility of reconciling this property right with the documentation obligation proposed by the VDW in the area of research and development through public funds. State institutions that provide public funding for the research and development of AI can impose certain conditions of use on the AI developer as a prerequisite for access to funding. One such condition could be the deposit of AI-relevant information, including certain algorithms. On the other hand, such a deposit requirement should be designed in such a way that it does not inhibit the willingness to invest

---

[96] See e.g. Article 9(2), TRIPS Agreement.

in the development of AI. This danger would exist if commercial competitors had unrestricted access to the deposited information. One could consider granting access against payment of a fee that compensates the AI developer for his efforts. Alternatively, access could be limited to non-competitors, such as researchers at public institutions. In addition, certain exceptions could be formulated, which would give a governmental control institution the right to publish deposited information, e.g. to avert a danger to certain public interests. The WTO TRIPS Agreement provides for a similar possibility in the protection of pharmaceutical test data, if necessary to protect public interests (cf. Spennemann/Schmiedchen, 2007). However, there is no explicit exception in the TRIPS Agreement for AI and trade secrets. Here, a multilateral consensus seems necessary. It would have to be precisely defined to what extent there is a public interest in a filing obligation for AI-relevant information and under what conditions this can restrict the recognised right to the protection of trade secrets.

(2) The following options for protecting products generated autonomously by AI are already being discussed (BMWi, 2019, p. 29):

- Leaving the *status quo*. In line with the idea expressed above about the importance of idea exchange and collaboration in the ICT/software sector, the independent development of innovation through strong AI could completely dispense with protection through exclusive rights. However, it should be borne in mind that other legal systems are already quite capable of protecting strong AI through intellectual property rights. For example, neither in the US nor in the British copyright system there is a requirement of a natural person as the author and a personal intellectual creation as the object of protection. Unlike in Germany, therefore, the products of strong AI can certainly be protected as the *copyright of* the person who created the conditions for the product to be produced (even if this is directly attributable to an independent AI).[97] Less extensive protection could prove to be a competitive disadvantage for Germany. AI development cannot be reduced to the free exchange of ideas, but also needs investors. It therefore seems disproportionate to exclude any intellectual property protection in principle. Rather, it should be possible to decide on a case-by-case basis the degree to which one wants to rely on exclusive rights or open collaboration. An *open source approach* would offer this flexibility. However, such an approach presupposes the existence of an intellectual property right, which can then be disposed of openly or exclusively, as the case may be.

---

[97] E.g. Sec. 9 (3) UK Copyright Designs and Patent Act 1988; similar in approach 17 U.S. Code § 101, according to which the commissioner of a work (as opposed to the actual author of the work) can be the person entitled to protection.

- <u>Creation of a legal personality for AI</u>. This option would allow for clear patent and copyright protection of autonomous AI and its works. The existence of legal persons illustrates that legal personality is not necessarily linked to human existence per se. On the other hand, legal persons are created by an association of natural persons. In this respect, one could be of the opinion that legal persons also carry certain aspects of human dignity because they represent a will expressed by a totality of natural persons (BMWi, 2019, p. 7). It seems doubtful that a comparable reflection of human dignity can also be assumed for AI. This question would first have to be investigated through a societal debate.[98] In addition, important liability issues would have to be regulated. For example, an AI endowed with its own legal personality would have to have possibilities to answer for violations of rights caused by it through independent liability and payment of damages. Because of the complexity of these considerations, the creation of an AI legal personality does not seem desirable to the author at present. Less drastic but equally suitable measures exist to protect the products of strong AI.
- <u>Adaptation of the legal system</u>. In line with the British or US model, the property right in products of strong AI could be awarded to the natural person who, as the last human link in the chain of command, has ensured that the AI can create such a product. This would mean abandoning the traditional requirement of a personal-intellectual creation in the area of copyright. This could prove difficult to implement in terms of legal policy (BMWi, 2019, p. 26).
- <u>Creation of a new ancillary copyright</u>. Less drastic than the adaptation of the German legal system (see above) appears the creation of an ancillary copyright for the results of strong AI. The institute of ancillary copyright is used in German law when a performance is to be rewarded that does not constitute a personal-intellectual creation. In particular, anyone who takes a simple photograph is granted such a right to the image, which expires 50 years after the image is taken or appears.[99] Since a photograph is not taken directly by a natural person, but by a photographic apparatus, copyright does not apply due to the lack of a personal-intellectual creation.[100] The results of strong AI could be protected accordingly.

Finally, it should be noted that the further development of AI depends not only on the protection of intellectual property rights, but also on the willingness of users to provide (share) data on which AI can be developed. In the European Data Strategy of February

---

[98] See in particular the discussions by Stefan Bauberger in Chapter 4 Machine Rights.

[99] See § 72 UrhG.

[100] In contrast to a simple *photograph,* a photograph is considered to be a *photographic work* protected by copyright if it fulfils heightened requirements of creativity and expression. In that case, it is not the function of the photographic apparatus but the creative use of the same by the photographer that is in the foreground of the assessment. Such photographs are therefore regarded as personal intellectual creations. Rights to such photographic works do not expire until 70 years after the death of the author.

2020, the EU Commission expresses concern about considerable deficits in the sharing of industrial data (see above). The applicability of intellectual property rights to data is problematic. This raises the question of an appropriate legal framework for data sharing (see below).

9.2.2 Applicability of Intellectual Property Rights to Data

AI and data are closely linked. Without data, on the basis of which AI can be developed, AI is inconceivable. Legally, however, they must be kept apart. While the concepts of technical invention (patent law) and creative formulation (copyright law) seem quite obvious for AI algorithms, these connections are not obvious for data. Accordingly, there is currently no law in the EU that establishes a specific property right in data (cf. van Asbroeck/Debussche/César, 2017, p. 22)[101]. Different variants of intellectual property rights come closest to such a protective purpose, even if particular problems arise from the fact that the system of intellectual property rights is not yet adequately adjusted to the new technical possibilities for the automatic compilation and analysis of data.

- **Copyright**: Protection of individual data as well as data collections or databases can be considered.[102] This is not the case with data that is compiled solely on the basis of logical constraints and without any personal intellectual creativity, such as sports results or temperature measurements (cf. van Asbroeck/Debussche/César, 2017, p. 70). Similarly, data or databases that come about through automated processes, e.g. AI, are not eligible for protection. The lack of a property right can mean, on the one hand, that the owners of such data are unwilling to make it publicly available, as they cannot expect anything in return and lose control over it. In such cases, the innovation potential of the data remains untapped. On the other hand, if such data is already publicly available, third parties do not need permission required by an intellectual property right to use it, e.g. in the context of scientific or commercial data analysis. The EU Directive on Copyright in the Digital Single Market, adopted in 2019, provides for exceptions to copyright for the purpose of analysing legally accessible texts and data (text and data mining), as text and data mining plays a crucial role in scientific and industrial research as well as in the development of AI.[103] However, rights holders can prevent the application of this exception to their works for the purpose of commercial research by means of an explicitly stated

---

[101] With further analysis of the national laws of some EU states. In Germany, it is proposed to derive a civil law property right to one's own data from existing provisions of criminal law and civil law (ibid., p. 57).

[102] See e.g. § 4 of the German Copyright Act.

[103] Articles 3 and 4 of *Directive of the European Parliament and of the Council of 17 April 2019 on copyright and related rights in the Digital Single Market and amending Directives 96/9/EC and 2001/29/EC.*

reservation. This reservation has been criticised in the literature as putting AI developers, commercial research institutions, and journalists in the EU at a disadvantage compared to similar actors in the US (Hugenholtz, 2019). There, commercial text and data mining may be practised regardless of the will of the creators. The ideal would therefore be the creation of a data protection law for non-copyrighted data that encourages the publication of data and does not set too high barriers for use by third parties, even in the area of commercial research.

- **Protection of databases**: In addition to copyright, the *sui generis property right for* databases introduced in the EU in 1996 comes into consideration here.[104] A "database" within the meaning of the Directive is a collection of works, data, and other independent elements.[105] The property right does not extend to individual data, but to databases in any form. Their makers can prevent the further use of the database or extraction of individual data by third parties. However, a prerequisite is that the producer has made a substantial investment in obtaining, verifying, or presenting the contents of the database. In this context, "content" is to be understood as an entirety that goes beyond individual data, or other elements. It has been pointed out in the literature that these requirements do not take into account the new technical developments in the field of AI and leave AI-created databases unprotected (cf. Van Asbroeck/Debussche/César, 2017)[106]. In the field of AI-automated data analysis, the content of a database is usually obtained by AI, i.e. automated. This does not require a substantial investment in the sense of the Directive, unlike perhaps the creation of the individual data itself.

- **The right to protect trade secrets** could give data holders a right to prevent the unlawful acquisition, use, and disclosure of the data under their control. The legal basis for this is the EU Directive on Trade Secrets already presented above. However, it also appears doubtful here whether data used in the context of AI analyses meet the protection requirements for trade secrets. To do so, they would have to have a commercial value, among other things.[107] This seems doubtful for individual data. Only in the context of a data set do individual data become valuable to show certain trends, behaviours, or developments. Data sets, in turn, are often shared among different actors in the context of an AI data analysis in order to abstract the data, compile it according to certain patterns, interpret it, and finally apply it as AI (van Asbroeck/Debussche/César, 2017, p. 123, Figure 5.1: *Knowledge pyramid*). If the different actors have not entered into a contractual

---

104 *Directive 96/9/EC of the European Parliament and of the Council of 11 March 1996 on the legal protection of databases.*
105 Article 1 of the Directive.
106 This is different from Schürmann/Rosental/Dreyer, 2019, who seem to affirm a database protection right for the result of the data analysis, i.e. the AI.
107 Article 2.1 b) of the Trade Secrets Directive.

agreement to keep the shared data secret from third parties, the shared data is no longer considered secret, which is a prerequisite for trade secret protection.[108] Contractual confidentiality agreements are therefore essential for the protection of data sets as trade secrets.

Alternative Approaches to Promote Data Sharing

As shown in the last sections, existing legal instruments offer limited protection for automatically generated data and may therefore be limited in their ability to encourage data sharing for innovation. Therefore, the creation of a law specifically tailored to machine-generated data has been suggested for some time. For example, the World Intellectual Property Organisation (WIPO) has initiated a consultation on the creation of a separate right for data, which is still in its early stages in 2021 (cf. World Intellectual Property Organisation, 2019).

An expert report prepared for the EU Commission proposes a non-exclusive right to data, which should be available to anyone who has demonstrably and legitimately processed or analysed certain data (see van Asbroeck/Debussche/César, 2017, p. 120). The non-exclusive nature of the right is intended to promote access to and exchange of the protected data.

In its data strategy presented in February 2020, the EU Commission does not address this proposal further. The EU data strategy announces a legal framework for a common European data space for 2020 and 2021 (cf. European Commission, 2020, pp. 12–15). The focus is to be on data sharing and the creation of corresponding rights. However, in this context the Commission seems to attach more importance to the concept of open data than to a redefinition of existing EU laws on intellectual property rights. Specifically for the area of publicly produced data, the Swiss federal government also adopted a "Strategy for Open Administrative Data in Switzerland 2019–2023" in November 2018 (cf. Swiss Federal Administration, 2019–2023).

Potential Conflicts Between the Protection of Economic Operators Through Intellectual Property Rights on the One Hand and the Protection of Users Under the General Data Protection Regulation (GDPR) on the Other Hand

Intellectual property rights of economic operators and rights of private users to their data under the EU General Data Protection Regulation (GDPR) differ in their objective. GDPR user rights are not designed for economic gains of the right holders, but concern the control of the data provided. Essential elements of this control under the GDPR are the reservation of consent, the right to be forgotten, transparency and purpose limitation of data processing, as well as data portability. These are outgrowths of the right to informational self-determination, which in turn is derived from the general right to privacy.

---

[108] Article 2.1 a) of the Trade Secrets Directive.

Despite these differences, overlaps between the two areas of law are conceivable. Private user data is used by platforms for commercial purposes and, as outlined in the previous section, may be subject to certain intellectual property rights such as a trade secret, at least in connection with the data of other users as data sets (without any commercial involvement of the users being envisaged). In addition, AI developed on the basis of user data may be protected by patents, copyrights, or trade secrets (see above).

Conflicts may arise here if the principle of transparency of data processing in Article 5 (1) (a) GDPR[109] is interpreted as entitling the user to disclosure of the data used for machine training or the algorithm used. If the latter are protected by a patent, reference can be made to the patent application, in which the technical details of the invention have to be disclosed in order to allow a replication by a skilled person after the expiry of the patent protection.[110] The ideas and principles underlying a computer program are excluded from the scope of copyright. Disclosure may therefore be required if it can be shown that an algorithm embodies a generally valid idea and not a creative expression of that idea. Such a demarcation appears difficult.[111] However, this is not decisive in practice: in the 2nd chapter of this book, von Gernler and Kratzer emphasise that the algorithm is usually known in broad outline and that it is more important to disclose the data used to train the machine. However, these may be protected by trade secrets, where secrecy is at the core of the protection requirements.

But also in cases where IPRs are not used to keep data secret, but on the contrary to share data with other actors, conflicts with Article 5 GDPR may arise. Article 5 (1) (b) GDPR imposes the obligation to use personal data for a specific purpose, i.e. not to further process it in a way incompatible with the purpose for which it was originally collected.[112] It is true that protecting data through intellectual property rights would create some legal certainty and thereby encourage data set owners to share them with other economic actors in line with the objectives of the European Data Strategy. However, this incentive to share data could be lost if at the same time there was a risk of a breach of Article 5 (1) (b) GDPR. Indeed, private sector representatives at UNCTAD *e-commerce week* 2020 stressed that the purpose limitation under Article 5 GDPR is a serious obstacle to further sharing of collected data (United Nations Conference on Trade and Development, 2020, p. 27, esp. 28/29).

---

[109] This provision stipulates that "personal data must be a) processed lawfully, fairly, and in a manner comprehensible to the data subject ('lawfulness, fairness, transparency')".

[110] Cf. Sec. 21 (1) 2nd, German Patent Act.

[111] The disclosure of the idea in the source code will usually require the decompilation of the protected programme. EU law sets certain barriers here. See above, Comment / Recommendations for action on the applicability of intellectual property rights to AI.

[112] This provision states: "Personal data must be [...] (b) collected for specified, explicit, and legitimate purposes and not further processed in a way incompatible with those purposes; further processing for archiving, scientific or historical research purposes in the public interest or for statistical purposes shall not be considered incompatible with the original purposes in accordance with Article 89 (1) ('purpose limitation')".

Comment / Recommendations for Action

The two conflicts between intellectual property rights and the GDPR described in the previous section will be addressed here.

1. Can the holder of a trade secret be required to give it up in order to comply with the transparency requirement in Article 5 (1) (a) GDPR?
2. Intellectual property rights in data are intended to promote legal certainty and thus the willingness to share data. Does the purpose limitation under Article 5 (1) (b) of the GDPR contradict this objective?

In order to solve the conflict described under 1, the purpose of the transparency requirement should be taken into account. According to Article 5 (1) (a) GDPR, personal data must be processed in a way that is comprehensible to the data provider. How exactly an individual needs to understand the details of this processing depends, according to the author, on the purpose of the data processing. If the purpose is to create a profile of the individual's commercial preferences in order to promote products or services more efficiently, it can be assumed that no essential legal interests of the respective individual are affected, especially since data is provided voluntarily in order to be able to access services that can be dispensed with if necessary, such as certain search engines. In such cases, a claim against a company for disclosure of its protected data records in favour of the transparency requirement seems disproportionate, as this would completely negate a recognised trade secret. Such a serious interference can only be justified if important legal interests are at risk and the individual does not have the choice to forego the data-based procedure. This is the case, for example, if data is used to assess creditworthiness, or in the case of sovereign examination of a claim by a government agency or a court (for example, in the context of expert opinions). In this context, the VDW has pointed out the need for transparency and traceability of AI-generated decisions (Schmiedchen et al., 2018, pp. 17–19). Neither EU nor German law currently explicitly states how a trade secret should be treated in the examples described above. The right holder will usually not be the state agency or a court, but a private entrepreneur who, on the basis of his protected datasets, develops automated procedures that sovereign decision-makers use. Consideration should be given to imposing an obligation on the right holder to disclose records in such cases. However, for the sake of proportionality of the interference, competitors should not be able to benefit from this, but access should be limited to explicitly designated experts who check for misuse or biased use of the data for training purposes, without pursuing their own economic interests. In this way, the core of trade secrecy, namely securing an advantage over competitors, could be preserved. On the other hand, the disclosure obligation should not be made dependent on the existence of concrete indications of an abusive or biased use of the data. It would be difficult for the person concerned to prove this and would severely impair the effectiveness of the disclosure obligation.

The conflict mentioned under 2. arises not only from the existence of intellectual property rights to data, but also from the sharing of data that is not covered by intellectual property rights. Affected are both actors whose business model is based on voluntary data sharing and those who need special incentives to share data, such as those granted by intellectual property rights, for the sake of legal certainty. Any incentive to share data, whether through intellectual property, *sui generis rights,* or open data models of the public administration,[113] could run counter to the requirement of purpose limitation of data collection. Thus, there is a risk of a conflict of objectives between promoting innovation on the one hand and data protection on the other.

For shaping the digital future in Europe, it is crucial to limit such conflict and to find a practicable balance of interests. Both goals have their justification in promoting prosperity and quality of life and should not be seen as opposites but as mutually reinforcing factors. Innovation becomes more attractive if it avoids disregarding individual rights. This can be achieved, for example, through anonymisation of data used or through the EU-wide recognised right to delete personalised data without calling into question the use of data sets in their entirety. Data protection that inhibits innovation in the long term indirectly benefits the strategic competitors USA and China. Rules established there, e.g. in setting new industry standards and exploiting private data, could set new global standards that contradict European values, but which Europeans would have to adapt to out of economic dependency if they themselves fell behind in the field of innovation. In more positive terms, a balance of interests between data protection and innovation could carry a specifically European model of society into the future. This balance of interests should strive to position Europe permanently alongside the USA and China as a global player and, through the attractiveness of the balance, to influence such countries that are in the second and third tier of digitalisation, such as Brazil, India, Indonesia, and South Africa. This requires a change of consciousness in the EU Commission as well as the European capitals, especially in Berlin, to understand and actively use the European social model not only as an EU-internal instrument for an "island of the blessed", but as a global power factor. Only in this way will Europeans succeed in permanently shaping digitalisation to enable and promote a dignified life in the tradition of European individual rights.

---

113 An example of this tension is currently provided by the use of patient data by health authorities in the context of the Covid-19 pandemic. The Swiss Federal Office of Public Health (FOPH) has been heavily criticised in the press for its refusal to publish case figures on new infections at the municipality or postcode level, arguing that analysis of these data could help determine locations of increased new infections. The FOPH had invoked data protection because in small communities, published data could quickly be used to infer individuals. As a compromise, it has been suggested that only data from the postcode level should be published, which does not allow such conclusions to be drawn (Skinner, 2020).

Chapter 10

Lethal Autonomous Weapons Systems – New Threats and a New Arms Race?

Götz Neuneck

Science and technology had an important influence on arms procurement, strategy development, and warfare, especially in the 20th century. The industrial and scientific revolutions of the 19th and 20th centuries have also politicised scientific fields, culminating in World War II with developments such as long-range missiles, radar, operations research, or cryptography, some of which were decisive during the war. In the Cold War, the subsequent dangerous and enormous resource-devouring arms races between the two superpowers produced weapons of mass destruction that have not been completely dismantled or limited to this day (cf. Neuneck, 2009). The diffusion of these technologies ensures imitation and replication by other states (proliferation).

Three important scientific breakthroughs, in the late 20th and early 21st century have led to significant advances in the civilian sector, but have also found their way into military technology:

(1) nuclear technology,
(2) biotechnology, and
(3) information and communication technologies (ICT).

These technologies, including missile and space technology, are highly "ambivalent" and for a long time formed the core of the dual-use problem, which could be limited by international law, arms export controls, disarmament treaties, and ethical regulations, but not completely contained. Dual-use refers to the potential use of technologies, products, and scientific and technical knowledge for both peaceful and for warlike purposes. In the course of history, there have been many examples of scientific-technical knowledge being used in armament and war technology (cf. Forstner/Neuneck, 2018).

In the 21st century, the focus of peace and security policy is on the field of information and communication technologies (ICTs), because here revolutionary developments are taking place that are visible to everyone and are even reaching into the private sphere. Keywords here are the internet, smartphones, or autonomous cars. However, the ICTs developed in the broadest sense in the civilian sector are also leading to an increased digitalisation of armaments and war in the leading Western countries.

Particularly, in the context of the debate on "hybrid warfare", social dynamics triggered by information operations are being discussed that could become relevant to war. Today, this is understood to mean in particular a stronger networking but also autonomisation of weapon systems. The connection of sensors, various platforms on land, at sea, in the air, and in space, the rapid processing and forwarding of data enable new operational profiles, missions, strategies and threat scenarios. However, this also can trigger new arms races, enormous arms expenditures and concerns about wars that violate international law or the lowering of thresholds for the use of force.[114]

A leading example for this trend is the development, procurement and deployment of unmanned systems, popularly known as "drones". Today, drone missions are no longer the privilege of the USA alone, but have become a serious threat to international security (cf. Horowitz/Schwartz, 2020). New revolutionary developments in autonomous systems are to be expected, which will also enable new weapon developments, operational capabilities, missions, and war scenarios. This implies new challenges in the field of national and international security and peace policy, i.e. for procurement, training of soldiers, handling in the field, protection against hostile systems, proliferation, and for arms control, arms exports, and ethical regulations for the prevention of war. In view of considerable investments in research fields such as pattern recognition, sensor technology, big data, robotics, cyber technologies, and artificial intelligence, a permanent arms technology impact assessment is necessary to understand the possible damage potential and the effects on peace and security more precisely.

The debate on the purchase of combat drones for the German Army (Bundeswehr) illustrates the diversity of arguments regarding possible mission profiles of combat drones. While proponents merely argue that the drones would only protect their own armed forces, opponents speak, among other things, of a lowering of the threshold for war. However, weapons systems are not procured for narrowly defined scenarios, nor would a national ban automatically raise the war threshold or mean a lower risk of violence. In reality, the fundamental question is how future decisions in automated wars are to be made in accordance with international law and whether preventive rules can contain destabilising developments.

In the first section, the current possibilities, uses, and problems of currently existing unmanned weapons systems are presented, while the second section is dedicated to the question of future developments in the field of autonomous lethal weapons systems. The final section addresses the question of how to limit or prevent possible destabilising developments (cf. Alwardt/Neuneck/Polle, 2017 and Grünwald/Kehl, 2020).

10.1 What Are Drones and LAWS and What Developments Are Taking Place?

Unmanned moving platforms have been around since the 1940s. The evolution from the first large ballistic missile, the V-2, to modern intercontinental ballistic missiles has been going on for a long time. Cruise missiles, on the other hand, use turbines, operate in the atmosphere, and have become part of the modern military arsenal, as have torpedoes. These weapon carriers can only be used once and enable heavy payloads to

---

[114] See also: Frank Christian Sprengel: Drones in Hybrid Warfare: Lessons from Current Battlefields, Hybrid CoE Working Paper 10.

be transported with pinpoint accuracy over long ranges. The modern arsenal has been joined by unmanned aerial vehicles (UAVs), which land or take off automatically, can be steered remotely and can be used several times. Depending on their equipment, they can be deployed for long periods, enable flight times of hours to several days, and do not put pilots' lives at risk. In the public debate, the term "drone" is usually used to describe anything that flies unmanned. Depending on the range and technology, different categories of drones can be distinguished. Drones with a fairly simple design have a low flight altitude and short range, while tactical drones are intended for medium flight altitudes and longer ranges.

Initially, drones have been introduced in the military for reconnaissance and surveillance purposes, in many forms. The dual-use potential of this development becomes clear when one considers the various types of civilian drones used for recreational purposes or by the police, fire brigade, agriculture, science, etc. In each case, they can be controlled remotely by an operator on the ground, whether by means of a TV-link or pre-programmed. It stands to reason that remote control was first applied to aircraft, but unmanned land or water vehicles are also gaining in importance. Partial autonomy comes into play in some existing drone systems, as functions such as automatic take-off or landing can already be delegated and performed autonomously.[115] [116]

Armed Drones: Pandora's Box Is Open

Technically, it is obvious that these "passive" flying objects can also be actively armed. When they are armed, they are called "Unmanned Combat Aerial Vehicles" (UCAV).[117] The step from reconnaissance to combat drone is obvious for the military, but non-state actors are also showing interest in these systems, as they supposedly enable "surgical strikes" at safe distances. If a country has a developed aviation industry, in-house developments, licensed developments and imports from other drone manufacturers are possible.

The first combat drones were used in the 1973 Yom Kippur War; since then, these technologies have been expanded, especially by the USA and Israel. The US Predator or Reaper combat drones were first used by the US in Afghanistan in 2001. The USA remains the trendsetter in this field to this day. The MALE drone programme (Reaper, Predator) was massively expanded and led to thousands of missions in Pakistan, Yemen, Afghanistan, Iraq and Syria and Somalia alone. According to an analysis by the Bureau of Investigative Journalism (BIJ), the USA carried out at least 14,040 drone missions in the four countries mentioned in recent years. The target here was often non-state actors in the context of anti-terror warfare.

---

[115] One also speaks of MALE systems: Medium-Altitude, Long-Endurance.

[116] The Bundeswehr has been using the "Heron" reconnaissance drone leased from Israel in Afghanistan since 2010 and in Mali since 2016.

[117] A distinction must be made between, among other things, multiple-use "combat drones", which operate from a safe distance, and single-use kamikaze drones ("loitering munition"), in which drones fly over a combat zone, detect a specific target and destroy it.

A distinction must thus be made between attacks by the US Air Force in the context of combat operations or secret missions carried out by the CIA intelligence service ("targeted killings"). The latter strategy is highly controversial under international law and is still rejected by many states. Such missions require not only the aircraft itself, but also a prepared infrastructure, i.e. depending on the range: ground stations, encrypted data transmission, secure landing sites, a space component for data transmission, etc.[118]

Tab. 10.1: State ownership, procurement, and use of advanced combat drones[119]

State of proliferation of combat drones		
Possession	In-house production	China, Georgia, Iran, Israel, North Korea, Pakistan, South Africa, Turkey, Ukraine, USA
	Imported / leased	Egypt[c], Azerbaijan[b], France[a f,] United Kingdom[b], Iraq[c], Italy[a], Kazakhstan[c], Netherlands[a f,] Nigeria[c], Saudi Arabia[c], Spain[a f], Turkmenistan[c], United Arab Emirates[a c]
Previous missions with the use of weapons		Armenia, Azerbaijan, Great Britain, Iran, Iraq, Israel, Nigeria, Pakistan, Russia Turkey, USA, United Arab Emirates[c]
Ongoing procurement	Development	Germany[e], France[e], Greece[d], Great Britain, India, Pakistan, Russia, Sweden[d], Switzerland[d], Spain[d], South Korea, Taiwan, United Arab Emirates
	Import / Leasing	Australia[a g], Germany[b], India[b] Jordan[c], Poland[b], Switzerland[b]

The following applies: [a] U.S. makes, [b] Israeli makes, [c] Chinese makes, [d] development as part of a consortium, [e] development as consortium leader, [f] previously unarmed combat drone, [g] uncertain information or not yet determined.

In recent decades, more and more countries have invested in the purchase or development of their own combat drones (Tab. 10.1). A list from 2017 lists 35 countries with their own combat drones, with the USA, China, Israel, and Turkey being the main exporters. While only the US, UK and Israel had armed drones in 2011, the import of armed drones has increased dramatically since then. Between 2011 and 2019 alone, 18 states have acquired armed drones, 11 of which use Chinese products.[120]

In 2019, 10 states had already used combat drones, often in a way that was illegal under international law. The US' use of drones as part of the "Grand War on Terror" against non-state actors (Al-Qaeda and IS) was sensational, as was the assassination of Iranian Major General Qasem Soleimani in Iraq in January 2020. Other states have adopted

---

[118] The Bureau of Investigative Journalism's Drone Warfare Project lists between 8,858–16,901 deaths in these four countries between 2010 and 2020, according to publicly available sources. 10 %-13 % were civilians. https://www.thebureauinvestigates.com/projects/drone-war

[119] Data sources include: *Who Has What: Countries Developing Armed Drones*. North America Foundation (NAF) World of Drones website. As of January 20, 2020. Available at: https://www.newamerica.org/in-depth/world-of-drones/4-who-has-what-countries-developing-armed-drones/

[120] Under President Obama, the USA restricted the export of armed drones on the basis of the 1987 Missile Technology Control Regime and only supplied them to the UK and France under certain conditions. Under Trump, this regulation was relaxed and deliveries to Taiwan, the UAE and India were made possible.

these patterns, with Turkey carrying out drone attacks against the Kurdish Workers' Party at home, Nigeria against Boko Haram and Iraq against the "Islamic State". Saudi Arabia and the UAE are using combat drones in Libya and Yemen. Drone operations in the war in Syria and Libya by Turkey (e.g. in Idlib) and Russia have encouraged Azerbaijan to conduct similar operations for offensive purposes in the 44-day war against Armenia in 2020 in coordination with tanks and artillery. Sensitive US targets (UAV hangars, CIA buildings, etc.) have also been attacked in Iraq by "kamikaze drones" of Shia militias supported by Iran. Meanwhile, advanced attack drones are considered a serious threat by the US military[121] (cf. Arraf/Schmitt, 2021).

The Pandora's box of the use of combat drones is thus open globally. Current developments show that the USA is no longer alone in the combat drone market and has competition from China and Turkey. Today's drones enable precision strikes even against individual tanks, make the battlefield more lethal, for example in Libya, Syria, and Armenia, and change the use of combat drones in favour of the attacker, as aircraft are freed up for other missions.

In the German debate on the acquisition of combat drones, proponents argue that the Israeli drones to be acquired, the "Heron TP", are to be purchased for the protection of the Bundeswehr. However, it has just been shown that combat drones can be used particularly well offensively against weaker opponents and also have a psychological effect. They are particularly well suited as "intervention weapons" in asymmetric conflicts where air superiority is present (cf. Ehrhart, 2021).

Although drones are referred to as "surgical weapons" due to their accuracy, which is technically correct at first, since the target can be fixed or hit more precisely than cluster munitions, civilian casualties are nevertheless by no means excluded. The crucial thing in alleged "surgical warfare" is always target planning in favour of avoiding civilian casualties, because collateral damage is always possible with any munition. An evaluation of publicly available sources shows that 10 to 13 per cent of US attacks also hit civilians. Local proximity of combatants in areas inhabited by civilians is also accepted by combatants to cause civilian deaths, but initially does not change the technical definition of "collateral damage" of today's weapons systems and the effort required for target planning under international humanitarian law.

The combined use of cruise missiles and drones caused significant damage in the attacks on Saudi Arabia's Abqaiq and Khkurais refineries in 2019, significantly disrupting oil supplies. The Houthi rebels in Yemen, backed by Iran, claimed the attack. However, the attacks were not from the south (Yemen) but from the north (Iraq) or east (Iran): 25 unmanned aerial vehicles hit Aramco's oil tanks and processing facilities in two waves in a well-coordinated and pinpoint manner on 14 September 2019. Apart from the surprise effect of this "coordinated multi-attack", the stationed air defences (Skygard, Patriot) were also ineffective. In the future, such scenarios are very likely in war zones.

---

121 Baku used Israeli "Harop" kamikaze drones to attack Armenia's radar stations and take out air defences, and Turkish "Bayraktar TB2" drones to take out air defences and armoured vehicles.

It should also be noted that it is, in principle, not so complicated to intercept today's drones, once detected, as they fly slowly, manoeuvre little or cannot defend themselves. However, air defence is not prepared for this. It now stands to reason that considerable efforts will be made to develop future systems that are harder to detect, can evade or fly faster. The technological arms race will thus be further fuelled, as the next section will show.

Besides the USA, European states are developing drones with stealth technology, but also other states such as China, Pakistan, India, Russia, Turkey are developing advanced armed MALE drones (see Table 10.1 and next section).

10.2 What Future Developments Can Be Expected?

As shown, the combat drones in use today are primarily remote-controlled, but some also have semi-autonomous functions such as automatic take-off or landing. However, the development of fully autonomous systems is at the top of the glossy brochures of drone developers in technically advanced countries. In the civilian sector, too, the development of self-driving cars or automatically landing aircraft, drones, and helicopters is being promoted with heavy investments. The superficial reasons given for greater or full autonomy are always to relieve the burden on humans and expected cost reductions. In the military sector, it is argued that autonomy is necessary in the event that no data connection to the aircraft is possible or in order to shorten reaction times in combat situations. In the battlefield of the future, more "autonomy" is expected for various functions, including target planning.

This indeed raises the ethically relevant question of whether machines should and may decide on the killing of an enemy. Internationally, this has led to a debate, in which different parties and schools oppose each other. Related to this is the question of whether certain destabilising developments can already be identified, from which prohibitions must then be derived, for example through arms control regulations or bans under international law. First, however, it must be clarified how autonomous weapon systems (AWS) are to be more precisely characterised and technically realised.

The further development of modern warfare is closely linked to the concept of Revolution in Military Affairs (RMA) (cf. Neuneck, 2011). RMA can be understood as the combination of weapons technologies, military doctrine, and the reorganisation of armed forces, so that the nature of warfare to date is fundamentally changing. Today's drivers for the advancing RMA discussion are:

1. the structural changes in the international order, i.e. the preservation of the qualitative military-technical superiority of the USA and the arms competition with China and Russia;
2. the high investments in research and development (R&D) expenditures and military expenditures of the USA and and their scientific-technical, industrial support;

3. the dramatic development of information and communication technologies, and the

4. integration of various technologies in force structures, training, and deployment.

The Trump administration has coined the term "return of great power competition" in various key documents. This refers to the competition between the USA and China and Russia, into which the scientific and technological rivalry between these states is also increasingly being drawn. For example, the 2018 National Defence Strategy states: "The security environment is also influenced by rapid technological advances and the changing nature of war: "advanced computing, "Big Data" analytics, artificial intelligence, autonomy, robotics, directed energy, hypersonics, and biotechnology" (Department of Defence, 2018, p. 3).

A Central Question: What Are LAWS and What Control Are They Subject To?

Lethal autonomous weapon systems (LAWS) are unmanned weapon systems or carrier systems that have weapons on board and are used for combat purposes. In certain operational phases, they are subject to only limited or no human control and are capable of operating in a targeted and partially autonomous manner in a complex dynamic environment. The autonomy of action is handed over to the machine itself. So far, the international community has not been able to agree on a uniform definition for AWS or a usable delineation of automated, semi-autonomous, and autonomous weapons systems. Table 10.2 is an attempt to characterise and distinguish the different terms, with the technological concepts building on each other. The distinction between partial and full autonomy poses particular difficulties.[122]

Tab. 10.2: Concepts for increasing autonomy that build on each other

Term	Use	Examples
Automated	System follows pre-programmed commands without variations	Patriot, landmines, close protection against ships
Semi-autonomous	Certain phases of an operation take place fully autonomously	Brimstone (UK); Long-Range Anti-Ship Missile (LRSAM); various systems in planning and development
Fully autonomous	Adoption of human cognitive abilities in goal planning, tracking, etc.	

The lack of a uniform definition represents an obstacle with regard to regulatory considerations and possible arms control policy approaches to AWS. However, since research

---

122 The Pentagon had already presented a definition for an AWS in 2012 in a guideline: "A weapon system that once activated, can select and engage targets without further intervention by a human operator. This includes human-supervised autonomous weapon systems that are designed to allow human operators to override operation of the weapon system." Department of Defence, Directive Number 3000.09. Subject Autonomy in Weapon Systems, 2012.

and development are only just beginning here, it is difficult to estimate what future capabilities AWS will have at all (cf. Alwardt/Neuneck/Polle, 2017). Permanent arms technology impact assessments within the framework of preventive arms control with regard to R&D programmes are therefore necessary.

AWS do not yet exist or cannot yet be clearly distinguished from increasingly automated weapon systems with autonomous sub-functions. When examining future AWS, it makes sense to also include all those unmanned weapon systems that are in the planning or development stage that promise increasing automated or (partially) autonomous functions as well as capabilities and could be combined in parts or as a whole into an autonomous weapon system or an AWS system network in the future.

In view of the expected developments, another problem becomes apparent here: the increased "autonomisation of war", in which more and more sensors and weapon systems are connected to take over semi-autonomous functions on the battlefield. Also conceivable here is cooperation between one or more AWS and operators, i.e. the combination of manned and unmanned systems, in US jargon *"manned-unmanned-teaming" (MUM-T)*. This can bring a range of new capabilities and military advantages (such as longer endurance, higher speeds with much faster reaction times, and better environmental analysis for target selection).

In the USA, there are various research institutions and universities in the field of robotics and AI that are engaged in R&D relevant to AWS (cf. Boulanin, 2016). Both the Pentagon and the armed forces have drawn up corresponding "roadmaps", in which robotics and autonomy are described as key factors for further developments and procurement. In addition to research fields such as machine learning, big data, manufacturing techniques, robotics, the focus is also on miniaturisation, swarm behaviour, and autonomisation.[123] According to the Pentagon, nearly $15 billion was spent on R&D, procurement, and operations of unmanned systems from 2016 to 2018. In 2019, the Pentagon established the Joint Artificial Intelligence Center (JAIC) to advance basic research, technology development, and military integration.

The fear is that the strong research efforts also in the civilian sector (keywords: artificial intelligence and robotics), the future generation of "autonomous platforms", e.g. by integrating automated and fully autonomous functions, could increasingly yield new capabilities and more effective weapon effects. This could also lead to new destabilizing scenarios. Technically, these changes could take place in particular at the level of communication, computing capacities, and software development, which is why they are particularly difficult to detect or control.

In the USA, however, there were also civil society protests: thousands of Google employees opposed the company's participation in the DoD project "Maven", which was to

---

[123] In October, the US military tested 103 Perdix micro UAVs (290g, 30 cm wingspan) that have been released from fighter aircraft and can communicate with each other. Because these networked systems can work together to decide whether the mission purpose has been fulfilled, such as reconnaissance of air defence systems, analysts assign an AI capability to the micro UAV. DoD Announces Successful Micro-Drone Demonstration, Release No. NR-008-16; 9 January 2017.

develop AI-supported methods for image analysis to identify, classify, and track persons. Concerns about the "lack of human control on the violent use" of LAWS have been also articulated in an open letter by 85 German computer scientists, which work in the field of AI and robotics.[124] This demonstrates impressively that some civilian researchers are unwilling to work for military applications that are used for violent purposes. Instead, they advocate international regulations.

The aforementioned R&D areas of AI, robotics, etc. are transformative technologies today, the main impetus for which is emerging in the civilian or commercial market, but which will also be taken up by the military sector. In some states, this will feed into the development of military weapons and doctrine, as well as transforming the respective aspects of future warfare. In an increasing number of countries, interest is already emerging in the future use of automated or autonomous weapon systems with advantageous military capabilities. These transformative technologies are seductive and problematic at the same time, which is why a precise armament impact assessment is necessary based on agreed principles.

Development is not focused on drones or aircraft alone. Armed unmanned ground vehicles (UGVs) or armed unmanned surface vehicles (USVs) and underwater vehicles (UUVs) have so far only been developed by individual leading industrialised countries and have so far only been deployed on a very limited, militarily irrelevant scale.

The US has long been the trendsetter in the development and deployment of LAWS. Key strategy documents emphasise the importance of autonomy, robotics, and artificial intelligence in new weapon systems for air, underwater/surface, or land operations. The so-called "Third Offset Strategy", has budgeted $18 billion for R&D over the last five years.[125] Key areas here are autonomous learning systems, human-machine decision making. Specific UCAV programmes of the US Air Force are XQ-222 "Valkyrie" and UTAP-22 "Mako". The US Army has presented a "Robotics and Autonomous Systems (RAS) Strategy", in which various robotic systems (for reconnaissance, transport, and combat) are to be integrated into US Army formations between 2035 and 2040. In the UK, the defence sector is working on autonomy and stealth programmes for UCAV (Taranis), France is leading the nEURon technology consortium to build a European combat drone, also involving Italy, Sweden, Spain, Switzerland, and Greece. Initial test flights have been conducted. These demonstrators, as well as unmanned escort missiles, are part of the French-German-Spanish Future Combat Air System (FCAS) project, which will focus on both a sixth-generation manned multi-role combat aircraft and unmanned escort aircraft, as well as new weapons and communication systems. Other examples from Israel, Russia, and China could be added.

---

[124] Offener Brief: Initiative für ein internationales Abkommen zu Autonomie in Waffensystemen, sponsored by the "Gesellschaft für Informatik, 1. November 2021, https://gi.de/meldung/gi-mitglieder-unterstuetzen-initiative-fuer-ein-internationales-abkommen-zu-autonomen-waffensystemen-1

[125] The first offset strategy is the introduction of nuclear weapons, bombers etc. in the 1950s. In the 1970s, microprocessors, stealth etc. made new conventional weapons possible. See: Robert O. Work; Shawn Brimley: 20YY. Preparing for War in the Robotic Age, Center for a New American Security, January 2014.

What Are the Expected Uses of LAWS?

Due to their capabilities and resulting advantages, LAWS will be particularly suitable for military operational scenarios that are difficult or too dangerous for human operators due to environmental conditions. Examples include heavily defended terrain on land or so-called *anti-access/anti-denial spaces* that require very fast reaction times or a high degree of manoeuvrability for "air combat". Characteristic here is a dynamic environment without a permanent communication link. Covert operations behind enemy lines are also likely. With regard to propulsion, navigation, and communication, other technologies are required on land or in the air than, for example, underwater, in the mountains, or under rapidly changing weather conditions.

The deployment of future LAW systems suggests changes in some areas of warfare, from which different security and peace policy problems and technological risks can be derived:

Due to the dwindling human influence on concrete processes in combat operations, the associated increase in difficulty of assigning responsibility under international law, and the danger that the principles of international humanitarian law can no longer be adequately taken into account, the risk of the use of force increases in principle. Furthermore, a lowering of the inhibition threshold for the use of force and the use of LAWS in the context of "anti-terrorism" actions or for the targeted killing of people could prevail.[126]

Due to the high level of technology, LAWS are also more susceptible to external electronic interference such as jamming or spoofing or possible system manipulation (*hacking*). Unmanned systems already show a considerably increased risk of accidents and failure due to technical errors. Future machine learning combined with AWS and the independent addition or extension of their programming to an "extended artificial intelligence" also entail a risk of unpredictable behaviour and possibly "unpredictably" acting AWS.

Misguided or unreliable LAWS may have destabilising or escalating consequences in a crisis. The accelerated pace of warfare associated with LAWS in automated warfare, and thus a greater burden on operators in decision-making processes, can lead to a sharp reduction in necessary pilot deliberation time, which in turn can lead to an unintended escalation in a crisis. Moreover, LAWS in crises could also induce more proactive military behaviour and more dangerous operations. On the other hand, the risk of escalation to full-blown war may be lower in a combat operation with drones because information can be obtained more easily on the ground and rational decisions can be made without fear.

The introduction of LAWS in a regional context may also have implications for regional or strategic stability: At the regional level, a military adversary would most likely react to a qualitative superiority in the field of LAWS by stronger armaments

---

[126] This argument applies not only to LAWS operations in war scenarios, but also in the context of "internal security" by police and security forces.

or new military strategies. Regional armament induced by LAWS could thus have a destabilising effect on existing power constellations in the medium to long term and increase the risk of escalation in a region. Finally, LAWS could also enable new military deployment scenarios based on their potential capabilities (e.g. in the area of maritime warfare or in *A2/D2 situations*[127]), which ultimately can lead to a shift in the balance of power in a region and thus, in turn, to new armament efforts.

Of particular importance would be scenarios in which nuclear weapons are integrated or involved, be it that new attack options on nuclear targets are pursued by means of conventionally armed UCAV/LAWS, be it that new nuclear-equipped unmanned delivery systems are deployed to undermine an opponent's second-strike capability within the framework of nuclear deterrence and thus endanger the strategic stability of deterrence. In this case, further nuclear disarmament would no longer stand a chance; nuclear modernisation or even rearmament would be the likely consequence.

Due to the increase in efficiency and capability potentially associated with them, LAWS represent a future means of further developing state weapons potential. An increase in the military efficiency and striking power of the initiating state can therefore lead to competitors also increasingly procuring LAWS in order to either maintain the respective balance or to change this in favour of one side. The consequence would be induced offensive/defensive armament spirals and associated new operational doctrines. The danger of a regional or global arms race in the LAWS sector increases, combined with a considerable increase in military potential and the resulting risks. A technological race of drones and anti-drone measures is also to be expected.

The majority of AWS hardware and software are, in principle, dual-use technologies that are largely freely available today and will be subject to few arms export restrictions. The development of very powerful LAWS will probably remain the preserve of a few key states for the time being. However, other states could develop less powerful LAWS or arm commercial autonomous systems or replicate existing or captured systems. In June 2021, a UN report caused a stir, stating that the combat drone "Kargu-2" of the Turkish company STM had directly attacked soldiers with autonomous control in the Libyan civil war (cf. Cramer, 2021). It is not known whether any damage was done. The incident also illustrates that verification of the autonomy function is a significant problem if the drone is not physically available (cf. Kallenborn, 2021).

This path is generally also open to non-state actors. The degree of autonomy of a system is largely determined by the programming, whose algorithms and programme components are for the most part not of decidedly military origin and are therefore difficult to control in proliferation. Furthermore, increased competition for the export of UCAVs can also be observed between some key states (e.g. USA, China, Russia), which poses the risk that existing export restrictions could increasingly erode and the proliferation of AWS could be accelerated.

---

[127] A2/AD means anti-access or anti-denial warfare. Defensive weapons or tactics such as land-mines or air defence are used to prevent an adversary from invading or traversing an area of land, sea or air.

10.3 How Can Future LAWS Developments Be Limited?

The preventive containment, limitation, or complete prohibition of LAWS is in principle possible through ethical or (international) legal prohibitions, manufacturing, or proliferation provisions, arms export control regulations, or arms control treaties. In recent years, there has been a broad international debate and diplomatic as well as civil society initiatives on LAWS. This raises the question of whether it is possible in today's international relations to preventively limit weapons systems that do not yet exist or to ban their use altogether, so that states do not even enter into their development and procurement or limit their use in the event of war. This is the task of preventive arms control ("ius contra bellum") and international humanitarian law ("ius in bello"), each of which is based on different factual situations, principles, and instruments. It must therefore be clarified whether the international community and states can agree on new prohibitions, verification measures, and sanctions.

So far, none of the existing arms control treaties directly prohibit, limit, or regulate LAWS in particular, simply because the corresponding technological developments were not even possible at the time they were drafted. However, both arms control law and international humanitarian law have developed a broad portfolio of principles, criteria, and measures (e.g. verification) that are in principle applicable to LAWS. It should be emphasised that it has been possible to ban certain types of weapons in five cases: Chemical Weapons (CW- Convention), Biological Weapons (BW Convention), Anti-Personnel Landmines (Ottawa Convention) and Cluster Munitions (Oslo Convention) as well as Blinding Laser Weapons (CCW).

Arms Control and Arms Export Control:

Disarmament and arms control primarily contribute to risk and threat reduction by being oriented towards the balance of power of certain weapon systems and actors as well as their verification capabilities. Proven criteria are non-proliferation, crisis stability, and arms control stability, i.e. avoiding an arms race. They refer either to complete bans or the setting of upper limits of defined weapon systems or expected damage effects. One focus of the treaties established during the Cold War (see, for example, Table 10.3) was the limitation of weapons of mass destruction and their quite visible delivery systems such as missiles, bombers or tanks. The verification of these treaties is primarily focused on easily identifiable delivery systems.

In the future, subjects of restraint or prohibition could also be autonomously acting weapon systems, which would thus in principle become part of the arms control process. The treaties listed in Table 10.3 provide for consultative commissions in which LAWS can be included in the treaty provisions. The 1991 Treaty on Conventional Armed Forces in Europe (CFE) covers the comprehensive limitation of five main conventional weapons systems in Europe, in which LAWS could also be integrated. Against the background of the general arms control crisis, the CFE Treaty has been suspended since 2007, but a new edition could draw on procedures and knowledge that have been devel-

Tab. 10.3: Arms control treaties and their applicability to LAWS (Alwardt/Neuneck/Polle, 2017)

Arms Control Treaty	Frame	AWS part of the contract?	Verification	Status
CFE Treaty	multilateral	Probably AWS in general*	Yes	suspended since 2007
New START	bilateral	Yes, autonomous UCAVs or UUVs; if they have characteristics of a strategic system and are intended for nuclear armament.	Yes	in force (duration until 2021; extension until 2026)
INF Treaty	bilateral	Disputed whether UCAVs are to be equated with cruise missiles, if applicable.	(Yes) has been considered implemented since 1991	Terminated
CW-Convention and BW-Convention	UN-Framework	Yes, AWS in general, if they are intended to participate in any way in the use of chemical or biological weapons	Yes (CWC) No (BWC)	in force
* If AWS meet the definition of one of the major weapon types; the definitions there do not exclude unmanned systems (probably a matter of interpretation).				

oped and verified for years. This also refers to the broad set of instruments of the OSCE or the Vienna Documents, in which various specific, risk-reducing "confidence- and security-building measures" (CSBMs) have been negotiated and successfully applied.

The prohibition of a certain degree of autonomy of weapons systems to be defined will hardly be realisable (or verifiable) on its own, since autonomous functions are essentially determined by electronics and software. It would be easier to regulate AWS either by specifying concrete conditions of use or by ensuring *"meaningful human control"*, although it remains open what the characteristics of "meaningful human control" (MHC) are based on. The British NGO "Article 36" has elaborated and deepened the concept of MHC. International Humanitarian Law (IHL) (see next section) provides principles and criteria for this. For example, the practical application of Article 36 of Geneva Additional Protocol I in relation to the testing, development, procurement, or introduction of new LAWS could have a preventive effect.

However, LAWS could also become part of preventive arms control and future (regional) arms control regimes (e.g. in Europe and South Asia or in the strategic context between e.g. the USA, Russia, and China) and thus contribute to crisis stability and international security. In the event of resuming or further development of the CFE Treaty, future military technology developments such as LAWS would also have to be taken into account. In addition, international efforts in the field of non-proliferation of AWS could be intensified, even if this is made considerably more difficult by the dual-use problem.

Many autonomy-relevant technologies arise in the civilian sector, cannot be banned, and therefore spread quickly. Only the integration into a LAW would characterise such a system as a weapon.

Strengthening arms export regimes with regard to AWS systems or important (identifiable) key technologies would be another important step here to prevent or curb the proliferation of dangerous military technologies. National arms export control is also guided by international regulations and contractually agreed restrictions on the supply of strategically important military-relevant technologies. Examples of multilateral arms export agreements are the *Missile Technology Control Regime (MTCR)* of 1987[128], the Wassenaar Arrangement of 1994, the Arms Trade Treaty of 2013, or the UN Arms Register on Transparency and Confidence Building of 1991. LAWS could be included in national export control regimes. However, they are not part of international arms control law, as they are unilateral supply agreements of certain groups of states.

International Humanitarian Law: the CCW Conventions and the UN GGE

International humanitarian law (formerly: international law of war) primarily includes provisions of international law that aim to protect civilians, buildings, and infrastructure as well as the natural environment as best as possible from the effects of hostilities in the event of war or international armed conflict ("ius in bello") (cf. Geiss, 2015). The four Geneva Conventions from 1949 and two additional protocols from 1979 are central here.[129]

Central principles for the use of weapons are the *principle of proportionality*, the principle *of distinction*, the *precautionary principle* and the principle of *avoiding unnecessary suffering. The so-called* "Martens Clause" states for all other cases which are not regulated internationally that civilians and combatants are protected by the principles of international law "as they result from the usages established between civilised nations, from the laws of humanity and the requirements of the public conscience." In 1980, the Convention *on Certain Conventional* Weapons, or the CCW, was adopted in Geneva, entered into force in 1983 and has so far been signed by 125 states.[130]

As contained in the full title of the Convention on Certain Conventional Weapons, CCW, "Convention on prohibitions or restrictions on the use of certain conventional weapons which may be deemed to be excessively injurious or to have indiscriminate effects", the aim of the CCW is to assess new conventional weapons to determine whether their use could "cause excessive suffering or have indiscriminate effects" and whether

---

[128] Here, a group of 35 states sets common export standards for ballistic missiles, cruise missiles, and UAVs on an informal and voluntary basis. The MTCR website is available at: http://mtcr.info/.

[129] Geneva Conventions (1949) and Additional Protocols (1977). Available at: https://ihl-databases.icrc. org/applic/ihl/ ihl.nsf/vwTreaties1949.xsp [28.09.2017].

[130] See: United Nations Treaty Collection, Chapter XXVI Disarmament, CCW (with Protocols I, II and III), Geneva 1980: https://treaties.un.org/pages/ViewDetails.aspx?src=TREATY&mtdsg_no=XX-VI-2&chapter=26&lang=en.

they should therefore be prohibited or restricted in declared wars or armed conflicts (cf. BICC, 2013). Five protocols regulate and prohibit different types of weapons and ammunition, such as (1) undetectable fragments in firearm ammunition (1980), (2) mines and booby traps (1980, amended 1996), (3) incendiary weapons (1980), blinding laser weapons (1995), and explosive remnants of war (2003).

Within the framework of "humanitarian disarmament", there have been civil society groups and non-governmental organisations for decades, which are also directly supported by individual states and have made it their task to achieve better protection of civilians in the event of war and in the aftermath and to ban dangerous weapons systems. The "International Campaign to Stop Killer Robots" aims to preventively stop and prevent the development, production and use of LAWS. The International Committee of the Red Cross (ICRC) has also issued several statements on LAWS[131] (see ICRC, 2019). The ICRC has proposed standards for the type and quality of "meaningful human control" in order to comply with IHL principles. International NGOs do not oppose autonomy in weapons systems per se, but call for the prohibition of certain weapons systems that can select and fire on targets without human intervention. States like Bolivia, Cuba, Ecuador, Egypt etc. or the European Parliament support this position. There is also, with varying intensity, support from academia, religion, the "AI community", and occasionally industry.

Since 2014, initial expert discussions on legal, technological, and military aspects of LAWS have taken place in the framework of the CCW in Geneva. As of November 2017, these discussions were transferred to a UN *Group of Governmental Experts (GGE)* mandated to "consider and agree on possible recommendations on options related to emerging technologies in the field of LAWS in the context of the objectives and purposes of the CCW". The development of a common language basis with regard to the concept of autonomy and practised transparency with regard to development and armament efforts in the field of LAWS would be a first important step. Moreover, not only existing proposals on transparency and confidence-building should be discussed, but also the circumstances under which LAWS could be prohibited by Additional Protocol I of the Geneva Conventions or another rule of IHL should be examined. Initially, the debate focused on the technological criteria of autonomy and then shifted to the operational context of a LAWS deployment. The UN GGE's final reports show very slow progress. In 2018, the GGE succeeded in establishing 11 "guiding principles". In 2019, it was agreed that these principles could form a basis for possible recommendations, and in 2020, it began to explore where common ground could be found.

While the CCW context is primarily concerned with the conformity of LAWS in relation to the principles of IHL, efforts to preventively control the arms of AWS or to

---

[131] For the landmine campaign, these include the International Campaign to Ban Landmines; the International Network on Explosive Weapons (INEW) and the International Campaign to Abolish Nuclear Weapons (ICAN) Stephen D. Goose; Mary Wareham: The Growing International Movement Against Killer Robots, in: Harvard International Review, Vol. 37, N°. 4, 2016, pp. 28–33.

integrate them into conventional arms control measures in the future have been lacking. AWS can so far only be implicitly covered by some arms control agreements (e.g. CFE Treaty), confidence- and security-building measures (e.g. Vienna Document), or export control regimes. This is only possible for the most part because the weapons definitions there, which are aimed at manned systems, can also apply to certain unmanned systems. However, these agreements either have only a very limited scope of application, are of a voluntary nature or are not (any longer) practised in conformity with the treaties, so that they cannot have the necessary effect to effectively counter the security policy implications of LAWS.

In his book "The Culture of War" (1995), the military historian John Keegan comes to the following conclusion:

> "The peacekeepers and peacemakers of the future have much to learn from other military cultures, and not only from those of the Orient, but also from primitive ones. Underlying the principles of voluntary limitation (...) is a wisdom that must be rediscovered. And it is even wiser to contradict the view that politics and war are only steps on one and the same path. If we do not firmly contradict this, our future (...) may belong to the men with the bloody hands"(Keegan, 1995, p. 553).

Part III
Political Regulation of Digitalisation

Introduction

Heinz Stapf-Finé

This third part of the book is dedicated to the political consequences that result from the first two parts of the book. The VDW assumes that the development and use of digitalisation, networking, and artificial intelligence requires active policy-making. We will present this in key policy areas such as education, health, environment, economy, labour, and social affairs.

As an extremely topical example, it should be emphasised that, for example, under the Corona pandemic conditions, the digitalisation pressure on schools and other educational institutions such as daycare centres for children and universities has increased enormously. Discourse on this usually deals with the question of accelerating the use of hardware and software. What is needed, however, is a primarily pedagogical debate on whether, when, and how digital technologies can be used sensibly to ensure learning success in the sense of "finding out" instead of "cramming". This example already makes it clear how strongly the effects of the use of technology depend on the social forms of organisation, in which the respective technology is used.

This is also clear in medicine and health care. Ethically, the non-use of digital technology, if it is superior to humans in diagnosis and treatment, would be unacceptable. On the other hand, digital technology can serve to control and monitor the (correct) actions of patients or those working in the health professions and lead to a loss of autonomy of action (in addition to higher health insurance costs).

How and to what extent can digital technologies be used to solve social or political problems?

This question also arises urgently in the context of digitalisation and sustainability. First of all, digital technology is energy-hungry, and the use of digitalisation (e.g., blockchain, streaming, and delivery services) is associated with high CO_2 emissions, even if globally active companies promise climate neutrality in the near future in their high-profile corporate responsibility campaigns. So, on the one hand, digital technologies can help solve environmental and energy problems, but on the other hand, they can also make them worse. Comprehensive technology assessments and, above all, democratic control of technology development and deployment, and thus ultimately of ownership, production, and distribution are essential.

This is made clear in the chapter on production and trade, which highlights the changes in the global economy and world trade under the conditions of the fourth industrial revolution. On the one hand, negative megatrends such as climate change, population growth, and growing social inequality could also be combated with digital tools. Rising productivity would also make it possible to increase human welfare. In reality, however, most emerging and developing countries are falling further behind in their economic performance in digitalisation-driven structural change, and international inequality is increasing significantly. This trend has been further accelerated by the Covid-19 pandemic and has hit South American countries, particularly, with great force.

The increase in productivity through digital solutions and AI applications will not be without effects on the quantity and quality of work in the various sectors. Even if at the moment the effects on (also cognitive) routine activities (e.g. accounting) are still being strongly discussed, there is an increasing threat of employment losses also in areas that previously did not appear to be rationalisable, such as medicine and law. Currently, there are also no indications that the fourth industrial revolution will lead to rising labour incomes or better working conditions; this seems rather unlikely at present, so that the need for political regulation becomes clear here as well.

Moreover, the questions raised in this context also make it clear how important it is to include all working people in the protection of social security. However, the unconditional basic income discussed in this context seems to be not only an undercomplex answer to complex questions, but also to drive the desolidarisation in society. Improved social protection in the face of increasing digitalisation also requires a social debate on how the digital dividend can be distributed more fairly.

The consequences of digitalisation in essential, possible fields of application discussed in the third part of the book thus forcefully show the necessity of comprehensive technology assessments for the protection of the individual, society, and the environment and make it clear that different scientific disciplines must conduct a joint discourse in social responsibility. It also becomes clear that only some of the technological impacts are inherent. The way in which digital technology is used within the framework of concrete social norms and rules, in the operational, social, and political context, determines whether the further digitalisation, networking, and development of AI benefits or harms the majority of people. The yardsticks here are people's individual self-determination and their fair participation in the wealth of the earth.

Chapter 11

Education and Digitalisation – Technology Assessment and the Demystification of "Digital Education" in Theory and Practice

Paula Bleckmann and Brigitte Pemberger

11.1 Introduction

> *"Schools, universities, but also educational institutions from kindergarten to lifelong learning are under increasing pressure to digitise quickly and extensively. It is therefore not surprising that educational institutions are currently inundated with a growing number of funding and advisory initiatives, but also training and further education offers. However, by far the majority of these activities are related to digital devices and the skills for (effective) use of hardware and software. Aspects of critical debate are frequently reduced to topics such as data protection or ethics, cyberbullying or fakenews."* (Hartong, Förschler, Dabisch, 2019)

It can already be considered a step in the right direction if there is at least a critical examination of the topics mentioned at the end of the quote. Under the pressure of the rapid implementation of online contact options in the pandemic-related lockdown, even these limited concerns were often neglected (cf. section 1.1). In the following chapter[132], however, with a view to the more distant future, precisely because of the new dynamics that have arisen with the measures to contain the pandemic, a number of fundamental and critical theoretical considerations are brought together that have so far been too rarely considered in the current political and academic discourse on "education in the digital age" (section 2) and especially also in the context of media education and teacher training in the digital age (sections 3 and 4). The focus here is on school age. However, many of the considerations are likely to be transferable to educational processes from kindergarten to adulthood.

11.1.1 Polarisation of the Discourse Due to the Pandemic Lockdown

The lockdown situations with closures of educational institutions such as schools and universities have in many cases exacerbated an already existing polarisation in the discourse on education in the digital age: while on the one hand politicians, practitioners, and experts ascribe to the lockdown the function as a kind of catalyst for developments that are desirable or already long overdue, on the other hand various experts document

---

132 The present chapter was written on the basis of a revised version of the first half of a paper by Bleckmann and Zimmer (2020), supplemented by remarks on the concept of analogue digidactics (Bleckmann, Pemberger, Stalter and Siebeneich, 2021), initial results of the MünDig study (Kernbach, Bleckmann, Tetzlaff, Pemberger, 2021 submitted), as well as reflections from the transfer initiative www.unblackthebox.org, which was created with Bleckmann's participation.

and criticise the acceleration and exacerbation of developments that are questionable or already proceeding at too high a speed. The first school of thought is represented in Germany both by the "Hackathon #wirfuerschule" and particularly succinctly by one of its organisers, Verena Pausder.

> *"Corona was an ideal tutor for the schools [...]. We have a digital pact to finance digital infrastructure, but the funds have hardly been called up. Very few teachers are trained. That means that now we were all forced to deal with the subject. And that reduces fears, lowers the inhibition threshold and perhaps also shows us that we are better than we thought."* (Pausder, 2020)

The other stance is found in Lankau and Burchardt:

> *"Instead of optimising school and teaching through digital transformation for metrics and technology, the focus must be back on the individual, community and humane learning processes. Digital technology can be one tool among many. Education, however, is relationship: the human being becomes a human being through the human being."* (Lankau, Burchardt, 2020)

From an analysis of international documents on the positioning of the "edtech" industry, i.e. business actors active in the field of hardware and software for educational institutions, Williamson and Hogan conclude:

> *"The business plan adopted by the edtech industry is summarised as 'support now, sell later', where businesses are expanding their services now in the hope they might lure schools and parents into long term subscriptions once the pandemic ends".* (Williamson, Hogan, 2020)

We have argued ourselves that it is above all important not to make the crisis the yardstick for future learning outside lockdown conditions:

> *"The pandemic seems to have removed the burden of proof: There appears to be no more need to show that digital education has better long term outcomes than other types of new or traditional learning arrangements. It is enough to know that it is better than nothing. It IS much better than nothing. It is like a straw that we hold on to because it is the only straw we think we have. That's easily justifiable. Only let's not forget to let go of the straw once the immediate danger of drowning is past."* (Bleckmann, 2020)

11.1.2 A Contradiction: Evidence of a Potential!?

Jesper Balslev (2020) derives the following conclusion from a summary of empirical evidence on the actual effects in connection with the analysis of international policy documents on the topic of "educational digitalisation" from the last three decades: Although the empirical evidence shows negative effects of the use of digital media on learning performance in numerous cases, more rarely also positive and in many cases neutral, the demand for increased digitalisation of educational processes has persisted for decades. Although the demand is not yet "evidence-based", a high POTENTIAL of digitalisation is considered to be proven: *Evidence of a potential,* is the title of the paper. In order to be able to eliminate this inconsistency in the future, an approach is needed that always compares the innovations through digitalisation with what could be achieved with modern pedagogy without its use: *"A more rigorous approach would be to compare analogue and digital interventions more systematically. [...] My analysis is that there is a technological bias at play, witnessed by the absence of analogue control-groups"* (Balslev, 2020, p. 154).

This *technological bias has been* reinforced by the pandemic circumstances. To counteract this, we again explicitly take Balslev's suggestion into account when presenting results from the MünDig[133] study. It is not designed as a control group study, but would, among other things, enable a detailed description of such *analogue control groups,* since it also covers analogue learning scenarios for the promotion of "digital competences" (section 5). In section 2, we adopt the interdisciplinary broad view of technology assessment (hereafter: TA) including analogue control groups and work with a comprehensive understanding of possible digital risks and digital opportunities for learning (cf. Zimmer, Bleckmann, Pemberger, 2019). In doing so, we follow, among others, Armin Grunwald (2020), who, as head of the Office of Technology Assessment at the German Bundestag, criticises a power- and corporation-driven determinism of technology and states: *"Enlightenment today means a digital maturity in which critical and uncomfortable questions are asked."* (Grunwald, 2020a)

Our questions would completely miss the intention of the authors, and fail being "unpleasant" in this positively critical sense if they were instrumentalised to legitimise an attitude of blanket rejection of digital media use in the classroom. After including the results of a long-term, transdisciplinary technology assessment according to the current state of knowledge, digital media would, probably be used more sparingly in everyday school life than seems politically desirable at present. In our opinion, however, they would enrich teaching in a much more sustainable way, because they would be used where and only where their use was considered beneficial after weighing up all the advantages and disadvantages. In contrast, an experienced or actual "forced digitalisation" often leads to educators at all levels of education and training working with digital media, but mainly using them in a way that minimises their workload, in order to meet the external requirements (cf. Zierer, 2018, more details in section 3).

11.2 Designing Human-friendly Technology Environments Instead of Self-optimisation and Forced Digitalisation

11.2.1 Counterproductivity in the Philosophy of Technology

As early as the 1970s, the philosopher of technology Ivan Illich coined the terms conviviality (Illich, 1973) and counterproductivity (Illich, 1973). In his works, he shows the dangers of a *counterproductivity of* technological development using the example of the systems of school, medicine, traffic and initially also media. He understands this to mean that a modern production system is first conceived, realised, and financed in order to provide benefits to the users. Development then continues in the same direction, but its overall effect is reversed from positive to negative: The proliferation of technology

---

[133] A nationwide online survey (www.muendig-studie.de) conducted as part of the research project "Media Education in Progressive Education (Montessori and Steiner) Institutions" funded by the Software AG – Foundation.

leads to a situation where only a few privileged user groups benefit from the development. Long before the internet and smartphones entered everyday life, Illich criticised *"a news system whose flood of information undermines meanings and swamps sense, growing dependence cemented by awareness"* (Illich, 1982, p. 135). He advocates rather than expecting individuals to self-optimise in order to, remain healthy and productive despite counterproductive developments in society as a whole, society should limit itself to so-called "convivial tools[134]", i.e. to design technological developments in such a way that it would be easy for the individual user to use them to support their goals without harming him or herself (Illich, 1982, p. 135).

Almost 40 years after Illich, the American sociologist of technology Sherry Turkle (2011) describes counterproductive effects of digitalisation on human social life in everyday life, i.e. not in educational institutions. Another five years later, an analysis of longitudinal data on the everyday lives of US adolescents reveals a picture referred to as the "smartphone turn" (Twenge, 2017): at about the same time as the widespread use of smartphones from around 2010 onwards, the respondents' self-report of life satisfaction declines, while there are increased reports of loneliness and depression, and other negative trends. (among other things).

11.2.2 Problem Dimensions Time, Content, and Function

According to a study by the German DIVSI (Institute for Internet Trust and Safety) (2018) entitled "Euphoria is past", approximately a third of respondents in Germany aged 14 to 24 are afraid of being "addicted to the internet", and around 40 percent are afraid of a largely digitalised future. The average screen time of German children and adolescents exceeded the maximum time recommended by experts by a factor of two in 2014 (Bitzer, Bleckmann, Mößle, 2014). In the lockdown, the screen time of children and adolescents has further increased by a factor of 1.5 to 1.75 according to preliminary study results (Langmeyer, Guglhör-Rudan, Naab et al., 2020; Felschen, 2020), so that screen time is now likely to be about three times the recommended maximum. The increase is even more pronounced in disadvantaged social groups. Screen media use for school purposes goes on top of this, as the above numbers refer the use in leisure time only.

However, the question of the temporal extent of the use of digital screen media is only one of at least three relevant criteria that allow us to distinguish between use that promotes learning and development and use that impairs development. We consider the delimitation of and overlap between three problem dimensions and two modes of use to be important (cf. Bleckmann, Mößle, 2014). The problem dimensions include *content* (what does the screen show?), *time* (how much time is spent in front of the screen?), and *function* (e.g. is it used to fight boredom? Are dysfunctional moods suppressed? Is the screen used in the family structure as a babysitter, means of education/pressure, etc.? Are social contacts suppressed or supported?). In view of developments in the last

---

[134] This means: tools for human coexistence

decade, we would expand the systematisation and also ask: what user data is collected, what profiles are generated (possibly as a sub-dimension of "content")? How long are the uninterrupted time windows of "leisure" (possibly as a sub-dimension of "time")? The modes of use are distinguished between "foreground media exposure" (the learner deals directly with the medium), "background media exposure" (a device runs in the presence of a learner) and "technoference" (a reference person such as a parent or teacher is distracted from communicating with the learner by a mostly mobile device[135].

11.2.3 Media Effects Research Between Private Leisure Use and Use for Educational Purposes

It is also important to distinguish between research results for extracurricular and school screen media use. In the balance, extracurricular or leisure use has a small but significant effect in inhibiting learning (Mößle, Bleckmann, Rehbein, Pfeiffer, 2012). Use at school or in the context of learning at school has very different effects, so far neutral with inconsistent findings, sometimes promoting learning, sometimes inhibiting learning depending on the form of use (Balslev, 2020, Zierer, 2015). There is a highly complex interdependence between the learning climate at home and the use and equipment of children with digital screen media in and out of school (see Fig. 11.1)[136]. "Tablets/laptops" (in the middle of the figure) are the most frequently used devices in Germany, but here they are representative of all digital screen devices used in the context of the educational institution, i.e. also PCs and smartphones. During the school closures in 2020/21 due to the pandemic, the place of use has completely shifted to the home, so that both online lessons and the use of digital devices for homework as well as "leisure use" take place in the private environment, which makes it difficult to limit screen media use.

On the one hand, children from disadvantaged social classes in Germany have on average "better" media equipment than their more privileged peers. As a result, however, they have significantly higher screen times, which are on average in the problematic range, and increased use of content that is not suitable for their age, which can partly explain their significantly poorer school performance (Pfeiffer, Mößle, Kleimann, Rehbein, 2008). On the other hand, the digitalisation of learning processes is propagated as an instrument to close the educational gap and to provide targeted support for students with learning difficulties. The example of the Romanian "learning PC lottery" raises awareness of the ambivalence of the availability of digital devices for use by socially disadvantaged children and young people, especially outside school (Pop-Eleches, Malamud, 2010). By winning a lottery ticket, disadvantaged young people could win a PC for free. Like their more privileged peers, they were supposed to be able to gain access

---

[135] See also Barr, Kirkorian, Radesky, 2020
[136] Figure taken from: Bleckmann, P., Allert, H., Amos, K., Czarnojan, I., Förschler, A., Hartong, S. Jornitz, S., Reinhard, M./Sander, I. https://unblackthebox.org/wp-content/uploads/2021/08/UBTB_One-pager_Physical_Health.pdf .

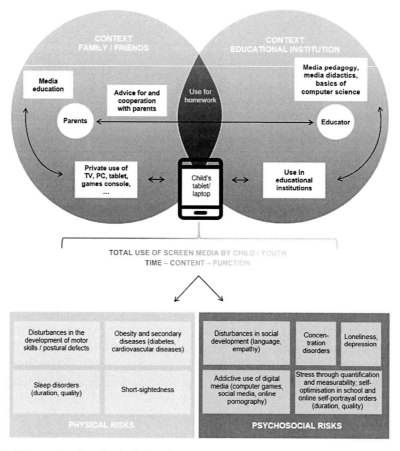

Fig. 11.1: Diagram showing the physical and psychosocial consequences of problematic screen media use in children and adolescents, taking into account the interdependence between, and factors influencing their use in private and educational contexts. (adopted from: Bleckmann, Allert, Amos, Czarnojan, Förschler, Hartong, Jornitz, Reinhard, Sander, 2020)

to computers and the Internet as a learning resource. In fact, the devices were mainly used for non-intended purposes, so that, on a longitudinal basis, the screen usage times of the "winners" increased and school performance deteriorated significantly.

According to the results of media effects research, the small but significant negative effects of screen media consumption are more pronounced the younger the users, the longer the times, and also for use without adult caregivers. Effects on physical (sleep disorders, obesity, delays in movement development), psychosocial (loss of empathy, concentration disorders, delays in language development) and cognitive development (measured by school performance) can be considered proven. A somewhat older literature review of more than 200 individual studies and about 40 meta-analyses and reviews, which is currently not available in comparable quality, summarises these results clearly (Mößle, 2012). In addition, there are now initial studies on the effects of *background me-*

dia exposition, according to which it has a negative impact on parent-child-interaction (Radesky, Miller, Rosenblum et al., 2014; McDaniel, Radesky, 2018), which is why the phenomenon is also referred to as "technoference" (from technology and interference).

The Freizeitmonitor (German annual leisure report) 2018 indicates that the widespread dissatisfaction with the control of one's own media use described above is not a phenomenon limited to adolescents: In reality, media leisure activities occupy the top 5 places for the most time-consuming leisure activities. However, the wishes of the adults surveyed go in exactly the opposite direction: they want more "real leisure time": more time to relax, more time in direct contact with friends and family, more time for sleep (Reinhardt, 2018). In the perception of educational practitioners, prevention of problematic screen media use is a big issue. In a survey on subjective further training needs among educational professionals at kindergartens and primary schools, it even ranked first among the topics listed in the survey (Kassel, Fröhlich-Gildhoff, Rauh 2017). Practising educators who work with younger children predominantly see the impairments of child development that problematic screen media consumption causes in terms of time, content, and function, and would like further training to counteract this by supporting and counseling parents and by working with the children, directly. For older children and adolescents (whose teachers were not included in the survey), we would assume that the need for further training would now include the prevention of digital risks and the use of digital opportunities for learning processes on an equal footing. The existing further education courses are now increasingly addressing the topic of digital risks, but mostly not at the level of reducing the use of screen media at home or in educational institutions (setting-based approach), but by enabling the students, or in some cases children from kindergarten age, to protect themselves from the risks through individual competences. Hanses criticises what he calls coerced self-optimisation in the area of health promotion and prevention and calls for setting-based rather than individual-based approaches (Hanses, 2010, pp. 89–92). In our opinion, the same applies to "media prevention": Approaches that centre on the individual (often: pupils) and want to train digital-risk-avoidance competences would *run the risk of delegating the handling of counterproductivity to the individual and* transferring the responsibility for his/her own failure to him/her through a coercion to self-optimisation in the face of unfavourable environmental conditions. This asks too much not only, but especially from younger people. Here, too, a social divide is likely to be intensified. Many less priviledged families lack the resources to implement clear rules on the one hand and to offer non-screen alternatives on the other, both in order to limit the time and content of digital media consumption in everyday life. The explosive nature of this dynamic is exacerbated in times of lockdown, especially by the loss of leisure activities that promote balance and community.

11.2.4 Lack of Interdisciplinary Discourse on "Education and Digitalisation"

In a position paper, the study group "Education and Digitalisation" within the Federation of German Scientists (VDW) called for a technology assessment that should not only be interdisciplinary, but transdisciplinary, and mentions in key words several disciplines that have so far not been taken into account in the discourse, or only insufficiently so (Vereinigung Deutscher Wissenschaftler, 2019). It would go far beyond the scope of this chapter to go into detail on each of the disciplines and topics excluded or underrepresented in the political discourse. Some of the topics mentioned have already been addressed above, and a few more will be explicated below. For all others, a few references, which do not claim to represent the respective discourse completely, have been included in the table below in order to facilitate follow up beyond the key words. In addition, a helpful resource is the "alternative checklist" for a (self-)conscious digitalisation of educational institutions compiled by the transfer initiative "UNBLACK THE BOX" (Hartong, Bleckmann, Allert et al., 2020), on which brief explanations and references to further literature can be found for 12 different critical questions.

It is noticeable that the disciplines that have been more strongly included so far (left column of Table 11.1) tend to emphasise positive aspects of digitalisation in education, while on the right, less considered so far, more critical aspects of educational digitalisation and more generally of digital media use by children and young people are highlighted. Even if this is observable as a tendency, limitations and exceptions can be found: For example, the aspect of handling personal data is often addressed in political documents, possibly as a cover-up for omitting other critical aspects. Altenrath and colleagues (Altenrath, Helbig, Hofhues, 2020, p. 584) even go so far as to conclude from their document analysis that *"data protection and data security [...] are the only issues that are critically discussed in the overall context"*. Thus, a statement demanding that instead of the planned use of the cloud software MS 365 at schools in the German federal state of Baden-Württemberg to rely on open source solutions was supported by a broad field of 20 organisations, including teachers', consumer protection, and parents' associations, but also by the German Society for Computing Science (GI).[137]

The allocation to the columns experiences an additional differentiation or softening when a distinction is made between the specialised discourse of individual disciplines and its reception at the political level. For example, Alternrath and colleagues (2020) criticise, based on the careful and comprehensive analysis of funding guidelines and programming at the German and European levels on the topic of education and digitalisation, that due to the *"interpretive sovereignties caused by discourse and power politics"* (Altenrath, Helbig, Hofhues, 2020, p. 584), a certain understanding of the goals of "education in the digital age" comes to the fore: *"Digital competencies" ultimately focus on technical-instrumental skills in the operation and application of technologies as well as the ability to use media for one's own actions"* (ibid., p. 584). This narrow focus cannot be reconciled with the much broader understanding of the goals of many authors of media education, which is aimed more at "digital sovereignty"[138].

---

[137] See https://unsere-digitale.schule/ [20 February 2021]
[138] Digital sovereignty, or media maturity (Medienmündigkeit) as it is called by Bleckmann (2012) also includes the ability to recognise situations/tasks where the non-use of digital media is preferable. For

Tab. 11.1: Dominance of individual disciplines in the political discourse on "education and digitalisation"

Disciplines that currently dominate the discourse on "digital education"	Previously neglected disciplines that should be included for transdisciplinary technology assessment
• Media didactics • Media Education • IT development (especially applications with the aim of increasing and standardising the recording of learning outcomes as well as educational governance) • Data-based/quantitative educational research • More recently also: (Legal) expertise in the field of data security and data processing.	• Historical and philosophical educational research (e.g. Hübner 2005) • Educational inequality research (e.g. knowledge gap, second and third level digital divide, e.g. Deursen, Helsper 2015) • Sociology of education (e.g. generation of social inequality through higher surveillance density, "inequalities of dataveillance": Hartong, Förschler, Dabisch 2019; critical sociology of education, Hauser 2011) • Algorithms, AI and Inequality (Allert 2020) • General didactics and teaching research (Zierer 2015, Hattie 2015) • Media addiction research (te Wildt 2015, Turkle 2011) • Public health and prevention science (e.g. Bitzer, Bleckmann, Mößle 2014; Sigman 2017) • Attachment research (including effects of digital distraction of caregivers of young children: Radesky, Miller, Rosenblum et al. 2014; McDaniel, Radesky 2018). • Paediatric and developmental psychological media effects research (Mößle 2012, Spitzer 2012) • Environmental medicine effects research (including non-ionising electromagnetic radiation, cf. www.diagnose-funk.org; https://www.radiationhealthrisks.com/scientific-studies/) • Critical Data Studies (political education on data education, Hartong 2016, Selwyn 2014) • Ecology (energy and resource balances of the production and use of digital devices: Grunwald 2020; Brunnengräber, Zimmer 2020; Gotsch 2020; Held, Zimmer 2020; Sommer, Ibisch, Göpel 2020).

11.3 "Added Value" and "Technology Acceptance" Yesterday and Today

11.3.1 Limitations of Technology Acceptance Concepts

Rekus and Mikhail see the teacher as "the most important 'medium' for student learning" (Rekus, Mikhail, 2013, p. 236). According to their understanding of the term, media have always existed in the classroom. Media in a narrower sense came into the classroom with the introduction of the first textbook by Comenius, followed by media no longer in use today such as diorama or magic lantern. School radio, school television, language laboratories and computer rooms were set up in schools at great financial expense in the corresponding periods of the 20th century, with recurring expectations of improvements in the areas of learning efficiency, motivation, and individual support (Hübner, 2005, pp. 274–293). But in each case, the results of the accompanying research on the new learning media did not meet the high expectations, so that Hübner describes a common sequence model with three phases: *euphoria – stagnation – disillusionment.*

---

example, a shopping list can be written efficiently with pen and paper; a handwritten love letter may be valued far more by the addressee.

According to the model, we are (still) in the phase of euphoria with regard to the use of online media in the classroom.

In the course of each of these historical cycles, low technology acceptance on the part of teachers was criticised as problematic by the innovators, who wanted to see the new technology increasingly used in the classroom, because it inhibited their idea of "innovation". With hindsight, the teachers' scepticism often proved to be justified. The criticised low "technology acceptance" could even be described in retrospect as reflective and forward-looking in the light of the evaluation studies, which in each of the cases mentioned showed cost-benefit balances that fell well short of expectations. In some cases, the introduction of the new technologies even had counterproductive effects in the sense of Illich (1982): Thus, in most cases, instead of the hoped-for decrease in educational inequalities after the introduction of the respective teaching technology, an increase was shown.

We therefore consider the current models for measuring technology acceptance[139] to be interesting, but the interpretation and evaluation made in the publications (high technology acceptance is good, low is bad) to be untenable. It leads to the problematic but consistent circular immunisation process described by Balslev (2020): The mantra is: More use of digital media in schools leads to better learning performance. Studies that examine the use of digital media in real-world settings to date predominantly show no increase in learning performance. Well, this must be due to the fact that teachers are not sufficiently qualified and motivated to use them. Measures that only qualify without increasing technology acceptance do not lead to the goal, since only both together lead to knowledgeable use. As soon as professionals have acquired both in the future, the use of digital media will lead to better learning performance, q.e.d. According to Rekus, instructional media are "indispensable 'means' to support teaching and learning" and he distinguishes between "teaching aids" and "learning aids" (Rekus, Mikhail, 2013, p. 234).

However, no object is always a "medium" by itself, but objects are only turned into media through their task-related use in learning processes:

> "On the one hand, *media should help the teacher to control the learners and present the subject matter so that he can plan and organise effective learning processes. On the other hand, media should contribute to supporting the students in their independent engagement with the respective tasks, so that they can begin their learning processes motivated, successfully persevere through them and evaluate them at the end*". (ibid., p. 236)

To do this, however, they need a broad spectrum of skills, which is comprehensively described by the TPACK model[140] (cf. Schmidt, 2020, pp. 74–84). In our opinion, the model can be followed well if the "T" is not understood to mean only (digital) technology, as many authors do, but both analogue and digital media. It would then be more aptly called the MPACK model, "M" for media.

---

[139] E.g. Technology Acceptance Model TAM (Nistor, Lerche, Weinberger et al., 2014) for practising teachers or Anderson, Maninger 2007 for future teachers.

[140] Technological Pedagogical Content Knowledge, originally called TPCK model

11.3.2 Questionable Benefits of Digital Media Use in the Classroom: (In)Visible Learning?

In his meta-analysis of empirical educational research "Visible Learning", John Hattie (2015), to whom Zierer (2015) also refers, examined six factors that influence learning success (i.e. quantifiable learning performance in the form of school grades): the learners, the parents, the school, the teacher, the teaching, and the curricula. Even though all six factors interact with each other, the role of the teacher is central since both areas "teaching" and "teacher" depend on the attitudes and competences of the teacher (Hattie, 2015, p. 280). From the authors' point of view, a fundamental weakness of the Hattie study is the sole reliance on quantifiable "learning achievements", which is not well compatible with a humanistically oriented understanding of education. Do we want to maximise outcome (good grades)? How are good grades related to other desirable "outcomes", like if we want to help children become democratic citizens, critical thinkers, designers of a sustainable future? Even if only this abbreviated form of determining "learning output" is taken into account, the greatest influence on learners, according to Zierer and Hattie, is exerted by passionate teachers, for whom there is an interplay of subject competence, pedagogical competence, and didactic competence: "The successful teacher acts like a director. He/she always has the goals of the lesson in mind, checks the selected methods, and takes into account the prerequisites of the actors" (Zierer, 2015, p. 91).

Specifically on teaching, Zierer (2015) found that (new) media achieve only a very low effect and that this is fairly constant over the last 20 to 30 years, even when differentiating between computer-supported and web-based learning. Zierer uses the SAMR model (Puentedura, 2006) as an explanatory model. He sees the reason for low effects on the learning success of the students in the fact that the teachers would often only use the new media as a substitute (*substitution*) or extension (*augmentation)* for the traditional media. Therefore, the acquisition of new media alone is not sufficient. Only if the tasks were *modified* or *redefined*[141] would the effect of digital media be more positive (Zierer, 2018, pp. 73–81). Zierer also explores the question of the extent to which the SAMR model could be transferable to the use of the above-mentioned classical media (without screen, not digital). Interestingly, according to Zierer, positive effects on learning success can also be measured in experiential education or cooperative learning, so that he concludes that

> "*successful teachers see themselves as change agents and do not use methods for the sake of methods, but always against the background of the learning situation. Digitalisation in the classroom does not mean using new media because they are en vogue at the moment. Rather, digitalisation in the classroom means weighing up the possibilities and the needs of the learners and only using new media when they are the best choice*" (Zierer, 2018, p. 107).

In this context, one could also speak of the learning goal of a *media literate teacher,* if the definition of media maturity (Bleckmann, 2012) is extended to the needs of both

---

[141] And accordingly adapted to a new conception of tasks! [Note by the authors]

persons, namely those of the teacher and the pupils. "Media Maturity" can only be achieved by those who know their own long-term goals and needs, who can assess the different media with their opportunities and risks, with their potential to satisfy these needs, and who can translate these considerations and deliberations into decisions and actions in everyday life" (Bleckmann, 2012, p. 34).

Weighing up the different alternatives is called "selection competence"[142] and is a sub-dimension of media maturity (ibid., pp. 103–108). The teacher's task is therefore to ensure an appropriate balance between protection against digital risks and empowerment for their responsible use, depending on the developmental stage of the students (cf. Bleckmann, Mößle, Siebeneich, 2018). This selection competence as an informed and reflected weighing of alternatives is essentially nothing other than a "technology assessment in miniature", i.e. *TA in a nutshell* at the level of an individual teacher. It should be noted that Zierer and Hattie's analyses are opportunity-oriented, but not risk-oriented. They ask "What does this do for learning success?", but not "What harm does this do elsewhere?". We clearly recommend going a step beyond the classical understanding of media didactics as represented by Süss and colleagues (cf. Süss, Lampert, Trueltzsch-Wijnen, 2013, p. 171) by not only considering the possibilities of media in the context of teaching and learning (both formal and informal), taking into account the preconditions on the part of the learners as well as the "framework conditions" present in each case, but also additionally including the respective risks in the use of digital media in comparison to analogue media in the sense of all the subject areas mentioned above in the right-hand column of Table 11.1. This recommended expansion is not considered in Hattie's analyses, i.e. risks of digitalisation outside the risk of lower school performance were not taken into account, so it is to be expected that this additional weighing of opportunities and risks would have an influence on Hattie's derived recommendations.

11.4 Exaggerated Dichotomies and Asymmetrical Promotion in Business and Politics Hamper TA on a Large and Small Scale

The implementation of critically-balanced, media mature decisions for or against the use of digital media in digital education policy as well as in practice is currently made more difficult by various factors.

---

[142] See also the contribution by Link, J. in this volume (Part 1, Ch. 3, Path dependence). Selection competence, which functions as an essential "organ" in decision-making processes based on the perception of actual needs and options, can be influenced by path dependency, among other things. The impression of an independent choice then arises for the respective deciding person, although system-immanent processes have largely taken this away from him or her through (customised) "preselection". Online systems reinforce these effects, as the constant availability of "an expert" (peer group, best friend, online encyclopaedia, choice buttons, recommended play list, etc.) may reduce or "unlearn" the weighing of different alternatives as a partial dimension of media literacy.

11.4.1 Questioning the Dichotomy: Digital=Modern=Good, Analogue=Regressive=Bad?

A number of recent publications on "education in the digital age" make an unfavourable simplification, which we have marked in bold in Table 11.2 below. Very simplistically, Lisa Rosa (2017) distinguishes between two different models of thinking, one good and one bad: She denies analogue media the qualities demanded by Rekus when she compares school and learning in the print age with that in the digital age. She comes to the conclusion that in the book-printing age the (outdated) central model of thinking was about "cramming in" and in the digital age the (modern) central model of thinking is about "figuring out". This also roughly corresponds to the narrative in Dräger and Müller-Eiselt (2015).

Tab. 11.2: Four-field scheme for cramming vs. figuring out, with vs. without digital screen media

	Old fashioned / traditional Instructive "Cramming in"	New / modern (Co-)constructivist "Figuring out"
With digital media	Predefined learning paths and learning content +/- "mock personalisation": input, output control, individual calculation for new input possible. Example: learning app, educational films	**Independent appropriation of the world, individually or in groups, research, processing and/or presentation.** Example: Pupils create explanatory videos
Without digital media	**Predefined learning paths and learning content, "Nuremberg funnel"** Example: Frontal teaching with blackboard writing	*Independent appropriation of the world, individually or in groups, experiments, research, processing and/or presentation.* *Example: Open forms of teaching (e.g. Reinmann-Rothmeier, Mandl, 2001), action/experiential education, Montessori, etc.*

In contrast, Muuß-Meerholz (2019) notes that digitalisation does not

> *"automatically bring more progressive pedagogy into education. At the moment, it looks more like the opposite: With new media, old pedagogies are optimised. More input, more practice in the traditional sense. More decontextualisation, more learning alone, with a fixed outcome, with predetermined meaning. We are optimising and reinforcing the flaws of teaching and learning in the book-printing age."*

Muuß-Meerholz thus supplements Rosa's good-bad dichotomy into a four-field scheme, within which, however, he only very briefly addresses the field at the bottom right (modern, without digital media). We have therefore added our own considerations in italics to Table 11.2. In our opinion, the false impression is often created that the transformations of educational processes described in the SAMR model are linked to the use of digital screen media. Through open forms of teaching, through action, experiential and progressive education (e.g. Montessori and Steiner) approaches, independent world appropriation is supported without the use of digital media and thus learning is fundamentally modified (M) or even redefined (R).

Thus, when using digital as well as analogue media in the classroom, both "cramming in", i.e. traditional instructional pedagogy, and "figuring out", i.e. (co-)constructivist pedagogy, are possible. Furthermore, as indicated by the dotted line between the right and left columns in Table 11.2, the "cramming in"-"figuring out" dichotomy does not take into account that teacher-centred instruction (whether supported by digital or analogue media) occurs in many different facets and does not only include memorisation for exams, which is associated with the term "cramming", nor does (co-)constructivist world appropriation automatically succeed better in groups or alone. "Figuring out" can also be the goal of ostensibly teacher-centred "frontal teaching"[143]. In fact, the Hattie study (see above) finds a positive effect of teacher-centred teaching on the learning success of students compared to open forms of learning.

It is surprising that despite the obvious theoretical shortcomings, the dichotomies drawn by Rosa and Draeger do not receive the expected critical reception, but are much quoted and positively taken up in the public debate in Germany. Not surprisingly, but particularly bluntly, this happens in the advertising statements of media corporations, and not only in the advertising of individual products, but also in the overarching measures of "public perception management" (cf. Linn, 2005). However, the simplistic line of argumentation outlined above can also be found again and again in international education policy documents over the decades: "Digital technologies *support new pedagogies that focus on learners as active participants*" (Balslev, 2020, p. 146). This could be explained, at least in part, by the following second point, which further complicates media literacy decisions by teachers.

11.4.2 Asymmetric Promotion in Business and Politics

In the German, but also in the international economic, educational, and science policy space, there is currently an asymmetry that does not treat analogue and digital learning paths equally, but rather favours those practices, science, and further education in whose implementation digital screen media are used. Lankau characterises and criticises the growing influence exerted by large, mostly international economic players from the media sector, who derive financial benefit from an increasing use of digital media in education (Lankau, 2017, pp. 20–36 and pp. 100–110). Förschler also describes in detail the emergence of large intermediary actors in the education sector and their influence on education policy discourse. These do not represent corporate interests directly, but the corporations that are represented steer decisions and provide large parts of the funding, including the German "Alliance for Education". These intermediary actors in turn exert influence indirectly through policy advice, and directly through their own continuing education programmes and the funding of digitised educational practice (Förschler, 2018, pp. 30–48).

In addition, the education curricula of the German federal states increasingly include mandatory use of digital media in schools. The disbursement of millions for the Qualitäts-

---

143 See also genetic-socratic exemplary teaching in Wagenschein, 2008, pp. 115–118.

offensive Lehrerbildung (Quality Initiative Teacher Education) in the area of qualification of pedagogical staff for teaching in the digital age, the University (Hochschule) Forum Digitalisation, the Digitalpakt#D worth billions and other major initiatives contribute to the fact that practitioners often experience a "digital coercion". This is regrettable, as many of the initiatives mentioned are not so narrowly defined: The long-term goal of the initiatives is to make young people "skilled for orientation in the digital world" and not necessarily to push the use of digital media[144]. There can be analogue alternatives for this. Balslev asks: "Can we train programmers without the use, or with very limited use, of technology (footnote: csunplugged.org)? How would students perform in settings that focused on attaining the grammatical, mathematical, logical, and social skills that often constitute the background factors for much of digital professionalism?" (Balslev, 2020, p. 146). It is all the more gratifying that with this anthology, many more contributions represent a view that could be assigned to the right-hand column in Table 11.3 below. In it, we have summarised in key words for different levels the essential differences between a digital coercion in schools and teacher education experienced by practitioners (not necessarily intended) and the approach of technology assessment we favour.

Tab. 11.3: Digital coercion vs. weighing long/short-term opportunities/risks for planning the use of digital screen media in school and teacher education

Digital coercion	Weighing up long/short-term opportunities/risks (TA)
Digital.	Digital or analogue?
Digital from an early age.	Differentiation according to level of appropriation: "Analogue" experience of the material world as the starting point for the digital.
Use the opportunities of digitalisation!	Weigh up the opportunities and risks!
What do the pupils learn from the medium?	What do the pupils learn from the relationship with the teacher (role model), what about the medium? How do the pupils relate to the world (also to each other)?
What does the screen show (content?)	What, how long, what for? (3 problem dimensions: content, time, function)
What is the effect in the short term (days and months)?	Also: What is the long-term impact (years and decades)?
Focus on the added value of achieving a specific learning goal in the school context (often individual training of subject competence in a school subject or higher-level "media competence" of the students).	Focus on many different influences (added value/advantage and "less value"/disadvantage) at the level of subject competence, personal development and health of the students, self-image, health and role of the teacher, as well as higher political level).
Often underestimation of costs (only acquisition, not maintenance costs/staff training/obsolescence).	Inclusion of long-term costs.

---

[144] It is worth taking a look at the 24 tasks of the Media Competence Framework NRW (pp. 10/11), which pupils should have acquired by the end of the 8th or 10th school year (Medienberatung NRW 2019). Most of the 24 tasks can be taught without the acquisition and use of screen media, even though the visual language used in the brochure does not suggest this.

11.5 Research Practice in a Differentiated Way and Its Further Development in a Health-Promoting Way

11.5.1 Media Maturity Matrix: Which Medium at Which Age for Which Purpose?

In the MünDig study, professionals, parents and older pupils were asked about their attitudes in the field of "Education in the Digital Age" using an innovative online survey instrument that was administered to all three target groups in a slightly adapted version. Based on qualitative preliminary studies, it had emerged that the established survey instruments were poorly accepted at reform educational institutions[145] due to a lack of differentiation. For a total of 10 areas (learning tasks/purposes), respondents were asked in which age range which concrete example activities with media with or without screens would be considered useful by the respondents. The figure shows that parents consider certain activities in the area of "producing and presenting" to be useful already at kindergarten age, while they consider others to be beneficial only well beyond primary school age. It is noticeable that the activities for the promotion of media competence that do not require the use of digital screen media all have a lower average "recommended starting age" from the parents' point of view than the activities that are linked to the use of screen media. These are descriptive results from the MünDig parent survey (Kernbach, Bleckmann, Streit, et al. 2021). So far, only the results on the need for further training are available from the survey of professionals. They show a high need for further training, which on the one hand lies in the area of acquiring technical application skills for the use of digital media in the classroom, and in some other areas, and on the other hand indicates an equally high need for further training on possible courses of action to help children become "fit for the digital age" without the use of screen media.

11.5.2 Analogue-Digidactics

In this context, therefore, and not only for progressive education oriented institutions, the question is, not whether children should actively design, communicate, research, etc. with media, but whether they should understand the basics of information processing systems instead of just learning how to use them. The answer to the second question is clearly yes. It is important to know which medium or which methodological-didactical approach seems particularly suitable for which age. We consider a type of didactics to be particularly suitable that minimises the use of digital screen media and has the long-term goal of initiating an active, critically reflective approach to digital media that is based on a fundamental understanding of the concept of media worlds. In summary, we outline the didactic foundation of *analogue-digidactics*, which is intended to enable adolescents to aquire skills for the digital age by starting off in an "analogue" way, taking into account the learning prerequisites according to their developmental stage. Analogue-digidactics

---

145 The survey was conducted at Montessori and Waldorf schools and kindergartens.

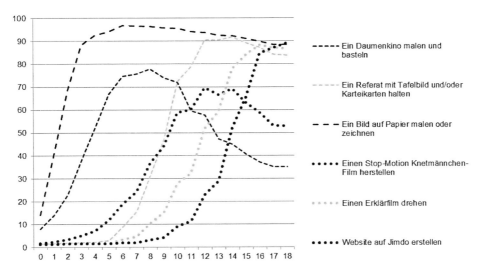

Fig. 11.2: Parents' survey MünDig study Area 1 of 10: Producing and presenting: Meaningful media educational activity by age (in %); case numbers: N=2187–2402

thus clearly distinguishes itself from digital-digidactics, which currently stands for teaching and learning using digital screen media.

In the course of the political demand for "digital education", it is important to remember that digital education begins long before children and young people work with, through, and on digital media. The foundations for this are laid with the acquisition of skills that are best learned in real life and much more difficult to acquire through "learning on the screen". In the best case, these are learning experiences that have a high degree of discovery character, which leads to an acquisition of knowledge that is based on hands-on experiences and insights and cannot be "taught" in this sense (cf. Stiftung "Haus der kleinen Forscher", 2018, p. 73). In the area of basic computer education in particular, there has been an increase in recent years in approaches that focus more on the structured thinking skills (computational thinking), mathematics, and language development that underlie programming and other skills for handling and designing information-processing systems than on technical use skills.[146]

The examples of analogue-digidactics for pedagogical practice in kindergarten and primary school were described for the first time in the research project "Conceptual development of the prevention programme ECHT DABEI"[147] (Bleckmann, Pemberger, Stalter et al., 2021, pp. 58–59 and 86–91) and focus on the following principles:

---

146 E.g. Curzon, Mc Owan, 2018; Stiftung "Haus der kleinen Forscher", 2018; Köhler, Schmid, Weiß et al. n.d.; Best, Borowski, Büttner et al., 2019; Humbert, Magenheim, Schroeder et al., 2019; Hromkovič, Lacher, 2019; Hauser, Hromkovič, Klingenstein et al., 2020; Initiative "CS unplugged – Computer Science without a computer": https://csunplugged.org/de/ [15 February 2021].

147 Before the renaming in 2015, the project was entitled "Conceptual development of MEDIA PROTECT" at the Alanus University of Arts and Social Siences in Alfter/Germany within the HLCA con-

- *Analogue before digital*: Get to know principles of media worlds first by using media without screens. Because "media education must build on the media evolution" (te Wildt, 2015, p. 308), screen time should not be increased unnecessarily.
- *Producing before consuming*: Putting the active creation of the analogue and later also the digital media worlds (not "pressing buttons" and "operating machines") in the foreground.
- *Transparency before "black box"*: Work with teaching and learning materials that allow maximum transparency of the mode(s) of operation. End devices such as tablets are therefore only suitable for children to a very limited extent because they have hundreds of internal processes that are not visible to the user, as in a "black box".

Practical projects of analogue-digidactics[148] can prove to be particularly valuable for pedagogy and media education because they can be realised at all educational institutions regardless of the respective pedagogical orientation. The intent is that the participating children and young people should experience themselves as actively active shapers, creators, and explorers, which has a strengthening effect on their autonomy and their highly individual development (cf. Antonowsky, 1997). All concepts of the practical projects of analogue-digidactics described to date can be "grasped" in the truest sense of the word. In an action-oriented experience, the children collect ideas for screen-free leisure activities "by the way", so to speak. The implementation of the practical examples requires neither an expensive infrastructure nor regular updates with new software versions. The transparency of the teaching and learning materials used invites children to get to the bottom of the question "How does it work?" in order to be able to critically encounter new technological developments later in adolescence and adulthood. The use of transparent media, such as a music box fed with (self-composed) punched paper strips to understand the basic principles of information-processing (digital![149]) systems, also has the advantage that students *learn that understanding is possible (!)*: The transparency of the chosen medium makes this possible – none of the processes involved are inaccessible to the fundamental ability to understand.

---

sortium (Health Literacy in Childhood and Adolescence) in cooperation with the Freiburg University of Education. Funding: Federal Ministry of Education and Research (BMBF).

148 See also the project "Analog-Digidaktik – Wie Kinder ohne Bildschirm fit fürs digitale Zeitalter werden" (Analog-Digidactics – How Children Without Screens Become Fit for the Digital Age) funded by the Software AG Foundation and the Waldorf Education Research Center from 2021 to 2023: https://www.alanus.edu/de/forschung-kunst/wissenschaftliche-kuenstlerische-projekte/detail/analog-digi-daktik-wie-kinder-ohne-bildschirm-fit-fuers-digitale-zeitalter-werden [26 June 2021].

149 See Chapter 2 by Gernler/Kratzer, whose well-founded discussion of digital technology, invoked in some circles as salvific, traces back to the bit as the smallest possible unit of information, as the carrier of the unambiguous answer to a yes/no question (corresponds in the case of the music music box to the question of sound/no sound, which is encoded on the punched tape as hole/no hole) and shows the relevance that knowledge of this fact should also have for decision-makers at the highest level. In the authors' view, the naïve overreliance on processes in the economy, society, politics, and the private sphere is linked to optimistic assumptions that are untenable on closer examination. In this context, too, the dissemination of analogue-digidactics seems promising.

In addition, analogue-digidactics has the following quality features: It is mobile and usable and implementable without electricity indoors and outdoors; free of advertising and unsuitable content; it is also inclusive, gender-appropriate, and ecologically sustainable".[150] So it is ultimately not surprising, but very gratifying, to read the concluding words of this chapter, coming from a fifth-grader in a group discussion commenting on a series of lessons carried out over several weeks. The lessons covered code "from ancient cuneiform script to the ASCII alphabet" and were carried out without the use of a computer[151], but with extensive use of a binary marble-adding machine, the Binary MAMA. This fifth-grader sees the computer "disenchanted" in the best sense of the word:

"I grasped that computers can't think, they can only calculate. "

---

150 In our opinion, the latter should be given much more weight in the near future in (educational) policy decisions for reasons of long-term sustainability, according to the recommendations of the demystifying reading of the "Jahrbuch Ökologie. Die Ökologie der digitalen Gesellschaft.", edited by Maja Göpel and colleagues (2020), much more weight should be given to the latter in (educational) policy decisions in the near future for reasons of long-term sustainability.

151 This example is only given here as an example. There are also concepts and practical reports on the "analogue-digidactic" teaching of other phenomena of digital media worlds. On the role of social media for the social class structure, the computer scientist and educator Corinna Sümmchen has been conducting "social media unplugged" with her classes for several years; see Sümmchen, 2019.

Chapter 12

Under Which Conditions Can 'Digitalisation' Contribute to UN Sustainable Development Goal 3 'Health and Well-Being'?

Johann Behrens

Introduction

The chapter summarises arguments from the VDW volume "Digital Promises of Salvation" (Korczak, 2020, Behrens, 2020) and is divided into three sections: The first section introduces the use of the terms (and devices/tools) digital and digitalisation in health care, with a special focus on the anthropological and technical significance of prostheses, diagnoses, and coordination. The second part follows on from previous chapters and clearly shows that much of what is attributed to the technology of digitalisation is not inherent to the technology, but derives from the social forms of organisation in which digital technology is used. Treating digitalisation as a productive force that asserts itself independently of the relations of production greatly obscures the point. The third part briefly explores some state attempts to regulate and embed digitalisation, closely following the main report of the German Advisory Council on Global Change (WBGU) 2019: "Our Common Digital Future" (WBGU, 2019).

12.1 What Does 'Digital' Mean in Healthcare? Prostheses, Training, Diagnoses, and Coordination

The term 'digitalisation' used in this book – fashionably common but highly narrowed – means the use of learning machines, neural networking, and deep learning. Learning tools are of enormous significance in healthcare, both in prosthetics and in diagnostics and coordination.

As Chapter 5 by Schmiedchen also states, prostheses are "an artificial replacement for a body part" (Kluge, 1999, p. 651) and have been in use for over 400 years. In their high functionality, which is the result of outstanding human developmental achievements, neither their non-digital form (e.g. glasses) nor their digital form (e.g. cochlear implant) shows their artificial intelligence, their empathy and tenderness, their cleverness, and their inwardness (= intellegere, Etymol., 2020, p. 585), but the intelligence of cobblers, tailors, or software developers alone is revealed. Very many compensate for their limited ability to run with a bicycle or automobile, use drills for drilling, typewriters for writing, or exoskeletons for lifting and carrying. Typewriters have recently been programmed as robots to recognise the patterns of our vocabulary and alert us to errors or alternative words. Paraplegic amputees hope that in the future they will be able to control their prosthetic hand with their thoughts alone, thanks to a chip in their brain, and that they will no longer be dependent on interrupted nerve pathways. Surgeons can already operate much more precisely and error-free with robots. The idea of a surgeon sitting with me while her robot operates on me is comforting, even though it might not be necessary. She can do her mail or read a good book while she is at it, if she only looks at me and the readings on her screens that control the operation from time to time.

For decades, people have used training automata that provide feedback. In language learning, automatic machines compare the pronunciation of a word with a standard and automatically give feedback. Such programs are not only appreciated by students; in a refined form that recognises individual patterns, they are also used by rehabilitation patients after strokes to regain lost language and other skills. Automata that can rephrase a person's statements into question form in a way popular in conversational psychotherapy caused quite a stir: "So you mean that ...?" (Behrens, 2019, 2020). This does not transform such automata into teachers or psychotherapists. Rather, they are technically supported self-therapy and self-learning. Questioning one's own statements and feelings is also possible in an analogue way, e.g. with the help of a diary, self-talk, or mental training. In order to function, training automata do not need any understanding of a language, a subject, or a person.

Trackers or wearables that record vital signs (Korczak, 2020, Behrens, 2020) digitally support permanent self-measurement and can thus trigger diseases known as cyberchondria (Heyen, 2016, p. 13; WBGU, 2019, p. 263, Behrens, 2020). However, this undermining of human self-esteem already outlined by Bauberger and Schmiedchen in this book does not stem from the digital devices used, but from a culture of competition, which is often also functional for the preservation of domination (more detailed in Part 2 of this chapter; see also Behrens, 2020). The de facto forced "voluntary" sharing of tracker data for individual advertising (big nudging) is not a technical problem either. Its cause is rather economic interests and old-established social relations that shape the use of digital technology (see part 2). The use of digital devices in combination with economic measures that are only relevant to advertising enforces a quantitative level of transparency that concretely threatens self-determined participation, idiosyncrasy, and privacy, and thus human dignity. Moreover, they are already undermining the solidarity-based financing of the health care system (WBGU, 2019, Behrens, 2019, 2020).

The danger of humanising digital automata is more difficult to assess. Without a doubt, it updates the bad habit of old homunculi horror stories of painting googly eyes on robots, as well as the completely misleading use of language of supposedly 'learning' machines and networks that are equipped with 'artificial intelligence' and supposedly drive 'autonomously' and make 'decisions'. Scientists are solely responsible for this completely misleading use of language (WBGU, 2019, p. 72). They encounter a situation in which almost all people anyway maltreat their machines (cars, washing machines, etc.) with curses, insults, and pet names that would not be appropriate for open machine systems but for self-willed living beings. In a randomised controlled pilot study of a hairy and 'talking' computer much used in rehabilitation and care in the form of a seal named 'Pabo', we endeavoured to clarify the effect and acceptance of this feedback automaton (Karner et al., 2019). The effect and acceptance corresponded roughly to that of a soft toy, bed bunny, or cuddly cushion, with which almost all people have comforting conversations with themselves from childhood.

All these prostheses and tools, including the particularly fast and persistent 'learning' machines that process information correlating to patterns, have in common that they only process input data and further develop originally given models and algorithms. Even if they continuously document patterns that no human being has ever seen before, they only process information. Machines are not intelligent in the sense of emotional and social intelligence (Etymol., 2020, p. 585). They do not live and do not want anything. They are not curious and do not seek sexual partners. They are not 'Homunculi' of which literature has been full not only since the Renaissance and Hoffmann's tales. All their (mal)functions have one cause: the people who originally set them up and put them into circulation and now keep and use them. The responsibility lies with them, not with the machines (see also Bauberger and Spennemann in this book).

Diagnostics and information-based coordination are, of course, also the basis of the prostheses described, but not all diagnostics and coordination lead to prostheses. Machines not only enormously facilitate the detection and transmission of vital signs and their reaction to them (cf. VDW, 2020, Korczak, 2020, Behrens, 2020). Above all, they show correlations in the documentation of patterns ('correlations') that the setters of these machines had never thought of. What we call 'digitalisation' today, i.e. the pattern recognition of learning machines and networks in Big Data, owes its origins to the health sector. After its beginnings with the itinerant healers of the island of Kos in the fifth century BCE (cf. Behrens, 2019), it was fully developed by the 'vital statisticians' of the 17th century at the latest. These used process-produced and, if necessary, self-collected data to recognise patterns, i.e. correlations, which were not previously conscious, indeed, which no one had suspected or asked about. John Graunt wrote the classic 'Observations' in 1665 full of previously unnoticed, i.e. invisible, regularities in population development. Quetelet (1869 and 1870) summarised such correlations, which could only become visible through vital statistics, in the construction of average men, an homme moyen, who does not even have to exist as a living person and yet says a lot about social patterns (see also Chapter 1.1).

It is such correlations of all possible characteristics that appropriately set-up machines can now calculate in immense numbers in a very short time. Since the inputters of the 'data' and the inventors of the original algorithms, which are then further developed by the machines, had no idea of many of these correlations beforehand and are now surprised by the formulas and models generated, the machines appear to be creative, autonomous, or even intelligent. Yet, they are merely tools of vital statistics that remain dependent on the quality of the data input to this day and require interpretation. An early triumph of vital or social statistics was the victory over plague and cholera in London through the vital statistics discovery of a correlation between certain wells and the number of plague victims in its vicinity. Many generations before doctors understood the epidemic in terms of cell biology and developed pharmaceuticals against it, the discovery of social statistics led to tentative cleaning of wells and thus to a decline in the epidemic. It is not surprising that recent empirical sociology, apart from Aristotle, is particularly fond of tracing itself back to these representatives of vital statistics and political arithmetic – indeed, the sociologist Nassehi attributes to them not only his subject, but right away modern society as a society that is digital at its core (Nassehi, 2019). This London case can be used to discuss three facts that are at stake in digitalisation.

- The first fact is quite trivial: data that can be correlated are constructs. They are conventional social indicators. Diagnoses are not dictated to us by nature, but are our constantly amended, tentative classificatory constructs (currently ICD tenth (!) version or ICF). In 1996, v. Ferber and Behrens have already explained that in the analysis of secondary data, which is typical for Big Data analyses, the users have to make an effort of ethnological data clarification (like ethnologists with a foreign tribe). Analyses can never be better than the quality of the data that goes into them.
- The second fact is even more exciting for the discussion of digitalisation: the pattern, i.e., the association or correlation, does not yet explain anything and cannot guide action. Only the experimental intervention, i.e. the cleaning of the wells in a few quarters at first, proved that the water could be causally interpreted (!) as the cause of the disease.
- The third fact to learn about Big Data analysis from the London case is the most important. The "average person" is the statistical construct of summaries of distributions in groups and rarely equates to actual people and groups. And almost never can the frequency of a characteristic in a group be used to infer a single individual in that group. For interventions targeting populations, this is acceptable. But in health care, I am typically interested in what benefit a treatment has for me personally, not what benefit it produces for the average of all others. The salvific expectation of digitised pattern recognition to automatically infer this internal evidence from the combination of a person's mass data is unrealisable (Behrens/Langer, 2004, 2021; Behrens, 2019, Korczak et al., 2020, WBGU, 2019).

The quality of the data also determines the other major achievement expected from digital technology for coordination in health care. Digital technology allows the sharing and processing of data, pattern recognition and pattern formation, evaluation of external evidence of others' experiences almost without loss of time (Behrens and Langer, 2004 and 2022). This not only facilitates documentation, and thus increases the time for 'talking' care and medicine in face-to-face encounters to build up internal evidence. It also enables the telemedical participation of specialists and improves the chances of coordination between different health care institutions.

Transparency in health care appears to be able to improve for many of those treatments that have so far been characterised by discontinuities in care and a lack of coordination between different treatment facilities. People fell and still fall into the gaps between health care institutions, each of which is only responsible for partial aspects. This is extremely threatening, often life-threatening, especially for the elderly, for those less able to speak their language, and for the poor (cf. the results of DFG-SFB 580 in Behrens and Zimmermann, 2017). Therefore, a patient card to improve information flows seems desirable to many. The hope for the electronic patient record, which all patients can bring from one health facility to another and which reduces the enormous risk of interruptions in care and the waste caused by duplicate examinations, was at the beginning of the hope for digitalisation in the health system in the 1960s and 1970s. None of the researchers involved in the large-scale state-funded projects could have imagined that more than half a century later a DFG Collaborative Research Centre would find the same interruptions in healthcare (cf. Behrens and Zimmermann, 2017) and that after 50 years the electronic health card would still be under discussion. What was left out at the time was that it was not the development of the technology that enabled the so-called 'connectivity'. The strength of the providers was underestimated, for whom it is precisely the prevention of connectivity that is in their own economic interest. In other words, the researchers relied on the development of the 'productive forces' and did not consider enough that the 'relations of production' shape the use and development of the 'productive forces'.

That is the topic of the second part. Here, however, I refer to the fact that no digitalisation can compensate for the weaknesses of the data that are entered as training data or data to be processed in the so-called learning machines, neural networks, and deep learning. Moreover, transparent coordination presupposes patients' trust in medical, therapeutic, and nursing care providers, which can neither be prescribed nor is always already given. Here is an example: many people with disabilities have given up looking for doctors they trust and have instead learned to compile their own medication and obtain the components that help them from different doctors depending on their preferences for different alternative methods. Transparency jeopardises this strategy (cf. Behrens, 2019).

12.2 Does 'Digitalisation' Promote Health? The Confusion of Technological Consequences With the Consequences of Long-Established Production Relations

After presenting the potentials of digital technologies for prosthetics, diagnostics, and coordination, the question arises whether these theoretical potentials are actually used for health promotion. As a benchmark of what "health" means, the following argument refers to the definition in § 1 of the Social Code IX, with reference to the Declaration of Human Rights of the United Nations and the Basic Law: The goal of all preventive and rehabilitative measures is self-determination and participation of all people in the life of society. Therefore, self-determination and participation constitute 'health' and, according to § 1 SGB IX and the ICF of the World Health Organisation, should be achieved through rehabilitative and preventive treatments and supports (Behrens, 1982, 2019,

Behrens and Zimmermann, 2017, WBGU, 2019, p. 39 also endorses this). Health thus refers to much more than the functional capacity of the human body.

Participation in the life of society always includes, first, sustainability because life is not possible independently of its natural and social foundations (also according to WBGU, 2020, goals 2 and 5 to 8). Second, participation means that the institutions in which I decide to participate or not to participate (e.g., a day care centre) are actually accessible to me. Participation can therefore only exist in a person's social space (cf. WBGU Charter, Goal 3, 2019, p. 398). Third, at first glance one can see the overlap between health and education: self-determination and participation in the life of society are just as much a goal of health promotion as of education.

As far as facilitating coordination through digital technology is concerned, the WBGU begins its report on 'Our Common Digital Future':

> "The future fate of the planetary environment depends massively on the progress of the digital revolution. ... Only if we succeed in orienting the digital upheavals towards sustainability can the turn towards a sustainable world succeed. Digitalisation otherwise threatens to be an accelerant of growth patterns that break planetary guardrails."

Without the facilitation of coordination and fine-tuning with the help of digital technology, sustainability is no longer even conceivable. But, as the next section shows, it would be completely wrong to rely on the development of digital technology.

An answer to the question of whether digitalisation promotes health is not possible, because the question is already oriented primarily towards technology and thus falsely appears as a consequence of digitalisation, which in reality is a consequence of much older social forms of organisation and relationships. The term 'technology assessment' also promotes this misunderstanding. What appears to be a consequence of digitalisation is not a consequence of a technology at all. This is the central thesis of this chapter: we know as little about digitalisation when we can construct pattern-recognising machines, neural networks, and deep learning as we do when we understand capitalism when we understand steam engines. It is not the "technology" of digitalisation, but its use in old familiar socio-economic relations that is to be assessed in its consequences. The term "technology assessment" actually hinders attention to these relations. Networking through learning, i.e. information-processing automata (which use non-linear and linear algebra for pattern recognition), can make an enormous contribution to the health goals mentioned; as tools of easier coordination in socio-spatial communal and global responsibility, faster diagnostics, better prosthetics (e.g. exoskeletons, pacemakers) and to time and resource-saving relief in favour of better and more sustainable human encounters.

However, it is by no means certain that these technically conceivable potentials will actually be realised. The reason for this is that networked learning machines seem to intensify social organisation problems that have existed for a long time. Precisely because societies have long been aware of these highly problematic developments and have become accustomed to them, they tend to react helplessly when they are exacerbated by the specific use of digital technologies. Yet the human ethical questions of digitalisation have been widely discussed since the debate on automation 70 years ago (cf. only Hannah Arendt 'Vita Aktiva' (2002), Hans Jonas 'Prinzip Verantwortung' (2003), and Jürgen Habermas 'kommunikatives Handeln' (2015 and 2019).

What is threatening, on the other hand, are the areas of social problems to which societies have become too accustomed:

a) Monopolistic expropriation: Although the technology was developed and financed publicly, algorithms, source codes and, above all, the personal data of users, without which a neural network would be meaningless, become the private property and trade secrets of monopolies whose profits are increasingly monopoly rents (Behrens et al., 1974, 2019). This has little to do with responsibility for the sustainable use of nature, including human nature. Consumer protection has long tried to assert the right to transparency and self-regulation against monopolised trade secrets (WBGU goals 12 and 17).

b) Advertising paternalism: The old ecclesiastical tradition of individualised advertising (the Vatican term for this is "propaganda") in the context of comparative individual self-observation ("confessional mirror") is continued and reinforced in unsustainable consumer advertising for self-optimisation including the optimisation of one's own children (Habermas). Self-optimisation can make people sick and lonely and undermine solidarity (cf. also WBGU Goals 4, 10, and especially 13).

c) Division and discrimination: The more (pseudo-) correlations are automatically discovered in mass data, the more risky many self-optimising people appear. Living 'at risk in the welfare state' quickly leads to intensifying spirals of (discriminatory) exclusion in the labour, credit, and housing markets (cf. also WBGU Goals 4, 13, 12). It is the weakness of much sociological work, which sometimes virtually recognises pattern recognition in process-produced routine data as the core of a 'theory of digital society' (Nassehi, 2019), that they do not want to admit how much most of the patterns found are based on so-called spurious correlations. Such ecological fallacies are far from being mere fallacies. They have a huge impact because discriminatory decisions in labour, credit, and housing markets are made on their basis.

d) Germany is a leading participant in the development of automatic, disguisingly 'autonomous' digital weapons systems, although the Federal President and Foreign Minister could have their correct criticism of these weapons systems lead to a national moratorium on this weapons development.

e) Social inequality does not only arise from unequal access to digital technology (cf. Global North-Global South – WBGU Goal 3). It also arises among its users (see part 3). The simple ethical principle formulated by Hans Jonas, that politics must be oriented towards the interests of the most vulnerable, is not promoted by subsidising capital-oriented digitalisation. On the contrary, digitalisation threatens to deepen social inequality and even lead to a digital divide.

f) Instead of using digital potentials for the integration of previously fractionated health care areas, digital and paper health information technology (HIT) systems that exist side by side and are separated according to professional groups repeat fractionations and thus apparently endanger patient safety (ZEFQ, June 2019). This shows once again that 'digitalisation' as a time-diagnostic term is as truncated, if not obfuscating, as the talk of industrial society instead of capitalist society. Even the WBGU overlooks (or at least pretends to) that the prevention of connectivity of process-produced data is not due to a technical problem, but rather to the economic interests of oligopolies and monopolies in sealed-off market niches. For them, preventing connectivity is rational in the short term.

g) Which occupational activities machines are therefore better at than humans, and when humans are ethically entitled to use these machines: While the machines of the industrial revolution relieved muscles, networked computers relieve some of the performance of the 'brain.' There are currently many very well-paid 'mental' jobs that require oversight, thoroughness, consistency, and powers of deduction. Whereas with the previous machines simple manual labourers had to fear for their jobs, now it is people who earn a lot of money with stock exchanges, taxes, bookkeeping, brokerage, building planning, statics, X-rays, company management, and many other things. This by no means automatically increases unemployment because, for example, capable people are urgently needed in elderly care. But it is not a gradual but disruptive development. While industrial development replaced muscle power in many areas, machine learning is replacing what used to be called 'mental' work. This is disruptively and unceremoniously hitting the top 10 % earning 'elites'. Instead of unskilled workers, financial advisors, and other highly specialised academics are now affected who will – at least in part – try anything to avoid having to move into elderly care even when people are desperately needed there. Perhaps they will try to ban the use of learning machines in their domains but still use them secretly. It is true that simple clerks are still needed to feed the machines with verified data, as before. However, these clerical jobs have never been carried out by the 10 % highest earners. But even care for the elderly – mentioned here as an example of all activities in which machines can hardly replace people – is not immune to changes with the help of digital technology. The Munich labour lawyers Giesen and Kersten (2017, p. 50) see only with digitally coordinated work realised what Karl Marx already saw for his time:

> "In the modern factory system, the automaton itself is the subject, and the workers are only as conscious organs attached to its unconscious organs and subordinated to the central motive force... As living appendages. ... Who works [in the platform economy] independently, who depends on it? Who instrumentalises whom here? People are the machines or the machines are the people?" (Giesen and Kersten, 2017, p. 50).

In fact: digital technology allows coordination and control of individual specialist nurses and doctors as well as therapists during their nursing and therapeutic home visits, adjusts route and work plans in case of delays, orders aids, records complaints, and allows the "telemedical" exchange of findings with distant specialists. As with Uber and Lieferando, nurses, doctors, and therapists can make house calls in their own cars or on their own bicycles and at their own expense as formally self-employed persons, for whom no social security contributions are currently due. But the example of home visits also shows that it all depends on who uses digital technology and how: Whether nurses and therapists use it to improve their networking with colleagues and clients in need of care, or whether the platforms or company managers use it to play the formally self-employed off against each other, to shorten home visits, and to minimise the cigarette breaks and small purchases between home visits. What has already been said for diagnostics also applies to coordination: learning machines, neural networks, and deep learning can never be better than the data they process and the algorithms they do it with. Machines are not subjects. Even if digital technology replaces the annoying supervisor of middle management with an ever-quiet machine voice and anonymises domination, it is humans who put the machines into circulation.

The same is certainly true in medicine for organ diagnostics. Today, laboratory machines are already superior in the diagnosis of body fluids, and pattern-recognising machines are certainly ideal for imaging procedures. Learning machines are inferior when it comes to building internal evidence together with patients to clarify their own participation goals and resources. However, this does not happen very often in medical diagnostics at present. Also, as far as prescriptions based on organic medical diagnoses are concerned, pattern-recognising machines are probably helpful. As far as specialist treatments are concerned, I was deeply impressed by the habilitation of a surgeon at the end of the last millennium. He proved that for certain complicated surgical operations, e.g. burns, a robot with a scanner was superior to a human surgeon and could therefore be used for ethical reasons. If this is the case, then for me it is just as true for surgery as it is for the court: the exclusion of learning robots, if they are superior to humans for the reasons cited, is unacceptable on ethical grounds. Patients and clients have a right to treatment and justice that is supported by learning machines. On the other hand, the activities of therapy and specialist care can only be supported or replaced by learning machines to a limited extent. Where they can be supported by learning machines, it would be unethical not to use them. After a stroke, a reduction of feedback training to regain linguistic and other skills from 300 hours to 5 minutes could be achieved when a machine processing the data into patterns was used for individual coding and feedback (source: Klaus Robert Müller, bbaw, 30 June 2018). What ethics would allow physiotherapists not to use this suffering-reducing learning machine in principle?

h) The renunciation of the human is clearly reflected in the naming of networked learning machines as "artificial intelligence" or the talk of "autonomous" machines instead of automatons (cf. WBGU, 2019, p. 94). The talk of artificial intelligence and especially the talk of autonomous machines instead of automobile machines already dangerously obscures responsibility: Of course, an automobile driving automobile is not 'to blame' for an accident, just as little as a surgical operation robot is not to blame for an operation error or a laboratory machine analysing blood or stool is not to blame for a misdiagnosis. Responsible for the error is the natural or legal person who put the machine on the market and the person who used it as owner or user, whether negligently or culpably (cf. Chapter 7). It is not that difficult to understand. Every instruction manual of a cleaning device informs about this, and all young driving students could recite their responsibility as buyer, owner, and driver in the driving test.

Unfortunately, this clear and actually simple fact is already obscured by the talk of 'artificial intelligence.' Even the WBGU writes in 2018, p. 4, that up to now "intelligence" has been a "unique selling point" of humans and the basis of human civilisation. Now we are letting technical systems imitate intelligence. This raises fundamental questions about ethics and human dignity." As the Senate's representative for people with disabilities for decades, I would like to strongly disagree with this. So-called learning machines, neural networks, and deep learning are in reality prostheses or aids such as glasses and hearing aids, donated organs, and assistive technologies, not competitors of humans in the struggle for human rights. Since the dignity of humans does not depend on their intelligence quotient, learning machines cannot, as the WBGU writes, ethically challenge the dignity of humans.

These eight tendencies away from using the given potentials of digitalisation for self-determination and participation in the life of society are widely known. It is almost cheap to respond to them with a call for socially sustainable 'regulation', or better, 'design', as politicians and civil society do.

12.3 Currently Discussed State Design Strategies: Opportunity for Impact?

In view of these dangers of digitalisation – which seem to emanate less from digital technology than from the social conditions in which digital technologies are used – states and their advisors discussed design strategies. Selected of these strategies and their obvious limitations are discussed in the following third part.

The internet has developed from a military to a free science and communication network to a nontransparent spying and advertising instrument in the hands of huge corporations and authoritarian states. To control this reversal, states have discussed or already developed various regulations to protect self-determination. They are all highly relevant to health as self-determined participation in the sense of § 1 SGB IX. Eight of them should be briefly discussed:

12.3.1 Is It Sufficient to Make the Use of Data Dependent on the Written Consent of the Data Authors?

The first protection strategy is the requirement that each person personally consents in writing to the disclosure of their data. Its effectiveness is questionable. However, as research reported in the VDW book cited earlier (Korczak et al., 2020) shows, this consent is overwhelmingly given without even accessing and thoroughly reading the user statements. The requirement of 'self-determined consent' is therefore presumably ineffective because consent to the disclosure of one's personal data is made mandatory for the use of a service or device. Self-determined consent is obtained through a situation that has elements of coercion for many consumers. However, the disclosure of personal data is not at all necessary for the provision of the service in question. Even the health data trackers (wearables) that millions of people bought at Christmas with their consent to share data do not need information from personal non-anonymised datasets at all if the consumer wants to know where they stand in comparison to others. Comparative data are publicly available from surveys with anonymised data. As a rule, anonymised data is also sufficient for science. Personal data are hardly necessary. All socioeconomic panels prove this. Therefore, the proposal to prohibit linking the use of a service or device to 'self-determined consent' to the use of personal data should be examined.

Demanding that data collectors be paid for sharing their data does not effectively protect self-determination either, because in the eyes of the corporations, the data collectors have already been paid for a long time. Corporations are already selling them the use of services in exchange for their personal data as a payment. The handover of personal data for advertising and similar manipulation purposes is the price of use, to which the data authors have agreed in writing.

Of course, it is a restriction of freedom of contract if citizens are prevented from paying for an app or other service with their most personal data for advertising and other purposes because they would not otherwise receive this service. However, such a restriction on freedom of contract is quite common in all democratic states. For example, voluntary consent to terms of use is not sufficient for many goods and services. It is not enough, for example, that when I use a car I agree to pay for any damage that may be caused by the car. In contrast: the TÜV and the police, not I, regulate and supervise which cars may be used at all. This is for my protection and the protection of third parties. The same applies to medicine and medical products. This should also be demanded for all digital products and services.

12.3.2 Test Health Promotion Apps Like Vaccines in Intervention Studies

Although extensive intervention studies on risks, side effects, and benefits are prescribed for the voluntary prevention measure 'vaccination', this does not apply to the health promotion apps offered in personalised advertising, which also aim at voluntary prevention. The risks, psychological and physical side effects, and the often questionable benefits of these apps are at least as explosive as those of vaccinations (for evidence, see Korczak, 2020, Behrens, 2019 and 2020). It is not without reason that exercise and dietary instructions notoriously call for clarifying with one's own doctor the suitability of the movements and diet for oneself. If this happens, what is the scientific basis for this claim if there are no intervention studies to build external evidence and practices of building internal evidence? As shown in the second part of the chapter, correlations are not enough, intervention studies are needed. At least since the invention of printing, markets, schools, and pulpits have been flooded with imperious advice on health promotion (see Behrens, 2019). 90 % of all magazines at the newsstand contain such advice. With the Internet, the offers tailored to the adolescent increases enormously. Since providers personalise advertising on the Internet, i.e. put information specifically at the back and thus make it less accessible to those being advertised to, enclosed public test reports are necessary. Therefore, health promotion apps are by no means harmless.

12.3.3 Preventing Pseudovoluntary Disclosure of Data That Someone Suspects Exist

Of course, in democracies, companies and insurance companies cannot legally access personal data, hacking is a criminal offence. But what if the provider of a job or a cheap private insurance asks for the health data of the fitness wristband or other electronic records before they finalise a contract? Formally, fulfilling the request is completely voluntary, i.e. self-determined. The conclusion of a contract is also completely voluntary. Of course, no private employer can be forced to sign an employment contract. And only statutory health and pension insurance companies are under obligation to contract, i.e. they are not allowed to refuse an applicant. Private insurance companies are completely free to choose who they insure and who they do not.

The problem is well known in the labour market and has long been exemplified for comparable cases: If a pregnant woman is asked by a potential employer whether she is pregnant, she is allowed to deliberately lie; without this lie, this deliberate deception being a reason for dismissal after employment. The labour courts are realistic enough not to rely on a prohibition of this question in a job interview. Only lying seems realistic to them. Consequently, one would be allowed to send falsified files when asked for fitness data or similar files. There are no corresponding regulations for insurance companies.

12.3.4 Disclose the Screening Patterns Instead of Protecting Them as Trade Secrets

The population in Europe has apparently become quite accustomed over decades to the fact that the criteria of the credit agencies (e.g. Schufa, cf. Korczak and Wilken, 2008) are non-transparent, according to which a mobile phone or a washing machine can be paid for by monthly instalments, or a flat is rented. This acceptance, which has been widespread for a long time, is a typical example of the thesis of this paper that it is not digitalisation, but previous practices that have made people defenceless against the supposed consequences of digitalisation. The demand of many economists and psychologists therefore makes sense that the creation and content of the screening patterns should not be protected as trade secrets but should be disclosed and made open to discussion. This is the only way to prevent statistical discrimination based on spurious correlations (see the second section). In the health care system, treatment that is decided only according to external evidence and not according to internal evidence (cf. Behrens, 2019) is a treatment error.

Under patent protection or trade secret protection for the screening results of learning machines (including neural networks), commercial enterprises do not publish their screening methods. Otherwise, the competition could use them. The processes guarded as trade secrets are a decisive step towards the monopolisation that every company strives for and that promises investors a return on investment. The consequence is well known: People are becoming more and more transparent as customers. The companies that collect, combine, and evaluate data, on the other hand, remain a black box and are becoming increasingly intransparent. That is how they get advertising contracts. In the interest of the customers, it should be the other way round: The companies that hold Big Data should be transparent, while the customers remain a black box. This is in line with Basic Law.

12.3.5 Preventing Supply Interruptions Through Information Flows (Connectivity)

Transparency in health care seems to many to be able to improve treatments, which so far have been characterised by discontinuities in care and a lack of coordination. People fell and still fall into the gaps between health facilities, each of which is only responsible for partial aspects. This is life-threatening for the elderly, for the less linguistic and less educated, and for the poorer (cf. the results of DFG-SFB 580 in Behrens and Zimmermann, 2017). Therefore, a patient card to improve information flows seems to be beneficial to many. However, for many people with disabilities, as already described, there are several arguments against such transparency (cf. Behrens, 2020).

12.3.6 Secure Jobs

The predictability of one's own remunerative employment is undoubtedly highly relevant to health. Such predictability is of great importance for the impression of having control over one's own life (locus of control, Behrens, 1999). Even people in high-salary jobs in research and teaching, who are among the 9 % of the best paid Germans consider see themselves as 'precariously employed' because of the fixed-term nature of their jobs, because their employment following their current well-paid job is uncertain. Starting a family, for example, does not yet seem feasible to them under the conditions of well-paid but temporary, and therefore precarious employment and is thus postponed. This is undoubtedly a health problem.

So-called digitalisation has some of the ambivalent consequences for jobs that have already been discussed in relation to automation, of which it is a subcategory: On the one hand, it facilitates and improves human work. Especially professions that claim to put the interests of their clients first can hardly afford, for reasons of professional ethics, to do without digitally enabled quality improvements through more precise and complete overviews, just to keep their jobs. On the other hand, digitally supported automation makes many jobs digitally replaceable, which may happen very suddenly. Some of these jobs are the best paid. Fortunately, those made redundant may find other job opportunities as there is a huge demand for care workers.

12.3.7 Breaking Up Oligopolies and Monopolies Under Antitrust Law

As important as the breaking up of near-monopolies under antitrust law is in many respects, it alone does not protect against being spied on and misuse of personal data for purposes of control and manipulation. If there are three dominant companies in the market instead of one, what should prevent them from handling personal data in the same way as the current monopolists? Therefore, restricting what can be consented to in the first place is the more direct route. It restricts the freedom of contract in one place, but it effectively protects the rights of consumers. The protection of competition in digital markets has at least two trivial challenges:

1) Since globalisation, companies are no longer bound to regions, making them more difficult to regulate.
2) Additional users incur almost no marginal costs for providers to supply in networks. Almost cost-free means that additional customers cause almost no additional costs for the providers. This favours the "the winner takes it all" principle and promotes centralisation and monopolisation tendencies.

The combination of 1) and 2) urgently raises the question that the WBGU formulates as "What would a globally networked competition law to contain economic power in the digital age look like?" (WBGU, 2018).

However, the problem does not lie in the superiority of the capitalist oligopolies and monopolies, but at least as much in the competition between the nation states. All oligopolies and monopolies are crucially dependent on states to protect and subsidise them constantly, otherwise they could not survive a day. If the competing states rely first on the economic growth of their oligopolies and monopolies, a "globally networked competition law" is secondary for them. Therefore, it is both a state failure and a market failure. Oligopolies and monopolies thus find a willing helper in the competition of democratically elected state governments, which prevents a worldwide competition law. For this, the term state-monopolistic capitalism was found in the 1970s, which has not lost its intrinsic truth to this day.

12.3.8 Protect the People by Implementing Your Own World-Class Digitalisation Industry

Germany and the EU would supposedly have to catch up economically and technically with the world leaders China and Silicon Valley in order not to leave the reigns of data to China and the USA; this is a common argument. In my opinion, it is wrong. If it were true, more than 160 states would be at mercy and unable to act because not all states on Earth can become world leaders! If they could only protect their people in this way, the situation would be hopeless. The protection of citizens must be ensured by national regulations that also lead to international, effective agreements to which all states adhere, whether they are economic-technical leaders of the world or not.

12.4 Conclusions

In the first part, this introductory overview has listed in the first part potentials for health promotion in prognostics, training, diagnostics, coordination, and their technical prerequisites, which go back to mathematical and information technology discoveries that are erroneously called learning machines, deep learning, neural networks, artificial intelligence, and autonomous machines. (see WBGU 2019 p. 72: There are no machines that 'learn', only those that automatically model, classify, calculate, and optimise patterns after appropriate primal inputs). For prognostics, training, diagnostics, and coordination, the aforementioned techniques, which are indispensable for sustainable use of the Earth, are at least so useful that doing without them seems ethically irresponsible. Doing without them already violates the Hippocratic oath. However, the patterns and suggestions for action that the machines calculate correlatively are dependent almost entirely on the quality and meaningfulness of the data that are entered as training data. Therefore, in the foreseeable future, machine-correlating pattern recognition will not be able to replace the classical intervention studies in healthcare when it comes to treatments. Big Data patterns can best stimulate intervention studies. That is already a lot.

In the second part, I tried to show that risks are listed as risks of digital technology, which in reality are not technology consequences at all but arise from much older socio-economic conditions. Even the word "technology impact research" is misleading. What the WBGU (2019), for example, contrasts on p. 340 as selected risks with the selected (technical) potentials it lists are, for the most part, not technical risks at all, but old-established social conditions. The fact that they are so old-established in business, school, church, administration, and advertising makes critical studies of 'digitalisation' somewhat helpless in the face of the use of new digital technologies by these old institutions. New digital technologies can considerably aggravate the consequences of old, familiar conditions. The term 'technology impact research' still expresses the earlier confidence in the development of productive forces that somehow prevail over relations of production.

In fact, the relations of production interact to determine the productive forces. The lack of attention to the existing relations of production is functional in capitalism!

Therefore, it cannot be assumed that regional alliances of nation states want to represent the interests of the world population against their own regional firms, unless there is effective political pressure for this through corresponding formation of countervailing power. In the third part, some general strategies are briefly listed, which offer starting points for this.

In the end, the most important point is to be noted: The technical term of autonomous machines that learn and operate with artificial intelligence obscures the most important issue of digitalisation for human dignity, which is responsibility. Those who call machines 'autonomous', i.e. making their own laws, and claim that machines make decisions autonomously instead of just automatically calculating and applying patterns and results, undermine responsibility and human dignity. He hides behind the broad backs of robots and other machines he has created himself.

Chapter 13

Reductionist Temptations: Artificial Intelligence and Sustainability

Reinhard Messerschmidt

13.1 State of Research, Current Discourse, and Status Quo

In line with the title of this chapter, it is first assumed that considerations on the potential and risk of artificial intelligence (AI) for sustainability actually address the entire topic of digitalisation and sustainability. Socio-technical systems such as AI can hardly be meaningfully discussed in isolation from directly related technical topics (such as Big Data and erroneous or biased training data or the explainability and trustworthiness of AI) and socio-political framework conditions (e.g. business models, values, legislation). In order to present the relevant elements of a paradigm shift towards sustainable development of AI systems, a compact overview of the current state of research and discussions of applications of AI is provided.

Only a few years ago, there were hardly any reliable scientific sources on the topic, but already a large number of different publications with political recommendations for action and research. Research conducted for the WBGU's main report "Our Common Digital Future" (WBGU, 2019) on the basis of 111 German- and English-language texts (Messerschmidt, 2020) shows not only a considerable range of recommendations already in 2017 and 2018, but also that beyond diverse technical and societal aspects, the pure word frequencies do not focus on AI, despite the already great hype at the time but on data, as the following Wordcloud representation illustrates (Fig. 13.1).

The situation has changed in the meantime, not least due to the WBGU's publications, i.e. the aforementioned main report (WBGU, 2019) as well as policy papers (WBGU, 2019a, 2019b) and factsheets (WBGU, 2018, 2019c). Their core narrative, beyond possible potentials of digitalisation for sustainability, to prevent it from acting as a "fire accelerant" of growth patterns that break through the planetary guard rails, has significantly shaped the current scientific, public, and political discourse. At least in some federal policy departments and the relevant communities, digitalisation and sustainability are now increasingly being thought of together. This is illustrated, for example, by the Federal Ministry of Education and Research's action plan "Natural. Digital. Sustainable." (BMBF, 2019) or the "Environmental Digital Agenda" of the Federal Ministry for the Environment, Nature Conservation, and Nuclear Safety (BMU, 2020). Parallel to this, relevant communities have emerged in recent years in various scientific disciplines at the interface of digitalisation and sustainability, and the growing interest in the topic is reflected in an increasing number of publications, an exemplary overview of which is provided below.

Fig. 13.1: Wordcloud created in software for qualitative text analysis (MaxQDA) on word frequencies in the text corpus studied – necessarily an excerpt without any claim to representativeness for the discourse, which was already difficult to delimit at the time. (Messerschmidt, 2020: p. 2)

Lange and Santarius (2018) and the anthology "What connects bits and trees" (Höfner and Frick, 2019) contains numerous contributions on the sustainable design of digitalisation. Furthermore, the topic was addressed in several articles in the magazine "FIfF-Kommunikation" of the FIfF – Forum InformatikerInnen für Frieden und Gesellschaftliche Verantwortung e. V. The third issue of the year 2020 deals with numerous highly relevant aspects with the focus on "Technology and Ecology", for example, from the environmental impact of corresponding devices and services (Gröger, 2020), corresponding environmentally compatible, life-cycle-based product design (Schischke, 2020) to software and sustainability (Betz, 2020; Hilty, 2020; Köhn, 2020) to sustainability goals for the operation and development of IT (Boedicker, 2020). Furthermore, the topic was also addressed several times in previous issues (e.g. Abshagen and Grotefendt, 2020).

The topic is also increasingly generating resonance in the public beyond relevant portals such as netzpolitik.org or heise.de. For example, in the guest article by the director of the Hasso Plattner Institute in Potsdam (Meinel, 2020) entitled "Only sustainable digitalisation can save the climate", it was pointed out that there could also be technological answers to the climate crisis "as long as IT does not make everything worse with its enormous energy consumption", which applies "especially to artificial intelligence". However, the topic is far from being sufficiently anchored across departments, society as a whole and in the mainstream of the specialised sciences relevant to product design. In the German Informatics Society, however, it has already been increasingly discussed at specialist conferences since 2019 and the 51st annual conference INFORMATIK 2021[152] is "all about sustainability" and deals with topics "such as green IT, resource conservation, use of intelligent technologies, and optimisation of systems" in four fields of action of the discipline ("ecological, economic, social, and technological") guided by the 17 UN Sustainable Development Goals of Agenda 2030. This thrust is also reflected in the BMU's two funding lines "AI Lighthouses for the Environment, Climate, Nature, and Resources" and 26 projects already funded, which illustrate the broad spectrum of possible fields of application for AI (Tab. 13.1).

Tab. 13.1: Project cover letters of currently funded projects of the initiative "AI Lighthouses for Environment, Climate, Nature and Resources". (Source: https://www.z-u-g.org/aufgaben/ki-leuchttuerme/)

Title/ Acronym	Brief description according to the project-executing agency's homepage
Funding line 1 "AI for environmental protection"	
AIR	AI-based recommender for sustainable tourism
AISUM	Detailed elaboration for the implementation of a platform for AI Empowered Sustainable Urban Mobility in the metropolitan area Berlin
AQUA-KI	Intelligent optical methods for the effective detection of microorganisms in water bodies
AuSeSol	Self-sufficient and self-optimising solar energy generation with integrated storage capacity
Cognitive Weeding	Development of a detailed concept for weed and weed management adapted to the crop type over cultivation periods with the help of artificial intelligence
DC-HEAT	In the "Data Centre Heat Exchange with AI-Technologies" project, the planning and operation of data centres in the Frankfurt area is to be designed with the support of artificial intelligence in such a way that the negative impact on the environment can be reduced and waste heat is used in the best possible way.
FutureForst	Development of a detailed concept for the use of AI in forest condition analysis and decision preparation for climate-adapted forest conversion
KI4NK	Development of an innovative idea concept for the promotion of AI-supported sustainable (online) consumer behaviour, taking into account provider and consumer perspectives.
PlasticObs	Detecting plastic waste in the sea from the air: The project aims to identify sources and distribution routes of plastic waste using aircraft and AI.
PRIA-WIND	Platform for ensuring species protection in wind power projects
Smart Recycling	SmartRecycling – AI and robotics for a sustainable circular economy
Unlikely Allies	The project networks experts from environmental protection and AI.
WindGISKI	Development of an AI-based geoinformation system for the socially acceptable selection of wind energy potential areas in the area of conflict between species, environmental and climate protection
Funding line 2 "Application orientation and foundation"	
AI4Grids	Using AI-based planning and operational management of distribution grids and microgrids, the aim is to achieve optimal integration of renewable generators and fluctuating loads as part of the energy transition.
CO:DINA	Creation of a transformation roadmap for digitalisation and sustainability in order to use digitalisation for the socio-ecological transformation.

---

[152] https://informatik2021.gi.de/

CRTX	Using AI-assisted spectroscopy and image analysis, the Circular Textile Intelligence (CRTX) project aims to achieve more specific sorting for second-hand use and fibre-to-fibre recycling to enable a continuous material cycle.
GCA	The "Green Consumption Assistant" is intended to help people consume more sustainably. For this purpose, it will display the concrete effects of consumption decisions and inform about more sustainable alternatives when searching for products in the search engine Ecosia.
I4C	Intelligence for Cities: AI-based adaptation of cities to climate change – from data to predictions to decisions
IsoSens	Development of an AI-based sensor to determine the isotopological composition of greenhouse gases for research into climatic processes
AI on the move	Use of artificial intelligence methods for optimised railway operations of the future
KInsect	To protect insects, systematic monitoring over a long period of time is necessary. The open source project wants to digitise monitoring and make it usable on a large scale with the help of AI.
BOX	An AI strategy for Earth system data to analyse, process and provide environmental changes.
Natura Incognita	A workflow platform for AI-based species identification
NiMo	The project "Nitrate Monitoring 4.0" uses intelligent systems to sustainably reduce nitrate in groundwater.
ReCircE	In the project "Digital Lifecycle Record for the Circular Economy", material cycles are made transparent and waste sorting is optimised with the help of artificial intelligence.
SustAIn	The project systematises and exemplifies the impact of various AI-based processes on sustainability.

As part of the SINTEG funding programme "Showcase Intelligent Energy – Digital Agenda for the Energy Transition", which was launched by the Federal Ministry for Economic Affairs and Energy (BMWi) in 2016, research was also conducted into intelligent ICT-based grids (smart grids) in five model regions. The ENERA joint project, which was completed at the end of 2020, included a field test for so-called software agents for the self-organisation of energy storage systems and their power feed-in to the grid using "distributed artificial intelligence."[153] In the area of land use, for which the WBGU (2020) has presented another main report, AI is already being used to promote sustainability – for improved monitoring, for example, through the evaluation of satellite data, or for new paradigms of small-scale agriculture with innovative use of AI and robotics, such as "pixel cropping" in the Netherlands.[154]

However, both for the decentralisation of the energy system and in agriculture, it is true in the sense of the "fire accelerant" narrative presented at the beginning (WBGU, 2019) that such potential can often only be raised if additional technical or political framework conditions are in place. Without legally compliant, interoperable smart meters and a widespread infrastructure for sector coupling (e.g. through bi-directionally chargeable e-cars and heat pumps in buildings), the use of AI to optimise the electricity grid is of comparatively little use. In the 14 projects funded by the BMEL (2020) as "Digital Experimental Fields in Agriculture", no fundamental paradigm shift is discern-

---

[153] https://idw-online.de/de/news766468
[154] https://wur.nl/en/project/Pixel-cropping.htm

ible as in the aforementioned Horizon 2020 project "Pixel Cropping" at Wageningen University, but the focus remains on increasing efficiency and ultimately intensifying the current industrial form. In this respect, AI is not a simple solution for complex problems with respect to sustainability. At best it could be used meaningfully after other problems have been solved. At worst, however, it can create new ones or exacerbate existing ones. Despite all the positive dynamics that undoubtedly exist, digitalisation in general and AI use in particular have so far tended largely towards the latter.

Accordingly, it would be naïve to believe that large IT corporations have recently discovered the "principle of responsibility" (Jonas, 1979) and would implement it in a contemporary way as responsible technology design (ÖFIT, 2021; Spiekermann, 2016) in the sense of "digital ethics" (Spiekermann, 2019). Unfortunately, responsibility has so far existed more as a PR buzzword, also since existing incentive systems and the concentration processes of the platform economy are tensely related to it. The IT security expert and blogger Felix von Leitner commented quite pointedly on the current situation: "If it says responsible anywhere, it is poisoned. There is no clearer indicator for 'is poisoned'. In any case, it is never responsible and certainly not for your benefit."[155] Accordingly, the positions of leading management consultancies and big tech companies linked in his blog post would merely be a patina for unchanged unsustainable practice – currently, the facts support this view.

The permanent collection and, to a large extent, AI-based analysis of behavioural data for the optimisation of advertising-financed business models, which is characteristic of "surveillance capitalism" (Zuboff, 2018), has considerable downsides, not only in terms of privacy, public welfare, and democracy (Nemitz and Pfeffer, 2020). It is true that the energy consumption of machine learning has been increasingly addressed in research and by the public in the last two years. However, the associated CO_2 emissions are as little anchored in the consciousness of developers and decision-makers at the corporate level or in politics and the public as the use for more intensive exploitation of fossil resources. Despite all the assurances of big tech that they want to achieve CO_2 neutrality within the next few decades, considerable scepticism seems to be called for here. This applies first of all to the obviously unsustainable use in fields of activity such as the extraction of fossil resources, which has been known for two years now – contrary to all the "ecological" rhetoric of the corporations:

> "'100 per cent renewable is just the beginning,' says a Google website. This is how the IT company likes to present itself. In 2017, it covered its entire electricity needs from renewable energies for the first time. Data centres next to wind turbines are part of the search engine company's self-image. Amazon and Microsoft also refer to their corresponding eco-commitment. [...] However, research by Gizmodo heavily scratches the eco-image of the Sillicon Valley corporations. According to the research, the big IT companies have built up countless partnerships with oil companies in recent years and set up entire departments whose sole aim is to provide services for the fossil fuel industry. The electricity purchased by the data centres should hardly be of any significance in comparison."[156]

155 https://blog.fefe.de/?ts=a15f7340
156 https://www.golem.de/news/oelfoerderung-wie-google-amazon-und-microsoft-das-klima-anhei-zen-1902-139655.html

Little is likely to have changed in this situation in the meantime, except that the electricity consumption of AI data centres is becoming more important in view of both the somewhat better information and data situation and the continued strong increase in the use of energy-hungry machine learning. This is all the more so since, thanks to the energy transition that is taking place too slowly but nevertheless globally, future oil production as well as demand are following a contrary trend.[157] In this respect, the data required for the application of AI are not "the new oil" in a double sense, because firstly, they are not a finite resource that can be wrested from nature and is consumed during use, and secondly, because oil definitely has no future in terms of sustainability.

Basically, the core message of the WBGU's main report "Our Common Digital Future" (WBGU, 2019) still applies to the relationship between sustainability and digitalisation in the broad sense and AI in the narrower sense, namely to avoid the latter becoming an accelerant of unsustainable growth patterns before any potential for sustainability can be realised. On the one hand, this is due to the energy hunger of Deep Learning, which has been increasingly discussed in the press and research for about one and a half years, but still far too little (e.g. Strubell et al., 2019; Dhar, 2020) – there is a great need for research, communication, and action here. A first step, for example, would be to systematically reflect on, collect, and optimise the CO_2 footprint of machine learning in application. A recent working paper by Stanford University (Henderson et al., 2020, pp. 15f.) not only addresses the fact that the location of the data centre can make a 30-fold difference with regard to the supply from renewable energies, but also proposes initial recommendations for action to various stakeholder groups. According to these, the systemic change(s) should be in research:

- cloud-based AI applications are in principle only carried out at data centres in regions with low CO_2 emissions, i.e. a high proportion of renewable energies;
- reporting based on standardised metrics for the most energy-efficient configurations possible be made widely available;
- further research into energy-efficient systems and the introduction of "energy efficiency leaderboards" for the online dissemination of best practices;
- source code and models are always published, provided it is justifiable for security reasons;
- energy-efficient configurations are anchored as standard in the usual platforms and tools by means of so-called "Green Defaults";
- climate-friendly initiatives should be promoted at conferences.

In the economy:

- training for machine learning should be immediately relocated to regions with low CO_2 emissions (see above) and documented accordingly (incl. so-called "default launch configurations");
- more robust tools for energy consumption/CO_2 emissions are introduced;

---

[157] https://edition.cnn.com/2021/02/11/business/shell-oil-production-peak/index.html

- energy-efficient operations are integrated as "default" into existing frameworks;
- source code and models (if necessary, only internally) are always published, provided it is justifiable for security reasons;
- energy-based cost-benefit calculations are contrasted with the gains from developing and training new models;
- reporting of model-specific energy metrics should be introduced.

Although this list is by no means complete and, for example, does not address the potentially necessary expansion of renewable energies in regions with low CO_2 emissions, it can certainly be seen as a first step from the community itself in the direction of more ecological sustainability – provided that this is followed by further steps and, above all, mainstreaming into the daily practice of the big-tech "hyperscalers". This could at least improve the energy balance, which has often been disastrous up to now, although the question of resources (Messerschmidt and Ullrich, 2020) is still not addressed. Moreover, there is a need for action not only at the technical level, but also at the legal level. Bietti and Vatanparast (2020) have recently pointed out that, ironically, a company that bears the name of the world's largest tropical rainforest contributes significantly to environmental degradation – despite Amazon's announced intention to become CO_2 neutral by 2040. In addition, it also contributes indirectly – not only by supporting the oil and gas industry, as mentioned at the beginning of the text, but also by supporting political candidates who deny climate change. Of course, this connection does not only apply to Amazon alone – the company name only served as a hook for the Harvard authors. They therefore emphasise that Amazon's practices are just one example of the environmental consequences of data-driven technologies, and that their share of global CO_2 emissions is expected to double from 4 % by 2025, according to THE SHIFT PROJECT (2019). While a relatively large amount of legal research addresses the consequences of AI for data protection and privacy, the ecological consequences are comparatively underexposed.

This finding is undoubtedly true. Nevertheless, with regard to sustainability, it is clear that both problems are interlinked – not only with regard to energy consumption through excessive data processing in "surveillance capitalism", but also with regard to a broad understanding of sustainability, from which a separation of ecological and social sustainability is not practicable.

13.2 Sustainability as a Guiding Principle – Also for AI

For a broader understanding of AI and sustainability, we will first briefly outline what is behind the multifaceted term, which has become increasingly anchored in everyday language and social consciousness at the latest with the 17 goals of the 2030 Agenda (SDGs). Regardless of the ambivalence of innovations known from technology assessment, it can be assumed, at least retrospectively, that real technological "progress" arises from value consciousness and thus human-oriented progress in socio-technical systems of digitalisation – such as AI – and cannot be meaningfully thought of without ethics (Spiekermann, 2019, pp. 29ff.). Normative foundations and clear purposes of technology design

(ÖFIT, 2021) are just as central here as an expanded understanding of values, in which the common good is at the centre and "not the return on investment or some automation index" (Spiekermann, 2019, pp. 65). This does not have to collide with sustainable, i.e. future-proof, entrepreneurial interests, because whoever implements digitalisation as an end in itself "destroys his own ecosystem of corporate values", the basis of which is just as much people as natural resources – without both, there would at best be room for posthumanist (night) dreams (as explained in chapter 5 by Frank Schmiedchen). In this respect, also with regard to AI, in addition to a human-centred approach, the safeguarding of stable ecological foundations of life is of particular importance within

Fig. 13.2: Digitalisation and sustainability goals visualised as a "wedding cake" model (WBGU, 2019c)

the many aspects of sustainability, since stable environmental conditions are the basis of all social and economic activities. The biosphere, as *primus inter pares,* is thus the basis of all sustainability goals – also and especially with regard to digitalisation and AI (Fig. 13.2).

In view of established scientific knowledge about planetary boundaries of the Earth system and its functions necessary for human existence, some of which have already been exceeded, the discussion and adherence to guard rails is indispensable. Although science can present well-founded proposals, the political decision on setting guard rails should be subject to a democratic,process (WBGU, 2019, p. 38). Since the

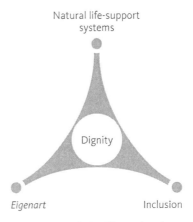

Fig. 13.3: WBGU's "Normative Compass" (2019: p. 30).

necessary demarcation of boundaries is subject to scientific uncertainties, the socially acceptable risk must be weighed up politically. Here, "only the democratic constitutional state is in a position to bring about the corresponding weighing within the framework of democratic procedures and to legitimately establish it in a generally binding manner" (SRU, 2019, p. 104). In this sense, the SDGs also make it clear "that a policy geared towards compliance with planetary impact limits makes it necessary to regulate the material flows of our economic and social systems" (SRU, 2019, p. 107). However, this is not a zero-sum game in which fundamental dimensions could be realised at the expense of others (Fig. 13.2).

In terms of human dignity, sustainability is only meaningful through the interaction of the preservation of the natural foundations of life and human participation and Eigenart (WBGU, 2019, p. 42). This means 1) adhering to planetary guard rails and avoiding or solving local environmental problems, 2) ensuring universal minimum standards for substantial, political and economic participation, and 3) recognising the value of diversity as a resource for successful transformation and a condition for well-being and quality of life. This guiding principle is already anchored at the national level in the sustainability strategy (Fig. 13.3), which, however, has not yet been effective enough in day-to-day politics.

The development of digitalisation and AI to date, however, seems to be fundamentally at odds with this guiding principle, as was shown at the beginning. Here, as in all R&I policy, the central challenge is to "shape policy in such a way as to prevent trends that undermine the ecological foundations of humanity. At the same time, social and economic development towards sustainability must be promoted" (SRU, 2019: p. 107). For AI, this means nothing less than a fundamental paradigm shift – on a social as well as a technical level.

13.3 Paradigm Shift Towards Sustainable Development and Application of Socio-Technical AI Systems

Behind the phrase "data waste" (Bietti and Vattanparast, 2020) used earlier in this text and the issues presented in the field of AI and sustainability lies a fundamental socio-technical controversy. Solutionist visions that technological solutions can address social and political problems without social and political engagement prove to be fundamentally flawed here as well. Ultimately, they follow the same unreflective logic of progress as their predecessors, who created the problems they are now supposed to solve. In this respect, "data waste" cannot be avoided solely through more sustainable or "green" ICT, because at its core it is a problem of democracy and distribution, which should therefore be addressed not only by experts and politicians, but also by the democratic public.

Thus, other models of ownership, production, and distribution would also have to be considered in a broad understanding of sustainability – in short: the (political) question of what "Our Common Digital Future" (WBGU, 2019) should look like. This, like the future of humanity as such, "depends on our ability to understand and intervene in society in order to minimize the impact of climate change: halt and reverse damage from social inequity; [...] [to] maximize the benefits of new technologies while minimizing their detriments" (Ito, 2018, p. 229). According to Joichi Ito, former director of the MIT Media Lab, in order to be able to do this with regard to socio-technical systems such as AI, a systemic understanding including possible leverage points for intervention at the level of laws, markets, technology architecture, and norms (Lessig, 2009) is just as necessary as changes at the paradigmatic level (Meadows, 1999). Contrary to hopes often raised at present, with Ito (2018, p. 72), "More computation does not make us more 'intelligent', only more computationally powerful." – This is especially true with regard to AI, because despite all the PR rhetoric about "smart", none of today's processes have intelligence in the human sense.

However, human intelligence is doubly challenged by them – on the one hand, to use the increasing computing power sensibly, also with a view to its ecological and social costs, and on the other hand, to always keep in mind the epistemological limits of current methods. This applies in particular to Deep Learning, not only in view of the average 3.4 % of incorrectly labelled training data (Northcutt et al., 2021), but also because this method has so far too often tended to arrive at the "correct" conclusion on the basis of false premises. Whether it is image recognition using rails as indicators for train detection, waves in the water for ships, or image tags for supposed horses, current research on the explainability of AI makes it clear that common evaluation metrics are blind to such circumvention strategies, which is why "the current widespread and sometimes rather unreflective application of machine learning in all industrial and scientific domains" should be fundamentally questioned (Lapushkin et al., 2019). From such a perspective, the prominent promises of autonomous driving or of AI capable of answering research questions appear like naïve and at the same time exuberant quantification of a complex world that is only accessible to such a methodologically still quite immature procedure to a certain extent. Cathy O'Neill has aptly summed this up in "Weapons of math destruction" (2017). The limits of such a reductionist understanding of AI become problematic not only when supposedly autonomous vehicles "overlook" obstacles with fatal results, but also when they, including all distortions in the data, penetrate the social world to the point of biometric total surveillance on such a massive scale as is already the case today.

Even if we grant computers intelligence within the framework of a minimal definition, according to Weizenbaum (1978, p. 300), the "most important basic insight [...] is that at the present time we know of no way to make computers intelligent, and therefore at the moment we should not assign computers any tasks whose solution requires intelligence". In the current digitalisation discourse, the model of reality is often confused with the reality of the model (especially with regard to Big Data and AI), which was and is problematic even in the case of a lack of reflexivity in the application of statistics (Bourdieu, 2004). The accompanying short-term overestimation of potentials and risks could lead to the paradoxical situation that the AI hype obscures the view of Big Data as the "elephant in the room" with all the sustainability risks in surveillance capitalism (Zuboff, 2018). This affects both the social sphere, for example through the crisis of digital publics (Nemitz and Pfeffer, 2020) or discrimination through distorted data and methodological intransparency (e.g. Wachter and Mittelstadt, 2018), and the ecological sphere through the currently predominantly applied data-intensive and thus resource- and energy-intensive procedures of machine learning. In addition, behavioural data-based advertising-financed business models, which ultimately aim to maximise the time on device of all users in terms of attention economy, appear fundamentally unsustainable (Daum, 2019).

In this respect, it seems plausible that Marcus and Davies (2019) argue for a "new start" of AI on a technical-paradigmatic level, as current Deep Learning could prove to be a dead end despite all the progress made in recent years (which is essentially based on the increase in computing power and data). The "data hunger" of current methods proves to be just as limiting as their lack of trustworthiness, transparency, and explainability as well as a differentiation between causality and correlation, the poor integration of already existing knowledge and reaction to unforeseen events in dynamic, complex environments (Marcus, 2018, pp. 6ff.). However, this by no means means throwing out the "baby" AI with the bathwater, but rather developing more robust procedures in the future that are also more sustainable and future-proof in a broad sense due to the elimination of the described shortcomings. For the next decade of more robust AI procedures, Marcus (2020) therefore proposes a hybrid, knowledge-based approach on the basis of cognitive models – whether and to what extent this path is purposeful may be clarified by the specialist community and ultimately only time will tell.

In addition to such technical questions, the ethically highly relevant question of the goals of the application of AI and the values on which it is based (Spiekermann, 2019) remains at least as relevant and, in view of the negative effects described, a changed paradigm is also needed here. Even if "European values" are often propagated at the European level with regard to digitalisation and AI, practice has so far fallen short of these in the absence of consistent implementation, for example with regard to the regulation of AI and facial recognition or its application in the military sector. This also applies beyond AI for tracking behavioural data with a view to the delayed e-privacy regulation. Similarly, an adaptation of the mechanisms of the platform economy to serve the common good, including the reduction of existing concentrations of market power and data stocks and the new conception of the democratic state as a platform (ÖFIT, 2020) or a "European Public Sphere" to "shape Europe's digital sovereignty" (acatech, 2020), remains primarily visionary. The realisation of another initially visionary concept of an open ecosystem for more digital sovereignty in the economic sphere – Gaia-X – now appears to be "capsizing inevitably" in view of the inconsistent demarcation from surveillance capitalist BigTech corporations, as the editorial of a leading IT journal recently headlined.[158]

In this respect, this article shares the conclusion of Bietti and Vatanparast (2020, p. 11), albeit based on a broad understanding of sustainability, because in order to prevent AI from essentially remaining a "fire accelerant" in the problems presented, it is necessary to question "whether we want technology companies, data-driven infrastructures, and the people behind them to have the power to shape the social and ecological conditions for our futures, and if not, who ought to be exercising such power, and what role law and public political engagement can play in shaping alternatives". With a view to the ecological dimension and the insufficient sustainability commitments of large tech companies in view of the situation described above, a keynote speech by James Mickens is applicable: "Optimistic facial expressions do not absorb carbon emissions." (Mickens, 2020)[159] A more sustainable paradigm for the development and application of AI can only succeed by embedding it in the "great transformation towards sustainability" (WBGU, 2011; Schneidewind, 2018). For the shaping of our common, digital future in all its complexity with today's and future machines, the following therefore applies:

> "While we can and should continue to work at every layer of the system to create a more resilient world, I believe the cultural layer is the layer with the most potential for a fundamental correction away from the self-destructive path that we are currently on [...]: a turn away from greed to a world where 'more than enough is too much,' and we can flourish in harmony with Nature rather than through the control of it".

Or in short: "resisting reduction" and "humility over control" (Ito, 2018a).

---

[158] https://www.heise.de/select/ct/2021/3/2028614113239018944
[159] https://www.youtube.com/watch?v=tCMs6XqY-rc

Chapter 14

Production and Trade in the Age of Digitalisation, Networking, and Artificial Intelligence

Rainer Engels

Introduction

This third part of the book has so far looked at the consequences of digitalisation on two essential dimensions of the so-called cultural superstructure (health and education) as well as on the connection with sustainability as it has been defined and discussed by the global community. This and the following two chapters will focus on the economic core. The comprehensive process of the current stage of digitalisation, networking, and the development of artificial intelligence is also referred to as the fourth industrial revolution. This began largely invisibly about ten years ago and has increasingly come to the attention of science, politics, and the public since 2015.

The fourth industrial revolution will be essentially complete in the most developed industrialised countries by 2035. According to the progress report of Plattform Industrie 4.0 (2020), 59 percent of German industrial companies with more than 100 employees already use Industrie 4.0 applications. 73 percent of companies would not only change individual processes in the course of Industrie 4.0, but entire business models. According to the study, the COVID-19 virus will not slow down this development but accelerate it – because digitalisation allows us to better deal with the crisis and get out of it faster. This is not just a German phenomenon: unlike the term "Industry 4.0", which is a German invention, all other major industrial nations have developed comparable concepts and are cooperating intensively with the German Industry 4.0 platform, albeit under different names in each case. The fourth industrial revolution is characterised by comprehensive digitalisation and networking of all production stages (machines, sensors, intermediate, and end products) over the entire life cycle up to self-learning self-organisation in a company (machine learning, artificial intelligence, Internet of Things). Even though this has already been done several times in this book, here is a definition: Artificial intelligence is the attempt to simulate rational or cognitive human intelligence on machines. Compared to the patterns of previous production design, the change is predominantly disruptive and not revolutionary in name only, albeit building on its precursors of electronics and automation. The impact will be profound and far-reaching and has the potential to fundamentally change the way production and consumption of goods and services are shaped economically, socially, and culturally across the entire value chain and product life cycle in many regions of the world.

Currently, companies and research institutions in the European Union (especially Germany and France, Denmark, Finland for e-health), the USA, China, Canada, Switzerland, the United Kingdom, Japan, South Korea, Singapore (in the field of smart cities), and Israel are technological leaders in various areas of the fourth industrial revolution.

In its 2016 research report, the Institute for Employment Research (IAB) expects Germany to lose around 1.5 million jobs by 2025, but this is offset by the expected creation of around 1.5 million new jobs as well. The employees who lose their jobs, however, are only partially qualified for one of the newly created jobs. Thus, disruptive change requires enormous adjustment efforts even in an economically successful industrialised country.

This contrasts with countries and regions of the world that have neither the appropriate technological basis, education, and university structures (especially in the STEM subjects), the necessary investment capital, nor integration into research and development networks. This is especially true for Sub-Saharan Africa, but ultimately for almost all developing and emerging countries. Even for most emerging countries (e.g. Brazil, Indonesia, South Africa, Egypt) it has become apparent that they can hardly keep up and, after a phase of catching up development, are once again being left behind. Everywhere in those countries, the consequences of development can lead to (further intensifying) existential crises.

Disruptive technological change is taking place in a global environment and with simultaneous megatrends. These include, above all, climate change and demographic development.

According to the overwhelming, almost unanimous majority of all climate researchers and without any exception of all climate research institutes, climate change is man-made and irreversible if the CO_2 increase continues as it has so far (cf. German Bundestag, 2019).

This can only be prevented if the temperature rise does not exceed 1.5°C. The necessary radical changes in the economy and society (decarbonisation and departure from the growth path), which are unprecedented in history, must be taken in the next 10 years if a sea-level rise of several metres is to be avoided, as well as the significant increase in life-threatening climate extremes, both of which would very likely cost the lives of millions of people.

Another central megatrend is population development. While demographic development in the economically strong and technologically developed countries mitigates the social costs of structural change to a certain extent, demographic development in many developing and emerging countries intensifies the problem pressure, in some cases drastically. Especially in the countries of Sub-Saharan Africa and South Asia, the massive growth of the population, characterised by an only slightly decreasing birth rate and the tendency to increase life expectancy, is dramatic in its consequences. While, for example, the population of Africa rose from over 228 million people in 1950 to 1.2 billion in 2015 (i.e. in 65 years), the United Nations expects it to reach around 2.5 billion by 2050. This means a doubling of the population in 35 years. Viewed in isolation, this development already poses a challenge to any socio-economic progress. In addition, destabilising effects result indirectly from the massive population growth. For example, severe youth unemployment is already generating social tensions and migratory pressures. Against such a background, the upcoming economic and social structural change of the fourth industrial revolution can have enormous negative effects, which can only be turned around positively by a wise economic policy and joint efforts of all actors. The potential innovative power of so many young people offers an opportunity for this.

However, the main cause of social disintegration is not population growth, but the unequal distribution of income and wealth (within and between nations), the oppression of women, the lack of universal access to contraceptives, and the lack of education, which in turn are a major cause of high birth rates. The growth of middle-income classes and social inclusion have led to decreasing population growth in all nations in the past.

Global urbanisation is progressing with unabated momentum. Already today, 50–60 % of the world's population live in urban conglomerations[160]. Cities and urban agglomerations are growing at a consistently high rate and new cities are being built. This trend, combined with the population growth in developing and emerging countries, leads to strong population pressure on urban areas and requires large investments in infrastructure and considerable social restructuring. However, strong urbanisation can also have positive effects: Investing in efficient infrastructure and initiating concepts for overarching spatial planning can promote sustainable urban life, which can then improve the conditions for investing in the establishment and expansion of digitally networked production and the development of local economies, including through "leapfrog" development (i.e. skipping stages of development) or even "moonshot" development (i.e. through visionary large-scale projects). Through the close coexistence of millions of people, critical minimum limits for effective demand are exceeded for the first time and interesting markets can emerge with forward-looking economic, educational, social, and infrastructure policies. An innovation and industrial policy oriented towards national and regional interests plays a central role here.

Climate change, population development, and urbanisation are the major megatrends that set the framework for the digitalisation of the economy and offer opportunities and risks that can be used or avoided through digitalisation.

14.1 Economic Activity in General

A common goal of the economy is to maximise social welfare while leaving no one behind in poverty. For some time now, the cause of poverty has not been an absolute lack of resources, but an unfair distribution within and between states. The binding nature of the 2030 Agenda under international law imposes national and international obligations on all governments to redistribute resources.

For redistribution to the poor to be equitable and democratically legitimate, it requires the accurate collection of wealth data. Unlike income data or commodity sales or com-

---

160 A distinction must be made here between the political boundaries of cities and the actual spread of urban conglomerations, which usually exceed these boundaries considerably. The people living there are then not included in the official statistics. Therefore, many more people live in urban centres than the statistics show.

modity trade flows, these data are not collected in most countries and for most categories of wealth. This is a political choice that is questionable under international law. Parliaments, tax authorities, and statistical offices must ensure the comprehensive availability of wealth data, especially financial capital. This is technologically possible without any problems. Of course, the availability of data is only one of the preconditions for redistribution to the poor, but this is the part of the political process that is greatly facilitated by digitalisation. Once again, international law in the form of the 2030 Agenda calls on the international community to end poverty. This is a strong argument. As an aside, however, the 2030 Agenda remains ideologically wedded to the old growth model, which makes real sustainability impossible (see also the previous chapter by Reinhard Messerschmidt).

Authoritarian governments use data to oppress their populations. Economic data are also misused for various purposes. At the same time, a globalised economy requires the free exchange of data – within fixed rules. Cybersecurity to protect personal and economic data is therefore indispensable and can also mean the exclusion of companies from certain countries (see the example of Huawei). On the legal framework for the digitalisation of the economy, I refer to the chapters by Christoph Spennemann.

Economically relevant and socially desired digital control tasks that affect human health, individual freedoms and choices, and life must not be decided solely by an artificial intelligence (AI). A human must always have the final say, AIs may only assist. Examples are triage decisions in the event of a disaster, personnel selection, creditworthiness assessments (in Germany called SCHUFA), allocation of funding. This also applies to autonomous driving according to the current state of technology.

Our current economic system is exceeding several planetary stress limits at the same time. The "new", digital way of thinking, if misused, can, like any technology, be a causal factor here, for example through accelerated growth due to increased efficiency, but it should instead help to solve these real problems (see chapter 13). Among other things, it requires the decoupling of economic growth and energy consumption. One example is the control of flexible electricity grids (smart grid), which would be inconceivable without digitalisation and networking. For decoupling, path dependency could be created through massive political intervention and used in a positive and constructive sense (see chapter 3). In today's prevailing digitalisation euphoria, which is gripping even the most peripheral corners of the planet, it is precisely the arsenal of methods of AI that is being credited with every conceivable – and unthinkable – miracle performance. In fact, they are possibly the most powerful tools ever made by our civilisation. Digitalisation and AI can indeed contribute substantially to creating a more humane and environmentally friendly world. A first step, however, would be to address the decoupling of growth and resource consumption in digitalisation itself. There is evidence that information and communication technology alone can achieve up to 20 % reduction in global CO_2 emissions from 2015 levels by 2030 (GeSI and Accenture Strategy, 2015, cited in WB, 2020). Of course, there should be no eyewash here regarding the CO_2 balance of digitalisation: Not insignificant shares of CO_2 emissions only arise from digitalisation and its exponential growth. The classic principle of internalising external costs has not yet been applied here either. Furthermore, the question remains to be answered as to where we actually want to grow, i.e. how we define progress in the future?

What could be more obvious than to apply these tools quickly and on a grand scale to the most pressing challenges this civilisation has ever faced? This is comprehensively elaborated in Chapter 13. The World Bank also sees the need to use digitalisation for a sustainable climate policy: it calls for a digital transformation to unleash the transformative power of digital technologies, both for emissions reduction and resilience across all sectors and to reduce the large and growing emissions of the digital sector itself.

More co-determination in the economy leads to less inequality, which in turn is more economically successful than the Anglo-Saxon model. To ensure the common good and the resilience of the economy, appropriate ownership models of productive assets are needed. Public ownership has recognisable advantages for many economic enterprises (such as savings banks, municipal energy, utilities and waste disposal companies, and public transport, as well as basic infrastructure, but also, for example, education and health care (COVID-19 has again made this all too clear)). With the collapse of the Council for Mutual Economic Assistance (CMEA or COMECON) under the leadership of the Soviet Union, however, the instrument of nationalisation has been thoroughly discredited. However, this often misses the point: what is crucial is that workers have a substantial say in their companies in order to build countervailing power. Here, the German co-determination law (especially if the rules of the Montan co-determination were also applied across the board) is a big step forward; in addition, there are various models of mixed economic systems (cooperative, public and private ownership models) that also lead the way here, also with a view to reducing poverty. In any case, digitalisation does not stand in the way of different combinations of market-oriented and public welfare-oriented instruments, as the recent past shows (renaissance of industrial policy, 110 countries have an industrial policy according to UNCTAD (UNCTAD, 2019)). It can facilitate processes, for example through stakeholder forums such as the German platform Industie 4.0, or through e-governance approaches.

14.2 Extraction of Raw Materials / Mining

Automation and robotics facilitate the extraction of raw materials without hard physical labour. At the same time, digitalisation allows for seamless traceability and recording of raw materials. Therefore, it is reasonable and expected for manufacturers of technology goods with corresponding raw materials to fully comply with individual and ESC human rights, in particular the core labour standards.

In the past, resource wealth has not led to an improvement in the living conditions of the local population. This is not primarily due to working conditions, but mainly to political systems and profit dynamics, in which large international corporations, in conjunction with local elites, get rich from raw material exports and the population and environment are exploited. In resource-exporting countries in Africa, for example, on average 3 % more people are illiterate and have a life expectancy 4.5 years lower than in

African countries that are not primarily resource-exporting. Women and children are also less well-nourished. The example of cobalt and coltan production in Congo shows how difficult it is to find solutions that respect human rights standards, even in the digital age. Automation in the raw materials extraction sector is already making work easier on the one hand and leading to job losses on the other. The simplified traceability of raw materials represents a great opportunity: If only certified raw materials with an assured origin are allowed to be marketed, this could at least put a stop to a large part of the worst practices. Use in small, informal mines could also lead to a reduction in child labour and slavery. The United Nations and international development cooperation (e.g. giz) are working towards this.

The integration of automation and technology, with the simultaneous implementation of the transformation agenda, could have a positive impact for the commodity-producing countries if the appropriate policies are in place. Policy-making here means advising partner governments in an altruistic way on how to shape their economic policies in a way that is more compatible with development, and in particular the country's raw materials policy, but also holding the private sector accountable in public-private dialogues and legal frameworks (analogous to the Supply Chain Act). However, a necessary prerequisite for this is that it is possible to effectively use the returns on raw materials for national development processes. In the future, investments in local people will play an even more decisive role.

Fossil fuels in particular, but also other scarce raw materials, are faced with the decision to be left in the ground, at least in part, in order not to overstretch capacities and to slow down climate change. Artificial intelligence should be used here for optimisation strategies, especially in conjunction with new investment patterns of institutional investors (e.g. (re)insurers, pension or sovereign wealth funds). It is difficult to predict at what technological speed the "peak" of important resources on earth will lead to a substitution by less scarce or renewable resources.

It is clear, on the other hand, that the still growing consumption of energy and raw materials (especially due to growth in emerging countries), with the help of digitalisation and artificial intelligence, makes the implementation of concepts for saving resources and energy (including recycling and the use of by-product flows) more efficient from an operational and economic point of view and thus easier to implement. There is still a lot of research to be done on the question of how exactly the raw materials needed for digitalisation, automation, and robotics can be included in the "bill". In addition, there are areas where automation/robotics or AI are needed to create access to the extraction of further resources (e.g. in "fracking"). This would also have to be included in a "sustainability balance", which would thus be less positive.

14.3 Production of Goods

Wherever complex production processes prevail, digitalisation, which has been developed into Industry 4.0 through networking and artificial intelligence, will prevail as well. This also applies to a lesser extent to services. According to the OECD, efficiency growth in the eight years from 2009 to 2016 was 20 % for high digital intensity industries compared to 5 % for low digital intensity industries.

Industrial automation is one of the world's most dynamic growth markets. For example, about 30 % of German mechanical engineering products are already software or automation technology. At the centre of this technology push is the automation of increasingly complex work tasks and new intelligent production processes that are networked with each other, exchange information, and are partially capable of learning (the Internet of Things in combination with Artificial Intelligence). In Germany in particular, this view is generally shared in the business community. However, this economic growth has so far hardly been oriented towards sustainability goals.

This is, as already mentioned in the beginning, not an evolutionary further development of computerisation and automation, but a revolutionary, disruptive development. The exponentially increasing number of sensors, the digitalisation of potentially all things and their networking with each other and with people will make it possible for the first time to control the entire economic process, including human processes, across the entire value chain and thus provide networked services from research & development to product design and marketing to production to recycling and disposal.

A very important factor in this context is the extent to which the digitalisation and networking of production affects global locations. Investments in robotics, for example, have a negative effect on the frequency and pace of foreign relocations from a development perspective (Jäger at al., 2015; de Backer et al., 2018). However, new technologies can have very different effects: communication technologies could promote global value chains, information technologies could shorten value chains (de Backer and Flaig, 2017). The digital networking of production (Industry 4.0) seems to create incentives to relocate production back to the home country due to its productivity and flexibility effects (Kinkel, 2019). Despite the growing relevance of AI, very little is known so far about its impact on the international orientation of companies (de Backer and Flaig, 2017). Manchanda, Kaleem and Schlorke (2020) structure the importance of AI for business as follows:

1. product complexity: AI enables companies to more efficiently produce sophisticated products such as automobiles, which contain a large number of complex parts and components that are all manufactured separately and assembled into a single unit at the end.
2. process complexity: manufacturers are now using AI by combining large volumes of data with computing power to simulate human cognitive abilities such as reasoning, language, perception, foresight, and spatial processing. AI is being used for predictive maintenance, production line inspection, and other tasks that range from the mundane to the frontline.
3. value chain complexity: real world benefits have been highlighted in a World Economic Forum survey of business leaders conducted during the COVID-19

crisis. Managers said, "the investments made in new technologies were now paying off". For example, they highlighted how Big Data, platforms, and the Internet of Things (IoT) had enabled them to quickly gather large amounts of information that helped them predict disruptions to value chains that would otherwise have adversely affected production. AI has helped them keep a constant eye on the value chain and allowed them to adapt more quickly, potentially ensuring the survival of some manufacturers.

Steffen Kinkel (2020) surveyed 655 manufacturing companies from 16 leading industrialised countries on their attitude to relocation. He found that the level of backshoring in relation to relocation (18/21 %) is remarkably high. He found that (re)relocation activities increase with company size and demonstrated that the intensity of relocation and the use of digital technologies both increase with company size. According to Kinkel (2020), research-intensive companies have a greater tendency to (re)relocate production activities close to their domestic R&D departments. The use of I4.0 and AI technologies also has a significantly positive effect on the tendency to relocate.

The OECD assesses that AI could bring about a real leap in global productivity – a pillar of the 4th Industrial Revolution – with significant impacts on humanity and the planet, similar in importance to water vapour and electricity. There is great potential to produce societal benefits for human welfare, but the risks of harm are also high. In the worst case, it could be used for nefarious and destructive purposes. Which path is taken depends heavily on political decisions.

In any case, this is already leading to enormously accelerated product development, individual customisation and the theoretical possibility of comprehensive control of the entire economy, regardless of whether the actors are state or private sector. Without effective democratic control, this is a highly dangerous development. It is not a new development but is taking place on a completely new level through the widespread use of personalised data. Companies are challenged to develop new business models in order to remain competitive.

For many, digitalisation, networking, and artificial intelligence lead to jobs that are less secure and less well-paid, and also less attractive in terms of content (especially in the platform economies in industrialised countries; in developing countries, platform jobs are sometimes also paid above average) and bring about a massive change in job descriptions through an increase in overall productivity, but especially labour productivity. Simple work, but increasingly also more complex and creative work, is being done by machines. Workers in digital industries are of course free to organise or to reject GIG worker contracts, but the enormous market power here requires special efforts to counter this with a functioning trade union system (this is shown by many prominent examples of US tech companies). State minimum standards must also be enforced in the digital industries, and new regulations are also needed in specific cases. At the same time, however, new, technically more demanding, varied, and very well-paid jobs are being created, although usually not in the same numbers. Unlike the transition from an agrarian society to an industrial society and the transition to a service society, the digital transformation affects all three sectors equally, so that – if no political counter-measures are taken – more jobs will be lost in most countries on a net basis. After the exaggerated forecasts by Frey & Osborne in 2013, there have recently been many more differentiated voices that give the all-clear, at least for industrialised countries, in the short and medium term. Working with scenarios gives us the opportunity to deal more openly with possible developments.

For emerging and developing countries, the situation looks bleak and in the medium term, a relevant number of previously promising emerging countries will suffer in par-ticular. In this context, Banga and te Velde (2018) look at the connection between AI and robots as an example. In the context of the productivity effect [achieved by modern technologies], it is important to note that progress in AI has led to the development of modern robots that are able to recognise structures, which allows them to replace work in a broader range of tasks, including more complex and cognitive tasks. There is already evidence that middle-skilled jobs (bachelor's level) are being hollowed out and a significant proportion of routine tasks are being replaced. Furthermore, according to Banga and te Velde, although lower production costs may create new demand and new jobs, profits are reinvested in the same technologies, resulting in a lower rate of job creation. And finally, jobs in other sectors created by spillovers would also be at risk of being replaced by automation – a clear path dependency, as described in the third chap-ter of this book. Shifts of labour from one sector to another are very difficult, especially in labour markets of emerging and developing countries, as can also be shown by the example of well-developed India, which is facing massive labour market disruptions.

Concerning platform economies, Graham, Hjorth, and Lehdonvirta (2017) propose four strategies to protect platform workers: Certification systems, unionisation of digital work, regulatory strategies, and democratic control of online work platforms. How re-alistic this is when work processes are AI-monitored and even human communication and interaction processes are AI-optimised remains an open question. One promising approach is to support technology cooperation on AI with developing countries. The BMZ-funded project "FAIR Forward – Artificial Intelligence for All" works with AI start-ups in five partner countries: Ghana, Rwanda, South Africa, Uganda, and India. Here, an attempt is made to strengthen a strategic partnership between Africa and Eu-rope, in distinction to the competitors, USA and China.

The ongoing process has the potential to counteract or reinforce the existing global-isation. This depends very much on how the three important economic nations (China, USA, and EU) decide. Under normal economic conditions, the international division of labour is the most efficient mode of production, provided that all factors of production are mobile. However, this is already not true for labour in normal times and COVID-19 has shown that the resilience of the economy in crises also depends on the security of supply and the openness of trade routes, so that a certain degree of regionalisation and even re-nationalisation can be rational (at least superficially). Digitalisation can support this. McKinsey (2020) estimates this effect (caused by COVID-19) such that – taking into account industrial economic decisions and national policy priorities – 16–26 % of global goods exports worth US$2.9 trillion to US$4.6 trillion will change country of production over the next 5 years if firms restructure their supply networks. This is still quite imprecise, but in the midst of the COVID-19 pandemic it is hardly to be expected otherwise. However, the magnitude exceeds exports from developing countries and will become highly relevant for them. Another figure reinforces this impression: 93 % of CEOs of lead firms in global value chains plan to increase resilience. A similar analysis is carried out by Zahn et al. (2020), who predict a drastic transformation of global value chains in the next decade.

The competitiveness of production locations is again changing significantly worldwide. Production in industrialised countries is becoming competitive again, as the location factor of labour costs is losing importance and factors such as political and energy security and especially market/customer proximity are becoming relatively more important. In this context, investors are often guided by current purchasing power and contrast it with the risks, which are still high in many emerging and developing countries due to a lack of legal security and democratic stability. The relocation of production to emerging and developing countries will thus be at least partially reversed (reshoring), as wage processing tends to lose importance. Short delivery routes (especially for products with comparatively high transport costs and for products with a short "lifespan" due to rapidly changing fashions and quality and safety considerations will become more important than reducing cost advantages, with risks remaining the same, especially in the poorer developing countries.

This trend will be reinforced by the requirements of the transformation agenda (Agenda 2030!) and the associated decoupling of energy input and goods output. The impact of increased critical purchasing behaviour and corporate responsibility in western industrialised countries is still open and also depends on the use of digital technologies for the traceability of supply chains, as this is the only way to ensure that human rights, social and ecological standards, and, at the same time, customer wishes are met in a transparent, verifiable, and ultimately justiciable manner.

However, customised production and automation can also make more regional production more efficient, so that it could be profitable in locations close to affluent customers, including in important urban centres of developing and emerging countries (especially coastal ones).

Also, according to UNCTAD (2017), it is not yet clear how advanced digitalisation will affect value chains, and whether these effects will always look the same. On the one hand, increasing digitalisation might require less presence of the investor in the target country. On the other hand, the company could produce more customer-oriented products & services close to the customer (e.g. through 3D-printing). In any case, new players are emerging that no longer sell products but provide services that are licensed. The involvement of local companies is therefore possible, but not mandatory – an exclusion of many through higher quality requirements is possible.

It is also still unexplored whether new business models that lead from one-off product sales to permanent services for the product will lead to more integration of external actors in value chains, or rather to in-house production and trade within the same company. In any case, the trend is away from products towards skills, services, and technologies whose underlying intellectual property rights are protected.

The increase in traditional philosophical or religious patterns of interpretation in the Western and Islamic world, but also in the P.R. China, which has been observable for years, have led to growing doubts about growth ideology and globalisation in unexpected coincidence with the environmental movement. Due to the above-mentioned tendencies, this attitude could be seen by companies as economically useful and lead to corresponding entrepreneurial decisions that intensify the process of reshoring and re-regionalisation beyond the productivity-induced level. The structural change also correlates with the fact that some leading trading nations are rethinking their global trade linkages.

Even if the digital participation of users from the global South is still low, especially in the commercial sector, and e.g. the use of robots in developing and emerging countries is only around 3 % of jobs, the growth dynamics are enormous even in developing countries and the indirect consequences are severe. Artificial intelligence and total networking, the convergence of technologies and of man and machine can strongly define future development worldwide either directly (e.g. for urban middle and upper classes) or indirectly. Nevertheless, emerging countries are also falling behind again in the process of digitalisation, and only a few industrialised countries can even competitively meet the high demands on technology and data.

One example is the semiconductor industry. This value chain is also dominated by East Asia, the USA, and Europe. Kleinhans & Baisakova (2020) argue that this value chain is characterised by deep interdependencies, an extensive division of labour between the countries mentioned and close cooperation throughout the production process. The semiconductor value chain is highly innovative and efficient, but not resilient to external shocks. Such a complex and interdependent value chain creates three challenges for policy-makers, Kleinhans & Baisakova continue: First, how do you secure access to foreign technology providers? Since all of the above-mentioned countries could disrupt the value chain through export control measures, foreign and trade policy plays a central role. Second, how do you create leverage by strengthening domestic firms through strategic industrial policy? Since no region will be able to have the entire production chain in its own territory, governments should support the domestic semiconductor industry to maintain key positions within the value chain. Third, how do you strengthen a more resilient value chain? Certain parts of the chain, such as contract manufacturing of chips, are highly concentrated and need diversification towards lower geographical and geopolitical risks.

Hallward-Driemeier and Nayyar (2018) see new technologies such as AI and changing characteristics of globalisation as a significant negative factor for industrial, export-oriented development in emerging and developing countries. In particular, the combination of productivity growth and employment promotion will not be able to deliver on its promises. The process of digital transformation encompasses production in the narrower sense as well as logistics and industry-related services. The increasing complexity of the value and logistics chain, simplified interface management, intelligent infrastructure, and the IT-supported merging of transport and goods flows will result in a substitution of production mobility by digitalisation and new logistics concepts and make investment concepts (e.g. just-in-time production) of the 20th century appear increasingly uneconomical. The COVID-19 pandemic has once again made this clear. Digital trade is taking an increasing share of total trade flows, US$4.5 trillion out of US$19 trillion over all trade volume in 2021, almost 25% already. This plays a significant role as digital trade makes it easier for consumers and producers to find each other and often helps to reduce transaction costs. This opens additional business opportunities for small and medium-sized enterprises (SMEs) in particular. It also facilitates global market integration. The conditions for the involvement of broad sections of the population in the production of digital services are at least not made more difficult.

For climate-neutral development, the mobility sector is the most important and least transformed sector in terms of CO_2 emissions to date, although the technologies are available. AI can achieve considerable efficiency gains here, in traffic control as well as in the energy integration of the electric-mobile vehicle fleet into the power grids as short-term storage media for buffering grid capacities. We currently have a climate policy-driven shift to electric drives in both private transport and local and long-distance public transport. At the same time, the pandemic is leading to a shift from public transport to bicycle/pedelec, car, and other individual transport. In total, CO_2 emissions have decreased enormously, but how this will play out after the pandemic has not yet been proven and will depend on how each individual country goes into the *"recovery"* (*green* or *business as usual*). The digital economy in the narrow sense still covers less than 10% of economies. Most disruptive changes are therefore taking place in other sectors.

Every year, countries lose billions of US dollars through tax avoidance by multinational companies alone. This money is missing for sustainable economic and social development. An important element in regulating this grievance is the creation of a common framework for measuring the digital economy, as the G20 is striving to do. However, as barriers to market entry are also lowering, business models are changing, sometimes dramatically, and digitalisation is leading to a growing willingness of societies to innovate. It is also likely that innovation performance will grow in at least some poorer developing countries, and completely new business models and/or, in some cases, entire markets will emerge. While industries are entering disruptive phases of upheaval more and more quickly due to increasing innovation speed and technology changes, and existing business models are disappearing, new ones will emerge, and different industries will grow together. At least in part, the start-up and beta culture will become the model for many companies in developing countries and has already become so in initial stages (e.g. Kenya). The globalisation of consumer trends, especially the continuing growth in demand for lifestyle/luxury/status goods, will tend to continue through e-commerce and platform economies and promote mono-/oligopolies (sometimes called superstars) that successfully act against growing market saturation with new sales and advertising concepts, among other things. Due to the still small number of wealthy customers in the poorer developing countries, however, they will benefit little in the short term from the fact that demand will be satisfied closer to the consumer in the future.

Digitalisation and automation in the economy, especially if one looks beyond the German/European horizon, thus shows considerable negative side effects, particularly for developing countries and most developed countries. Hence, the digital economy requires very consistent and thoughtful political management if the potential that digitalisation has to offer is to be sustainably exploited. That is, in a manner that transforms the economy in a climate-friendly way, and that creates a fairer economic system, especially without poverty.

Chapter 15

The Future of the Digital Work-Oriented Society

Christian Kellermann[161]

Introduction

Digitalisation has accelerated the transformation of labour. Even though work today is still prone to changes, there are certain periods during which the changes occur more rapidly. In this process, the implementation of new technologies plays an important role; however, this always takes place within the context of (and in interaction with) other social and cultural changes and circumstances. Digitalisation should also be perceived in the milieu of new forms of globalisation, demographic development, changes in education, and, last but not least, a shift in working people's values.

A qualitatively new technological stage has begun under the label of "artificial intelligence" (AI) as part of the implementation of certain digital technologies used within the labour context in society. AI is the collective term for algorithms that can process large volumes of data and that are capable of learning and finding complex solutions autonomously. AI has the ability to self-optimise and can be characterised by an immanent complexity and non-transparent approaches ("Black Box"). This distinguishes AI from other instruments (including digital tools) that have been used thus far. At the company level, AI systems can organise, manage, and control labour. These systems can also create a self-organising and optimising structure of business and labour relationships, as is done, for example, on labour platforms. The application of AI systems in the labour context thus leads to qualitative and quantitative changes which, in turn, require an adaptation of the regulatory framework as well as of labour relationships.

Particularly on the macroeconomic and industry-wide level, a question has arisen – in addition to the issue of the new quality of work – of what quantitative employment impact would occur if AI as well as other digital technologies increasingly penetrated into business processes. However, the discussion about the potential impact on overall employment is very varied and inconsistent in Germany as well as in many other countries. The predictions of what effects AI involvement will have on employment vary strongly: this is also related to the prospects of putting AI to use and its potential.

Will the automation potential lead to technological unemployment, or will AI bring about growth and employment for all and at a higher level? Or will it heighten the polarisation that the first waves of automation and digitalisation have already set in motion?

---

[161] Christian Kellermann is a senior researcher at the German Research Center for Artificial Intelligence (DFKI) and assistant professor at the HTW – University of Applied Sciences in Berlin. An extended version of this article was published together with Thorben Albrecht by the Friedrich Ebert Foundation (Albrecht/Kellermann 2020).

15.1 AI and the Technological Limitations of Its Applicability

The methods of AI were made possible by the connection between high-performance hardware, big data, and machine-learning procedures (algorithm-based processes). As explained in the second chapter of this book, the development in storage technology, processing power, and speed have outlined a *qualitative advancement* from what used to be a predominantly theoretical approach to application-oriented AI. Nevertheless, AI still comes up against limits of computability, not least due to an incomplete information situation. This is associated with the limited applicability of individual AI systems in practice.

There is also uncertainty when predicting the further development of AI, which does not prevent some from granting AI too much potential in the work environment and endorsing it without restriction. As of today, genuine AI applications in most businesses are either very limited or are still being projected for implementation in the future, albeit mostly for only occasional and partly experimental use. In Germany in particular, many SMEs are hesitant and only gradually beginning to develop ideas for more complex AI applications. However, the "Mittelstand Digital" programme and the German government's "Future Centers AI" provide concrete support for the AI-based innovation of SMEs.

This article therefore focusses on the development of specific areas of AI application in recent years to delineate the practical potential of the technology. Machine learning is a relatively well-developed and significant application area of AI in diverse production fields. With the aid of machine learning, cross-sectional tasks in enterprises are easier to control using AI, for example: (1) higher-level recognition of actions, which enables abstract recognition of comparable situations (pattern recognition) as well as (2) handling of extremely large data volumes. This means that machine learning makes it possible to use examples and observations to draw conclusions and make generalisations which, unlike memorising (immediate storage), can help propose and transfer solutions to various situations. Typical practical application areas include target-specific advertising and marketing, logistics, predictive maintenance, customer relationship management, and people analytics. For example, user data such as purchase or search behaviour across different platforms can help automatically produce similar offers. This is one of the reasons why the distribution phase of the supply chain (advertising, the Internet, and customer relationships) and machine learning are in an interdependent relationship.

In many other contexts, AI methods currently mark the technological boundary of digitalisation in the business and labour environments. There is still a big gap between appreciating the potential of the technology and the specific application of AI in the workplace. In spite of this, it is conceivable that AI – and along with it the "intelligent" digitalisation of the labour world in the foreseeable future – will increasingly and essentially change. In a way, this is already taking place. There has been evident development and application potential in the growing predictability of processes in such labour-related areas as production, services, administration, and agriculture. The precision level, however, has suffered mostly from the fact that AI as such is hard to gauge and measure, and many business and work processes (in which AI could be integrated) are similarly difficult to quantify, which results in their problematic formalisation.

Despite the lively debate about the potential of AI to fundamentally change our society and the working world, the specific effects of AI on the labour context are still under-researched, which is why the discourse tends to be anecdotal. Some of the anecdotes are presented by AI developers, whose aim it is to fully mechanise human intelligence: an ambitious and desirable goal for some, but an ambivalent and dystopian objective for others. Nils Nilsson, a pioneer in AI and robotics, defined this goal as "[the] complete automation of economically important jobs" (Nilsson, 2005, p. 69). Nilsson proposed that an "employment test" could help measure what share of human work might be acceptably performed by an AI system. AI systems would only need to pass the same qualification tests as people are required to pass to be allowed to do special jobs. Current versions of such tests have come to the conclusion that in knowledge-based careers, AI is becoming increasingly competitive and has a more vertical outreach in enterprises (Webb, 2019; Muro et al., 2019).

These and other tests, however, suffer from a whole range of methodological restrictions, which can strongly reduce their validity. In particular, the performed jobs and tasks are frequently described only in keywords and are quickly dubbed as "redundant" when contrasted with the alleged capabilities of AI. Another factor is that not all adjustments are implemented in businesses merely because they can technically be done. Putting technological innovations into practice, including the area of AI technology, needs special considerations and requires actual limitations to be put in place.

15.2 Use Cases in the Tension Between Profitability and Regulation

Several factors influence whether and to what extent AI is actually used in individual businesses. The first factor is the technological performance of AI. The second factor is what AI is allowed to do (what regulatory frameworks are in place: under what conditions, for what tasks, and with what requirements AI can be implemented). The second point includes ethical and regulatory limitations as well as the question of how much AI can be integrated into production and labour processes: i.e., its "integration capacity". Both of these factors establish a framework for a business decision on whether an investment in AI will be truly beneficial in a given case. The third factor (the issue of what the business calculation of costs and income will look like) is particularly hard to address in the case of AI. The decision about the actual implementation of AI in businesses thus depends on what AI can do and what it is allowed to do, as well as what it can bring to the business in question.

Regardless of the frameworks, many business contexts make it difficult to gauge what investments in AI will actually mean for production and labour processes, and whether they could help achieve the desired progress in production. Because of the fact that the application of technology changes as much as technology changes itself, such decisions are hard to make for many businesses. The speed and scope of AI development in the coming years is wide open. With individual implementations, it is difficult to estimate how fast the operational adaptation to AI use will be applied and how long it will last. Moreover, the regulatory framework is a fast-changing variable because it is only in its initial stages in many areas. Finally, it is difficult to assess whether the technology can be successfully integrated into the work organisation of a business and whether it can actually improve the overall processes in production or the provision of services. This depends very much on "soft" factors, namely, whether the interaction between AI and people (colleagues, customers, and so on) will succeed and is sufficiently productive. If this interaction fails, negative consequences cannot be ruled out.

The tension between moving technological boundaries, unclear regulatory framework, and operational functionality makes the decision-making process concerning making economic investments very uncertain. If a positive decision is nonetheless made and an investment in AI is carried out, it might be driven by a general technological optimism (a desire to present oneself as a "front-runner"), or it might simply be acting upon the advice of often exaggerating consultancy institutes advocating for a technological future. As a result, even expensive and disastrous investments are quite possible (note, for example, the experience with the "CIM Ruins" [Computer-integrated Manufacturing] in the 1970s and 1980s). On the other hand, the productive potential of AI might not be utilised at all because of general uncertainty and scepticism. In both cases, an informed discussion and impact assessment of the application of AI technology can help reduce the uncertainties affecting the decision-making processes in businesses.

What are the expected outcomes of AI application in the working environment or directly in the workplace? In which areas does (or could) AI play a role? These questions are not least relevant when one considers the fact that AI implementation will affect operational functionality (the "functionality of labour") and alter the qualitative conditions and the power relationships in the workplace and across an entire company.

Based on the current (qualitative) potential-focused analysis, it can be expected that AI implementation will likely affect cognitive routine tasks; in this sense, "routine" should be more broadly formulated to include progressive AI capabilities (see more on this in the next section). At present, such tasks already encompass such things as the processing of standard cases in finance, insurance, medicine, health care, and many areas of the law. In these areas, AI plays a role that goes beyond merely evaluating data, making predictions, and producing diagnostics: it is also actively involved in research. However, the discussion about pattern recognition in the medical field has shown that a full-scale replacement of human decision-making is not (yet) possible. At the same time, there has been enough evidence to prove that AI could be used in an assistant capacity provided this is done within reasonable limits (cf. chapter 12 on this and the next paragraph).

The social and care services provided to people are another field of potential AI implementation that demonstrates its limitations. Construed as the physical agents of AI, robots are often the barometers of the degree of the automatisation of a production line or service as a consequence of digitalisation and the related impact on labour and employment (cf. Dauth et al., 2017; Bessen, 2018). AI can control a robotic system, which in theory could carry out all kinds of assistance activities in a company or household. Image recognition, sensors, and actors (that is, all technologically complex construction units) are thus increasingly capable of fulfilling diverse human-like tasks. Nevertheless, personal services make it particularly evident that reasonable decisions and emotional tasks will still need to be largely performed by humans. One of the reasons for this is that AI cannot develop human emotions (simulations are possible, but so far can only be compared with human emotional expressions to a very limited extent). Another reason is that humans, as emotional beings, will accept reasonable information about themselves primarily from other people rather than from machines. For example, in the insurance business, where the assessment of insurance claims is fully automatised, the communication of rejected claims is normally carried out by people. Overall, we have to assume that activities which require emotional intelligence and empathy, as well as those that involve making ethical decisions, will largely remain in the human domain for some time.

A central question is therefore in which jobs AI is used to complement human activities and in which jobs it may rationalise people away. Depending on which concrete decisions are actually made in the company and at the workplace, this also influences the possible quantitative effects on employment in a company, an industry, a sector and ultimately on the economy as a whole (the global economic consequences are the subject of the previous chapter).

15.3 Intensified Polarisation in the Labour Market

The discussion about the quantitative impact of digital technology implementation on employment has been going on for several years now. Quite a few notable studies have come to the conclusion that the digitalisation of the labour environment will result in massive upheavals in the form of job losses. This also applies (perhaps most of all) to AI and the growing use of machine-learning systems.

Methodologically speaking, the reasons for such results usually include the comparison of profession profiles in labour market statistics as well as assessments of technology potential based on AI development or the numbers of registered patents. This aids the argument that in the next two decades up to a half of all jobs could be eliminated due to digitalisation, particularly as a consequence of using AI. Professions in transport, logistics, manufacturing, and services are among the most endangered. Low- and middle-income groups will be at the centre of the job-cutting process, and this is why technologically driven unemployment further catalysed by digitalisation would accelerate existing polarisation. In addition to manual routine activities, machine learning and mobile (lightweight) robotics could perform cognitive tasks without firm specifications and thus bring about upheavals in the middle and upper strata of the workforce (Frey/Osborne, 2013; 2017; Muro et al., 2019).

There is a consensus that digitalisation can potentially lead to the substitution of labour. The only question is how great this potential is and what counterforces there will be if they are necessary. The difference between a direct disruption in the labour market and moderating effects is usually made by a "net calculation" across society (Arntz et al., 2017; Arntz et al., 2018; Dauth et al., 2017; Fuchs et al., 2018). Jobs or activities eliminated by digitalisation are contrasted with new jobs in other areas. Growing productivity propelled by digitalisation brings macroeconomic competitive advantages, added value, and ultimately employment effects (McKinsey Global Institute, 2018; World Economic Forum, 2018).

A common assumption in the scenario of technology-driven unemployment is that primarily very simple tasks can be automated. Nonetheless, the higher the routine degree of an activity, the greater the potential is for its substitution. Profound research has clearly shown that no routine is like any other; after all, even the simplest routine tasks are incorporated into work processes as well as the whole organisation and cannot be simply broken up and structured anew. Many of these simple routine tasks are valuable precisely because they require practical knowledge and experience that is hard to transfer and formalise. In other words, it is something that AI has not been able to sufficiently reproduce so far. A static, isolated, and separable understanding of routine work often does not do justice to the tasks in question and in turn only exaggerates the realistic capabilities of AI. The capacity of human labour is, on the contrary, determined by means that are qualitative and context-dependent and cover a broad spectrum of activities (Pfeiffer/Suphan, 2015). Knowledge that is not formalised but "tacit" (Autor, 2015) goes beyond formal qualifications and involves such human senses as intuition, gut feeling, and emotions. It also comprises general knowledge and common sense, i.e., precisely the degree of understanding which AI and machine learning still have great lengths to reach. If the labour capabilities are not part of this perception (if that is even possible), it will quickly transpire that the pendulum has not yet swung to the AI side and that there is still a long journey ahead to achieve simple and informal human capabilities in the labour context.

The binary opposition between routine and non-routine tasks is therefore very limited and encourages a premature definition of digital technologies (including AI tools) as instruments of human replacement. It is evident, however, that the existing processes and organisation of work are becoming more efficient owing to AI: driverless transport systems, man-robot-collaboration (cobots), smart glasses, 3D-printing and additive manufacturing, digital assistance systems, enterprise resource planning, digital twins, and other innovations are increasingly becoming part of the company-level and industry-wide division of labour. Such systems have far-ranging effects on individual jobs and are accompanied, in many cases, by a concentration of workflow and workload (Dispan/Schwarz-Kocher, 2018). In addition to the qualitative aspects of these alterations in labour, this development also leads to possible quantitative changes. But drawing conclusions based on specific (negative) quantitative employment impacts does not do justice to the inherent complexity of routine tasks.

On the other hand, the potential for substitution is accompanied by positive quantitative employment expectations. Up to the individual burn-out limit, the compression of work is connected with a growth in productivity. Digitalisation and the implementation of AI require investments to be made, and these investments can mean either more or less employment. In an optimistic scenario, the implemented digital technology will lead to a technological upgrade and change the capital structure in a business. Rising overall employment is thus a consequence of the growing demand for a concrete type of capital. The increase in production demand in industries providing the inputs for this type of capital leads to growing employment in the economy as a whole.

Furthermore, investments in digitalisation technology alter the cost structure and thus also the relative competitiveness of companies. Businesses that cut their costs thanks to digitalisation can lower their prices and increase the demand for their products and services correspondingly, provided there is constant demand from other sectors of the economy (Arntz et al., 2018). Consequently, the output of the investing businesses increases and produces new income in the form of wages, profits, and capital income. An important aspect in this scenario is the competitiveness effect as well as the division of earnings into capital- and labour-related income. Rising productivity reduces the production costs for automated activities, which can lead to a growth in profits or an increased demand for work in non-automated activities. Corresponding simulations have shown that this is often followed by long-term surges in the overall demand for the workforce in the economy (Fuchs et al., 2018).

Looking at these and similar considerations more closely, it becomes apparent that they strongly rely on spill-over effects from one sector to another. However, they are much more complex than is commonly assumed. Many problems emerge; if capital productivity increases in relation to work, technological innovations are generally labour-saving, meaning that it is rather unlikely that growing productivity would result in increased average wages. If labour is replaced by technology faster than new labour is created, then technology replaces work; and no increase in labour demand in other sectors is necessarily created. Ultimately, digitalisation decisions made by a business will always be the result of a calculation of relative factor prices, i.e., relative prices for all necessary production factors. It is therefore not unlikely that digitalisation will be accompanied by sinking average wages, because the substitution effect leads to a decreased labour demand (Acemoglu/Restrepo, 2018). Ultimately, the outcome will be a growing inequality first on the labour market and then in the whole economy (Korinek/Stiglitz, 2017). This context also highlights the need for a regulatory framework or a redistribution of profits from innovations, because otherwise the overall outcome from technology-driven innovation can be negative for society as a whole when compared with the situation before the innovation (Acemoglu, 2019).

Even if higher wages become a reality as a result of digitalisation in an industry or the economy as such, this does not mean that potential purchasing power will materialise in new purchasing activities in either the same or other business areas. A lot depends on other factors, such as the way digitalisation creates a new structure of economic demand (for instance, via new purchasing patterns), how high or low earners profit from it, and what savings rates currently shape the economy. Until now it has been empirically observed that higher qualified labour profits from digitalisation and human activity are mostly complemented by it. There is flexible demand for the corresponding products and services provided by the workforce; at the same time, however, there is inflexible labour supply in these areas (i.e., a shortage of skilled workers). Activities performed by workers with lower qualification profiles paint a different picture. The demand for manual activities is relatively inflexible as far as their price is concerned; if the price of manual activities drops due to digitalisation, the demand for them does not rise correspondingly. We are now experiencing the preliminary stage of "Polarisation 4.0" (cf. Autor et al., 2017; Autor/Salomons, 2017).

The automatic interconnection between digitalisation, boosts to productivity, and a growing (macroeconomic) demand for labour, as implied by several involved actors, turns out to be rather dubious. It still remains to be seen whether digitalisation and the increasing use of AI will have positive or negative effects on labour and employment. A lot depends on what direction is taken and the relevant regulatory framework. A *laissez-faire* policy would not be a good path towards a digitised work-oriented society. On the contrary, uncontrolled digitalisation with AI on the top threatens to throw our work-oriented society off its current "balance". In the end, there will be (too) few winners and (too) many losers.

15.4 Contouring the Digital Work-Oriented Society

In the foreseeable future, however, it seems that we will not be addressing the fundamental question of whether people will be working or not. Technology-driven unemployment, being a potentially comprehensive and long-lasting state of affairs, principally involves a scenario that is itself rooted in technological progress, just like a new "pareto optimum" for labour and employment.

Nonetheless, in order to achieve and maintain a socially, economically, and environmentally balanced work-oriented society within such a highly dynamic process as digitalisation, it will not suffice to evaluate the impact of technology implementation and fine-tune the corresponding measures by means of regulation. The rapid development of core digital technologies, with AI and machine learning at the current technological frontier, requires a *continuous and holistic* approach to labour and the socio-economic dimension of technology (Kellermann/Obermauer, 2020).

It is important to continuously and carefully observe all technological developments in society. An open and independent technology impact assessment, which is focused on specific uses as well as social impact, thus remains a fundamental prerequisite for the tangible evaluation of the potential of technology, including its social impact. In this regard, the establishment of an AI Observatory by the German Ministry for Labour and Social Affairs is a step in the right direction.[162]

---

[162] https://www.denkfabrik-bmas.de/die-denkfabrik/ki-observatorium

In addition to technological advances, the ongoing transformation of labour, industry, and individual workplaces and labour contexts has to be continuously assessed. Representative evaluations of AI implementation in the labour context have been scarce in Germany as well as in Europe. One exception is a joint research project between the American IT corporation IBM and the German services union Ver.di, which has been commissioned to enable the application of Watson-AI for IBM customers (IBM, 2019). The project specifically pursues the question of what effects AI implementation could have for services activity.

For a broad-based, work-focused technology assessment, it is necessary to categorise and continuously record the different uses of AI in the work context. It is not primarily the technical differences between different AI systems or their level of technical maturity that is important, but the role that AI plays or should play in the company. Depending on the role, there are different requirements for the regulation and transparency of AI as well as for the framework conditions for the participation and empowerment of employees and their representatives in a company.

Institutions and regulations need to be created with which the use of new technologies can be coordinated with social norms and needs, both at the individual workplace and with regard to the role of work in the labour society as a whole. To this end, four levels of action are initially of particular importance: there is a need for

(1) an appropriate political-regulatory framework,
(2) transparency about the objectives and functioning of the AI or technology deployment in the specific use case,
(3) the participation of workers in the deployment and use of technologies; and
(4) the empowerment of employees to deal with AI in a constructive-critical way.

These four levels of action contribute to a new social framework that can anchor the disruptive-dynamic development of technology in a predictable and transparent institutional context. The social goal is not to lose control over technology through such an institutional and regulatory framework, or to give AI and digitalisation the necessary scope for development and application, but at the same time to be able to calculate the effects of the technology at the individual workplace, on affected occupational groups and on the economy as a whole, and to establish reliable monitoring mechanisms at various levels. Otherwise, there is a danger of increased polarisation in the labour society, of incomes and thus also of increased polarisation of society as a whole. If we do not manage to regulate whether and how the "intelligent" machine should work for and with people, this will ultimately also have fatal consequences for the normative question of how, with what and why we want to continue to work for each other based on the division of labour.

Chapter 16

New Social Question and the Future of Social Security

Heinz Stapf-Finé

16.1 Changes in the World of Work and Consequences for Social Security

Digitalisation and the spread of machine learning ("artificial intelligence") pose major challenges for social security, as do other megatrends (cf. BMAS, 2017, pp. 18–39) that go hand in hand with this, in particular growing social inequality, flight and migration, changes in lifestyles and family forms, and demographic change. (cf. also chapter 14).

Closely linked to digitalisation is globalisation, understood as the increasing importance and liberalisation of cross-border economic flows (Lang/Mendes, 2018, p. 38). At the same time, a worldwide information and communication network has been one of the motors of globalisation, and vice versa the expansion and modernisation of the network has progressed with increasing economic interdependence. Especially with the end of the West-East conflict, neoliberal globalisation (cf. Giegold, 2006, pp. 106–107), politically driven by the so-called Washington Consensus (cf. Stiglitz, 2006), gained momentum through an expansion of the free movement of goods, capital, services, and labour.

The distributional effects of globalisation have already posed major challenges to existing approaches in social policy. It is true that additional economic growth as a result of globalisation led to a certain reduction of poverty in many countries in Asia and also in other regions of the Global South. Since the growth effect in industrialised countries was for many years less pronounced than in the catching-up economies of Asia, there was thus a partial reduction in inequality between countries. Within all countries, however, inequality increased because it was mainly the economically better-off strata that benefited from the increase in income and the incomes of poor people did not rise significantly. This effect could have been avoided if the exchange conditions between raw materials and finished products had been more equitable.

Other challenges are the effects of the unbounded economy on the environment, climate, and working conditions. Tensions and conflicts are on the rise, leading to a large movement of refugees and migrants. UNHCR reports: "The number of people fleeing war, conflict, and persecution has never been higher. At the end of 2018, the number of people fleeing around the world was 70.8 million. This compares to 65.6 million people at the end of 2016" (UN Refugee Agency, 2020).

With increasing global interdependence and the spread of western patterns of consumption and living, an individualisation and pluralisation of life models has developed. This is evident in the change in gender roles and the increasing participation of women in the labour market. Although there are still traditional single-earner households, the family model with the man in full-time and the woman in part-time employment now predominates. But working single parents are also gaining in importance (cf. BMAS, 2017, pp. 32–33).

Closely related to this change in family forms is the demographic change due to the decline in birth rates. So far, it has been possible to compensate for this by increasing the participation of women and older people in the labour force. But soon the baby boomers will retire with corresponding demands on the social security systems. The shortage of skilled workers is often mentioned in the same breath as the demographic development. However, this ignores the fact that this has a lot to do with insufficient training efforts on the part of employers and often not particularly attractive working and pay conditions.

16.1.1 Effects of Digitalisation or Artificial Intelligence on the Labour Market

Social policy is facing a number of challenges that are not solely due to the fourth industrial revolution, characterised by digitalisation, robotics, machine learning, networking, and the Internet of Things. There is consensus in the literature about the fact that digitalisation, big data, and machine learning will lead to further structural changes in the world of work. Christian Kellermann analysed this in detail in the previous chapter. However, opinions differ on the strength of the impact: In individual surveys, the impact of digitalisation on employment is estimated to be rather high: according to a survey by HDI Insurance (2019), 60% of respondents fear job losses due to digitalisation in Germany. However, 72% believe that their own jobs are not at risk. A forecast by the Ifo Institute on behalf of the Chamber of Industry and Commerce for Munich and Upper Bavaria concludes that total employment in Germany will range between -4.8% and +5.5%, depending on the scenario. However, a polarisation of the labour market has been observed in recent years: There has been higher employment growth in occupations with low and high qualification levels compared to occupations with medium qualification levels (cf. IHK für München und Oberbayern, 2018, p. 2). The Institute for Employment Research (IAB 2015) examines the substitutability potential for occupations, i.e. the share of jobs that could already be done by computers or computer-controlled machines today. The highest potential is found in manufacturing occupations and is 70%. In all other occupational segments, the potential for substitutability is less than 50%. The least substitutable are activities in social and cultural service occupations. The impact on employment subject to social security contributions will be limited, according to the IAB. About 15% of employees face a high substitutability potential, compared to 45% with medium and 40% with low potential. The team of authors even considers positive effects on employment possible: "The computer-controlled machines have to be developed and built. Skilled workers are needed to operate, control, and maintain the machines. Skilled workers who can handle the new technology must be trained" (IAB, 2015, p. 7).

It is possible that projections are too imprecise with regard to the development of employment due to digitalisation. Muro, Whiton, and Maxim (2019) suggest looking at the effects of robotics and software on the one hand and artificial intelligence on the other. This is because routine or rule-based procedures would tend to affect low- or medium-paid occupations, while the possibilities of artificial intelligence tend to affect the better-paid occupations. Images such as robo-lawyer or Dr. Watson, which carry out legal tasks or medical diagnoses quickly and precisely, also fit this description (cf. Ramge, 2018). A comparison of job descriptions and AI patents, which incidentally was made with the help of AI, shows that bachelor's graduates will be more affected by AI compared to groups with lower-rated education. However, it is not yet possible to provide precise information on the degree to which individual occupational groups will be affected (replaced or supplemented by AI or new jobs created), making it difficult to estimate the impact on employment (Muro, Whiton and Maxim, 2019, p. 23).

Robotics or automation is the focus of a study by Oxford Economics (2019). For the USA, they found on the basis of longitudinal data that in the past industrial workers migrated mainly to the sectors of transport, construction and maintenance, and office and administrative work when their jobs disappeared. According to the econometric model used, these sectors are the ones that will be affected by further automation in the future.

Realistically, it can be assumed that one trend that started in the past will continue in the future: the reduction of lifetime working hours, mainly due to advances in productivity. While in 1800 a total of 153,000 hours of a lifetime consisted of working time, in 1900 it was 132,000 hours. In 2010, remunerised employment had dropped to 48,000 hours. An estimate for 2100 assumes 35,000 hours of average lifetime working time, with times for education and training gaining further importance (cf. IGZA, 2018, pp. 48–49). By reducing the working hours per week, it could be achieved that the gain in quality of life is distributed fairly.

16.1.2 Working Conditions: Platform Work and Hybrid Workers

A preview of the future of work is provided by the area of platform work, in which online platforms act as intermediaries between those offering and those requesting work. Roughly speaking, activities can be divided into cloudwork and gigwork. In the case of cloudwork, the work can be carried out regardless of location, i.e. completely via the internet, and if it is also irrelevant who carries out the work, this is referred to as crowdwork. With gigwork, the work has to be done on site by a specific person, such as with online-mediated driving services or food delivery services (see Schmidt, 2016, p. 5). Meanwhile, it can be assumed that the boundaries between these types of work are fluid: "Today, platforms on the net increasingly serve to generate orders in the off-line sector. Therefore, it is much more necessary to think cross-medially. The boundaries between purely offline and purely online are fluid." (Bertelsmann Foundation, 2019, p. 5). For some, the platform workers are considered a digital proletariat: "The new business models rely on an army of millions of more or less precarious workers who can be hired and fired in a very short time depending on the order situation. So, the platform operators rely on solo self-employed people and private individuals who do jobs on the side as extra income." (Schmidt, 2016, p. 3) By others they are considered to be privileged workers: "The platform worker in Germany tends to have above-average qualifications and be financially better off." (Bertelsmann Foundation, 2019, p. 6). What makes it difficult to organise the interests of platform workers is the fact that what they like about their work is the flexibility and freedom of design and that for most it is a part-time job. However, there is a need for action, as most of them complain a lack of social security and protection rights (cf. ibid.).

16.1.3 Consequences for Social Cohesion

Already in the past, there was a trend towards a growing inequality of income distribution. To such an extent that the OECD (2014) was concerned that inequality could inhibit economic growth, because economically worse off population groups invested too little in their education since they expected little return from it. The White Paper of the Federal Ministry of Labour and Social Affairs (2017, pp. 178–179) deals with a "new social question" in connection with digitalisation, which could reinforce already existing inequalities due to an increasing inequality of income and wealth distribution, the spread of the risk of poverty, the loss of work subject to social security contributions and the insufficient protection of certain groups of employees.

In addition, there are the effects of automation and AI on the development of regional disparities. While Muro and others expect smaller rural communities in the United States to be less vulnerable to disruptive technological developments with regard to the effects of AI, the Oxford study concludes that rural regions are more vulnerable in the face of automation. For Germany, the study determines that Chemnitz, Thuringia, Upper Franconia, Upper Palatinate, and Freiburg are among the most vulnerable regions. The least affected regions are identified as Hamburg, Darmstadt, Upper Bavaria, Cologne and Berlin. These are major challenges to social cohesion, which people in Germany are already concerned about. One of the main reasons for concern is that technological progress is perceived as a threat (see Wintermantel, 2017, pp. 3–7). Therefore, it will be crucial to distribute the expected digitalisation dividend due to rising productivity more evenly (cf. Oxford Economics, 2019, p. 7; BMAS, 2017, p. 180).

16.2 Social Security Development Paths

Considerations regarding the future development of social security must be based on a consideration of the development path so far because it cannot be assumed that radical system changes are politically enforceable. Moreover, it is possible that previous lines of development point in the right direction, as will be shown below.

16.2.1 From Categorical Insurance to Employment Insurance

The introduction of social insurance was the answer to the social question as a result of industrialisation. The systems of health, accident and pension insurance (originally:

invalidity and old-age insurance) introduced in Germany at the end of the 19th century had a pronounced categorical character and initially insured blue-collar workers, later also white-collar workers. In the 1920s, unemployment insurance was added due to high unemployment. In 1995, the fifth pillar was added in the form of long-term care insurance, which proved that the social insurance system, which was often declared dead, is extremely alive and also sustainable.

As the IGZA team of authors was able to demonstrate by looking at the social security systems of Austria, Denmark, Germany, France, Italy, Sweden, and the United Kingdom, the originally categorical systems have been extended to ever wider circles of the population. At the end of the 1930s, for the first time more than 50 % of the labour force was legally insured against old age, sickness, and accidents at work, and by the mid-1970s, 80 %. Traditionally, insurance against unemployment has had a lower coverage. "In 2017, almost all the labour force in Western Europe belonged to the insured group of health insurance, about 90 % were insured in pension insurance, 87 % in accident insurance and an estimated 75 % in unemployment insurance" (IGZA author team, 2018, p. 14).

In addition to the expansion of the group of insured persons, another Europe-wide trend is that benefits have been expanded over time. Even though this process has suffered setbacks in the meantime, since the mid-1990s a tendency towards benefit cuts and privatisation of benefits has set in. In addition, the introduction and extension of benefits to improve the reconciliation of work and family life gained momentum. In summary, there is a historical trend towards the expansion of social insurance to cover all workers. In the future, this trend should be continued to new forms of employment and the associated risks of phases of non-employment in connection with digitalisation.

16.2.2 Unconditional Basic Income as the Answer to All Questions?

Frequently, the demand for an unconditional basic income is raised in connection with the discussion about the effects of digitalisation on the world of work, although the discussion often remains quite superficial. In the foreseeable future, basic income is not a real alternative, not least because there is a lack of social acceptance for such a fundamental change in the system (cf. Wintermantel, 2017, p. 6; BMAS, 2017, p. 180), for which the financial requirements would be so high that it could put established social systems at risk. One of the main reasons against a basic income is that "one size fits all" cannot do justice to the diversity of social needs. Compared to a basic income amount that is the same for everyone, structured social security systems are much more targeted and offer tailor-made solutions for specific target groups. Moreover, the idea that a basic income could broaden the concept of work and create an inclusive labour society could turn into the opposite and deepen the existing social divide: a basic income for many and a privileged position in the world of work for a few (cf. BMAS, 2017, pp. 180–181). Possibly, basic income could even lead to a subsidy for the employer side in the area of low wages, which would be topped up by a basic income.

Nevertheless, the debate about basic income makes sense because it provides many suggestions on how the existing social security system can be made fit for the future. For example, with regard to access to forms of basic and minimum security and a better distribution of the productivity gains of digitalisation, so that a better-funded basic security enables real participation in social life.

In addition, the idea of a basic income could be incorporated into models of lifetime working time accounts in order to be able to secure phases of socially desired time off for education, for example. Section 16.3.4 deals with the corresponding models of the IZGA.

16.3 The Future Social Policy

There have already been adjustments to the social security system in the past, but there is still a need for reform, especially with regard to the following changes in types of employment and living, such as:

- switching between the different kinds of employment of dependent employment, self-employment and civil servant status;
- switching between remunerated employment and family and/or care work;
- discrimination against women and single parents;
- precarious situation of solo self-employed and small traders.

16.3.1 Employment Insurance

In its position paper on the Green Paper "Work 4.0", the Gesellschaft für Versicherungs-wissenschaft und –gestaltung (GVG) points out that the number of solo self-employed has increased by around 27 % since 2000 to over two million in 2014, even though the trend has been declining again since 2012. Crowd work is a field of activity in which solo self-employment is widespread. The GVG problematises: "The alternation of phases of dependent employment and phases of solo self-employment can, under certain circum-stances, create social security gaps." (GVG, 2015, p. 10).

Therefore, there is a need to further develop the statutory pension insurance into an employment insurance that covers all forms of employment that are passed through in the course of a lifetime and is thus an instrument for dignified protection at old age (cf. Bäcker/Kistler/Stapf-Finé, 2011). In the future, digitalisation may create the opportunity for a stronger adaptation of employment histories to different preferences. This also requires social security for all forms of employment under one roof in an insurance scheme for all employed people.

Certain forms of family and care work (voluntary social or ecological year, federal voluntary service) should be supported by state contributions from tax revenue, as should phases of training or unemployment. Solidarity-based compensation can thus be achieved with the help of tax revenue. In the case of pension insurance, an extension of funding to all types of income is not necessary; at old age, only the loss of earned income should be replaced.

The situation is different in the area of statutory health insurance and long-term care social insurance. In addition to the extension of coverage to all forms of employment, and thus the abolition of private health insurance as full insurance, all types of income should be subject to contributions and thus everyone should contribute to the solidarity-based financing according to their economic capacity. This would put an end to a situation that has led to good risks (well-earning, healthy people who usually live alone) increasingly being able to escape solidarity-based financing through the effect of contribution assessment thresholds and through special systems for certain occupational groups such as civil servants, architects, etc. The same is true for the private health insurance system. Corresponding proposals have been on the table for some time (cf. Pfaff/Stapf-Finé, 2004) and have demonstrated financial feasibility and conformity with the legal framework. As the review of the history of social security has shown, this would be the completion of a development that is already underway.

A further expansion of social security is also important in the sense of strengthening the so-called "automatic stabilisers", which have already contributed in the past to maintaining purchasing power in phases of economic crisis and thus to stabilising the economy in the short term.

16.3.2 Digital Social Security

The employment insurance pension and the citizens' insurance health and care could easily be linked to a model of digital social security (cf. Weber, 2019). The basic idea is that platform workers receive a certain percentage of the agreed earnings transferred to a personal digital social security account of the client. The account system could be administered by an international organisation such as the International Labour Organisation (ILO) of the United Nations or the World Bank and transferred to the national social security systems. Alternatively, a direct transfer to a national social security system would be possible, comparable to the transfer of sales tax. In pension and unemployment insurance, claims could be generated from this, which correspond to the amount of the paid-in (minimum) contributions, possibly increased by tax-financed subsidies to guarantee minimum security benefits (if it is possible to subject the platforms and internet corporations to fair taxation). Health and long-term care insurance benefits could also be generated from minimum contributions, possibly supplemented by funds from tax subsidies. The proposal also provides that platforms that want to circumvent these regulations in order to engage in social dumping would be restricted in their offer on the territory of the country.

16.3.3 Minimum Security

Traditionally, the German social security system is strongly related to maintaining living standards and less to poverty prevention. However, in the course of the introduction of Hartz IV, the fear of losing status has become widespread, as workers are usually referred to basic security after only one year of unemployment and have to accept any job regardless of its level and quality. As a result, a discussion developed about an adequate minimum income, which led to the introduction of the statutory minimum wage, and later the basic pension. The basic pension is more a recognition of the lifelong achievements of long-term insured persons with low incomes than a minimum pension that serves to prevent poverty. However, a minimum pension was introduced. In future reforms, the systems of minimum security in working life (minimum wage), in unemployment (basic security) and in old age (basic pension) should be better coordinated so that the receipt of minimum benefits in working life and in unemployment also qualifies for a minimum pension.

Inspired by the debate on an unconditional basic income, future reforms of the basic income system in Germany should examine whether the amount of the benefit actually corresponds to the socio-cultural minimum and thus enables participation in social life. If necessary, cuts must be made to the goal of work at any price in favour of good work. In addition, in the course of simplification for those receiving benefits, the path should be taken to bundle transfer benefits that are often received together, such as housing benefit, child supplement, and employment promotion measures, into one benefit. In addition, it should be possible for benefit recipients to earn some extra money through work without being threatened with a too severe withdrawal of benefits. This would considerably relieve the situation of people in the lower income bracket (cf. Bruckmeier/Konle-Seidl, 2019).

16.3.4 Life Stage Orientation and Employment Insurance, Flexible Transitions Into Retirement

If automation facilitates and partly replaces human labour in the future, the time gains from automation should be distributed fairly. On the one hand, this requires a further reduction in weekly working hours, which should be possible while maintaining the previous level of prosperity. In addition, due to a stronger hybridisation of life phases, a stronger life-phase orientation of social security is necessary.

An approach could be to include working time accounts in different variants. These accounts could be used to save working time credits that can be used for breaks in work life. IGZA (2018, pp. 66–73) discusses three variants: Individual working-time accounts, in which credits are accumulated that are retained even when the employer changes. The second option would be solidarity-financed models into which automation gains could flow. Another variant would provide start-up credits that could be claimed at the beginning of working life. It would also be possible to link this to a basic income, which could guarantee the remuneration from the average weekly working time for the entire working life.

Proposals for the further development of unemployment insurance into employment insurance go in a similar direction. However, it is not about covering time off, but rather periods of further education and training, which are likely to be taken up even more frequently than before with increasing digitalisation. One model currently under discussion envisages the introduction of a further education budget of 26,500 Euros for each eligible person over their entire working life. The group of persons liable to insurance includes persons subject to compulsory insurance under unemployment insurance, solo self-employed, marginally paid employees, recipients of benefits under SGB II or SGB III, as well as inactive persons who are expected to enter the labour market. Financing would come from taxes and contributions and would be paid for further education with recognised and qualifying degrees. In addition to further training, wage replacement benefits could also be paid during the period of further training. The model calculation by Hans and others (2017) shows that the introduction would be possible without additional taxes, spending cuts, or higher debts because the expected expenditures would be offset by additional tax revenues, additional revenues in social insurance, and reduced expenditures due to the reduction of existing or future unemployment.

The previous policy of raising the statutory retirement age has led to a precarisation of retirement. People who want or need to retire earlier do so at the cost of significant reductions in pension benefits. Or people who cannot make it to retirement age for health reasons are referred to the disability pension. The following trend must be counteracted: "The flexibilisation and deregulation of employment relationships, in connection with other social developments such as individualisation processes or the dissolution of traditional gender and family arrangements, has led to a destandardisation of employment histories, to an increase in 'atypical' and precarious employment relationships and to a spread of unsteady employment biographies." (Fröhler et al., 2013, p. 592) Making the transition to old age more flexible could make a significant contribution to better securing flexible employment histories in old age. A whole range of measures could be considered (cf. ibid., pp. 595–634), some of which already exist but which should be more easily accessible:

By paying additional contributions to the pension insurance to better compensate for pension reductions or partial pensions, which can be drawn more flexibly than before after the reform of the partial pension law in 2017. This makes it possible to draw a partial pension in addition to part-time employment in old age.

Not to be confused with partial retirement: this is a model of reduced working hours before retirement. The remaining working time until retirement is halved, and the employer tops up the reduced salary and pays additional contributions to the pension. There is no legal entitlement; the model tends to favour better-off workers. The lifetime work accounts discussed above could also be used for the age transition.

So far, flexible entitlements are only being discussed as a model: every insured person receives additional earning points every four (alternatively five or six) contribution years. An earning point corresponds to a pension entitlement earned from average earnings. The additional entitlements could be used to fill contribution gaps, compensate for periods of below-average earnings or offset pension reductions in the event of earlier retirement.

Those who enjoy the benefits of an occupational pension have the best chances of financially securing the transition to retirement. This is because drawing an early occupational pension is more favourable compared to paying additional contributions to compensate for pension reductions in the statutory pension. However, transferability depends on a number of conditions, such as the way in which the occupational pension is implemented. Another problem is that there is no legal obligation to provide occupational pensions, although there have been cuts in the pension level of the statutory old-age pension in the past, so that it only provides a significant contribution to maintaining the standard of living. Future reforms must therefore find a satisfactory answer with regard to a compulsory scheme.

16.3.5 Age- and Age-Appropriate Work

In order to be able to keep ageing workforces in employment for longer in the face of increasing demands in terms of adaptability, a wide range of corrective (operational deficiencies are remedied once identified), preventive (occupational health and safety concepts are taken into account in work design), and prospective measures (occupational health and safety concepts are used in the planning and revision of work structures) are necessary (cf. BAUA, 2017, p. 14).

Employees would like to see significantly more effort in areas such as ergonomic design of the working environment, inclusion in company training programs, mixed-age teams, health care programs, participation in company development and change processes, and lifetime working accounts. In addition, they would like to see the targeted use of older workers as trainers and advisors in the company, as well as part-time work. A little more than one third would also like to have the opportunity to change jobs within the company. However, human resources managers in some companies fall far short of these expectations and express significantly less need for action (cf. BMAS, 2013, p. 9).

Many companies do not even comply with the legal obligation to prepare a risk assessment to determine which occupational safety and health measures are necessary. In 2011, 50.9 % of companies stated that they carried out risk assessments, in 2015 it was 52.4 % (cf. Sommer et al., 2018, p. 4).

For the area of small and medium enterprises and the creative industries, it has been shown that special counselling and support services for employees who are precariously secured can be helpful. Therefore, cross-company offers for workplace health promotion should be expanded, which could also support employees in the digital economy (cf. Simon et al., 2011).

16.3.6 Distribute the Digital Dividend Better

"To the extent that automation in industry and services results in people being replaced by machines and software, the main question is how the resulting productivity gains are distributed". (BMAS, 2015, p. 44) There is agreement that the "robot dividend" (Oxford Economics, 2019, pp. 6–7) must be better distributed. But the path is arduous. The EU Commission's attempt to levy a tax on the business of internet companies has failed, partly out of fear that the USA might react with countermeasures (cf. Handelsblatt, 06 March 2019). So far, France has single-handedly introduced the GAFA tax (Google, Amazon, Facebook, Apple), which affects a total of around 30 digital corporations, including Airbnb, Uber, Instagram, Ebay, Microsoft, Twitter, and Wish (cf. Handelsblatt, 11 July 2019).

In addition to a taxation of data flows, a modified assessment of social security contributions would also be conceivable, as for example envisaged in the concept of citizens' insurance, in which all incomes would be subject to contributions. The proposal of a value-added levy, which would take into account not wages but the entire value added, and thus also capital gains as the basis for social security contributions by companies, goes somewhat further. Other proposals include changing the ownership structure in the digital economy by declaring people to be the owners of their data, for the use of which a fee would have to be paid. There are also proposals to give employees in the digital economy a stake in the companies, which could revive the discussion about democratising the economy. There is no silver bullet here, but it would be important to initiate a social debate on how a fairer distribution of the digital dividend could be achieved so that everyone benefits from it and not just a handful of internet giants.

16.4 Outlook: Towards a European Social Policy

A discussion about the future of social security cannot avoid the question of a stronger Europeanisation of the same. The decision to introduce a common currency without harmonising economic and social policy has so far led to social distortions. Countries in the European South can no longer make their currencies cheaper through devaluations; the only chance they have to remain competitive is to dump wages and social benefits.

But so far, social policy has largely remained the responsibility of the respective member states and the inclination to change this is not particularly great. This is also shown by the discussion on a European unemployment insurance. The proposal envisages that the employees of the member states pay part of their wages into a joint social insurance scheme and receive a benefit in the event of cyclical, short-term unemployment, which could be supplemented and topped up by the respective country (cf. DIW, 2014). Such an instrument would not only have a social policy effect, but also and above all a cyclical policy effect, as economic downturns could be cushioned (cf. Dullien, 2014, p. 4). No sooner had the outgoing Commission President called for a European unemployment insurance scheme at the beginning of 2019 than the Commission felt compelled to clarify the matter (cf. European Commission, 2019). The nasty word of the "transfer union" was doing the rounds again.

Globalisation, digitalisation, climate change, and strong refugee movements do not stop at national borders. That is why social security requires cross-border solidarity. It is time to formulate common standards for workers in Europe to end the competition for low wages and social standards. Furthermore, measures to coordinate social security systems must be modernised in the sense of improving common social standards. Cross-border mobility also needs to be enhanced and a common European response is needed with regard to the consequences of digitalisation for occupational health and safety (cf. BMAS, 2017, pp. 182–184). Investments in education could also be seen as an expanded social policy; the Corona crisis has shown like a magnifying glass how much catching up is needed with regard to more educational justice; joint European initiatives would also be conceivable here. A truly social Europe in this sense will also be able to reawaken diminished trust among the population.

Part IV

Responsibility of Science

Klaus Peter Kratzer, Jasmin S. A. Link,
Frank Schmiedchen, Heinz Stapf-Finé

The appeal to all scientists in the world to remain true to their responsibility in science and research but especially to think carefully about possible applications of their research has been the impulse for founding the Federation of German Scientists (VDW e. V.). In 1957, the main motive of technological assessments was primarily relevant for the applicability with regard to nuclear weapons. Since then, this methodology has been extended in the following decades to encompass many other research fields. In addition to peace and security politics, VDW activities include research and international negotiations regarding energy, climate, biodiversity, and biotechnology.

In 2017, the broad topic of digitalisation was added. Today, we as the VDW approach you, our reader, with this compendium to jointly think about potential risks of the ongoing digitalisation. The book contains numerous fundamental reasons for scientific support of our request for thorough technological impact assessments of further digitalisation, digital networks, and the development of artificial intelligence.

Repeating the fundamental message of the VDW, we demonstrate that scientific or technical feasibility do not automatically imply rationale. Of course, our discussion on how we want to live today and in future is by no means complete yet. Instead, this compendium should be handled as a scientific tool for education, self-education, and further discussions. The compact summary presented and the collection of various research results and research questions offer comprehensive arguments for further consideration. The diverse discipline-specific perspectives fruitfully support the ongoing societal discussion on potential implications of the self-reinforcing use of digital technologies and devices. It may even be considered the beginning of a new way of thinking. That way, potentially beneficial developments can be pursued and socially less desirable scenarios can be avoided. Accordingly, we invite you to discuss these issues with or without us, the VDW.

In conclusion, we would like to connect the many perspectives that this book offers and to construct a logical chain that connects all chapters.

Chapter 1 shows that collecting and using data for reasonable political and economic decisions is not a new phenomenon. However, the question, which data are collected how, by whom, for which reason, and in which context, demonstrates that actively created and indirectly generated dependencies (chapter 3) would self-reinforce and, without interventions, tend to reduce future alternatives.

Technology-philosophically, it is necessary to expand that argument: Even if there are no ethical standards for judging an abstract algorithm, the application of technologies is not value-free. There is an application context by embedding technology in societal or economic settings and the respective values and norms (chapter 4). However, these standards must be the result of a societal debate and not the random result of numerical or logical algorithms.

Our explanation of the technical background (chapter 2) indicates that euphoric interpretations of the current state of research on AI do not at all necessarily point to the creation of a singularity in the future. However, human innovation during the past six centuries suggests that things we consider improbable or even impossible today may become part of everyday life in the world of tomorrow. This is particularly true if you take into account, which protagonists push the current development of digital technology with what kind of financial and intellectual resources while deploying ethically questionable business models. This includes those who dream of becoming immortal by uploading their conscience into a data cloud (chapter 5). It is time that politics and a critical public oppose such trends, particularly once post-humanistic beliefs are promoted that unknowingly create dependencies (cf. chapter 3) that lead to a lingering replacement of humanity (chapter 4) by technological euphoria and dependence on technology. At last, such stance at least indirectly condones the incremental abolition of humankind or its destruction in a singular event (chapter 5).

Our analysis of machine rights (chapter 6) points out that there are already first attempts in the political arena to grant machines legal status. Regarding the question whether machines can be natural persons, we can calmly watch the trial of "Commander Data" in Star Trek and listen to Jean-Luc Picard's plea in the distant future (Snodgrass, 1989). Until then, our answer is: No! However, regarding the question whether machines can theoretically be a legal person, the answer is: Yes! However, immediately the following question arises: Is this wise? Which consequences does the status imply for responsibility or liability? If a machine was liable for future "decisions" (chapter 7), it would necessarily have to have a legal status before, at which the liability can be directed. At the same time, such legal framework would formally discharge the companies that create the machines or the people that use these machines from liability. The implications of such machine rights need to be assessed legally and technology-philosophically as early as possible (chapters 4 and 7).

In particular, military applications are an example of obvious liability problems and ethical questions, which require legal frameworks at an early stage (chapter 10). At least since Marc-Uwe Kling we have a slight idea of what it means if wars fought totally autonomously, utilizing weapon systems that trigger each other (Kling, 2020): 50 million casualties!

Therefore, it is important to define machines legally as tools of a person in order to make sure conscience and humanity are at least included in the evaluation of a given situation and the resulting actions (chapters 4 and 6).

Regarding the processes and products of the digital transformation, digital networks, and AI, technical standards are currently developed to minimise the risk of machines or algorithms causing any damages (chapter 8). This voluntary self-organisation of the economy develops routines and inspection schemes and establishes a comprehensive technical action framework. In these frameworks, existing standards are recorded and a harmonisation of standards at interfaces is moderated. Standardising organisations support and ease the introduction and sometimes indirectly even the further development of digital products and technologies. On the one hand, this sounds very efficient because adjusting standards once they are established is by far more difficult and economically more expensive than developing and applying joint standards from the very beginning. However, on the other hand, in the context of the digitalisation these fast-track standardisations have the tendency that what starts out as society-conform private bottom-up standardisations in the end results in top-down politico-economic regulation instead (cf. chapter 8).

Chapter 8 highlights that although these "technical" negotiations lead to compromises between the economic stakeholders involved, an open societal discussion regarding these topics is avoided nonetheless. The result is that standardisation organisations decide today what will be daily routines that will be performed without further thinking in tomorrow's world. It is even more problematic if businesses that are practically monopolists can unilaterally set standards and interfaces to consolidate their market position. For many people this creates the notion of inevitability with regard to specific digital applications.

On the one hand, the disruptive changes that the Fourth Industrial Revolution causes in economics, the labour force, and society in general (chapters 14 through 16) cause the "normal" distortions of fundamental structural change. On the other hand, such structural change creates unparalleled transformations of the geostrategic status quo. The technologically more advanced regions drastically increase their lead in comparison with the lesser developed countries. Substantial reshoring can be expected, which will generate extensive disruptions in international supply chains. The focus of industrial production on the three regions North America, South and Southeast Asia, and Europe will probably lead to a distinct redistribution of wealth from the lesser developed countries into the three regions with the highest degree of technology (chapter 14). It is unclear whether this leads to an expulsion of small and medium-sized businesses from the market.

In contrast, the re-regionalisation of production also has the potential to create more sustainable patterns of production. This could foster societal cooperation in the industrial countries (chapter 13). However, this requires economic and social politics that aim at maximizing the benefit of society, which necessitates the further development and adjustment of institutions as well as their structures and processes. From this, the following politically very relevant questions arise:

How can monopolies and oligopolies be avoided and a resilient and market-oriented economic system be secured and refined? Such economic system should consist of private enterprises of various sizes, public corporations, traditional and new collectives as well as other innovative kinds of business.

- How can digital dividends be collected and how can they be invested for the benefit of society?
- How can the remaining human workload be organised so that individual stress is minimised compared to today's working lives? Professional activities should be balanced with individual (psychological and spiritual) development, family life, societal engagement, continued learning, and creativity.

- What would be an effective and efficient social system given the boundary conditions of economics and the labour market defined above?
- How can the growth in the demand for energy of digital systems be curbed? Energy demand still increases despite growing energy efficiency due to the needs of mainly block chains, media streaming, data mining, and the training of so-called AI-tools.
- How can the resilience of digital economic chains and cycles be increased with regard to digital attacks? How can they be made transparent and secure?

In contrast to the numerous progressive possibilities, there is a qualified chance that humans become just a peripheral tool in the otherwise digital and interconnected production processes of an efficient and machine-oriented system (chapter 14). Phases of increased rationalisation that have already started are not going to be automatically compensated. Unlike previous phases of rationalisation, this one will predominantly affect people with intermediate education and those working in offices – broadly speaking (chapter 15). This means that for the first time since the 19th century mainly those positions are abandoned that are occupied by women.

The expected consequence of continued digitalisation, denser IT networks, and the implementation of AI is that without a reasonable political intervention societal disparities will increase progressively at the national level and very likely substantially at the international level. This in turn may cause or augment societal tensions and social conflicts.

Another challenge is the lack of transparency in AI systems. Particularly senior computer scientists lose the ability to verify the underlying calibrations of AI-tools. The tools become some kind of black box, which contains decision rules that cannot be analysed at all. It is theoretically possible to determine, which data are used to train the AI systems. However, it is highly unlikely that the source code, which utilises these data, will ever be disclosed. Thus, there is no control mechanism whether the individual computations and decision steps of the AI-tool actually correspond to the specifications and the intentions of the user.

Since neither the user of an AI-tool nor the computer scientist who codes the software completely understand the algorithms that are applied, it is highly unlikely that the AI-tool satisfies the specifications or the intentions of the user in every possible situation. Therefore, this can be considered to be an explorative procedure: You take a data set and supply it to an algorithm whose functionality cannot be verified. Then you check the results for feasibility. An insight gained from such procedure is that such algorithms can detect similarities in the data sets if the amount of data provided to the tool is sufficiently large. Any additional data set can then be classified and similarities to the already existing data sets can be highlighted.

So how can we assess the application of an AI-tool e.g. for medical purposes such as in radiology? Should the use of all available tools become compulsory for medical personnel – including the utilisation of AI-tools – since AI systems can scan new data from medical imaging much quicker and more precisely, given the complexity of the procedure? Are patients ethically entitled to be treated with AI-tools in order to obtain specifically tailored treatments, e.g. in combination with other new approaches such as mRNA technology? (chapter 12).

A key problem is that even AI-tools commit errors. Humans need to be able to correct them subsequently. They have to understand that an error was made and then induce some counter-action. The AI-tools have to be recalibrated or reprogrammed so that the resulting output is qualitatively satisfactory. It is impossible to successfully recalibrate an AI-tool as long as no human understands how a given AI-tool actually functions.

Since the beginning of the global COVID-19 pandemic, there is a particular urgency to increase the degree of digitalisation in the education sector (chapter 11). Apparently, there it is not asked whether this corresponds with or contradicts the pedagogical experience of childlike development. It is unclear at which age and to what degree the use of digital media is useful during classes at school. Maybe there are alternatives that are more effective or even less expensive. In addition, there is the possibility that a quantitative increase use of digital media leads to a qualitative decrease in learning success and maybe even has adverse health effects in children. Current experience shows that a prescribed digitalisation of schools without having satisfactorily answered the questions posed in chapter 11 and without prior sufficient media literacy of the school children and their parents will cause more harm than good in the long run.

There is a related problem, which can affect both adolescents and adults: If learned skills, proficiencies, and knowledge are replaced by digital devices and applications, they can be unlearned. Changes in brain activity indicate that e.g. the intensive use of navigation systems impair the human capability for orientation – people get disoriented much more frequently (cf. McKinley, 2016, pp. 573ff.). Returning to the example of the radiologist, this means that the regular use of AI-tools in body imaging reduces the proficiency of the medical personnel to interpret the medical image correctly (i.e. to detect tumors etc.). Consequently, there is a risk that the continued use of an error-prone technology reduces the skills and competence of health professionals. If a robust technological assessment came to such a conclusion, would it not be necessary for ethical (and economic) reasons to prohibit the use of AI-tools for this specific purpose?

We have reached a point, at which the precautionary principle should be applied. If the functionality of an AI-tool cannot be verified and its coding or utilisation can cause risks, which are unknown in likelihood, kind, and extent, then its implementation and use should be prohibited until reliable knowledge about the risk potential is available.

If you look at the development of autonomous weapons systems (chapter 10), e.g. the susceptibility to failure of AI-controlled armed swarms of small drones, which cannot be prevented, must lead to their ban and their declaration as weapons of mass destruction, based on the law of nations.

At present, there is only very limited fundamental research addressing technological impacts and security issues, particularly with regard to machine-machine-interaction and machine-environment-interaction. There is substantial public need for such research nonetheless, which needs to be funded. Research on risk assessments, societal interdependencies, and future steps of development needs to be designed and intensively promoted in a user-oriented way that accompanies the technological development.

The necessary funding of such research should support interdisciplinary and transdisciplinary basic research particularly with regard to possible technological designs. The overarching goal is to develop specific interdisciplinary rules and regulations for the process of technological design, which includes its programming. Cooperations and formalised transdisciplinary scientific communication in research and teaching can foster mutual progress and boost the public dissemination of the insights gained.

Decision makers and influencers need to become acquainted with relevant science-based information about the possible consequences of new digital technologies. It takes substantial effort to process such information and review it properly. Decision makers need to be empowered to reach well-informed decisions. We would like to contribute our share to induce the necessary debates and negotiations in order to effectively prevent or at least minimise the conceivable risks of a continued digital transformation. It has to be noted that the majority of research and development activities is not publically controlled and occurs in the setting of global competition. Moreover, the democratic control of military research is also very limited.

All research and development must follow normative ethical and legal principles. The same holds for the continued development of digital technology, specifically for AI. Codified rules are necessary, which include all pertinent legal aspects at all levels: voluntary self-commitments, administrative regulation, laws, constitutional amendments, multilateral agreements with mechanisms for implementation etc. This includes embargoes and moratoria. Norms have to be set, which are not only valid in the specific context of application but also throughout society. The precautionary principle should be legally expanded if this is necessary for a comprehensive averting of danger. Work on these legal issues needs to start promptly, since the design of internationally valid legal frameworks that contain useful means for implementation normally not only takes years but decades.

Effective public (and multilateral) structures and mechanisms for sanctioning are necessary to ensure that digital products and processes are verifiable at any time. This also applies to all phases of research, development, and implementation. In particular, market rollouts should only take place after sufficient safety evaluations, consisting of thorough tests in realistic scenarios. Supplementary technical expertise is necessary. However, neither this expertise nor business considerations may influence regulation. Furthermore, a technically informed, globally functioning, democratic control mechanism of research and development in this sector is required. This needs to be supplemented by ethical self-commitments in research, development, and implementation.

There is another basic rule for the development and implementation of any AI system: A fundamental requirement is the irrevocable coding of ethical principles, which remain operational in all conceivable modes. In case of situations, in which this is not or no longer guaranteed, all necessary actions always have to be possible in a timely and precautionary manner. This demand can have the effect that certain methods of AI can only be implemented in a limited way or maybe not at all.

The most important guideline should be that AI can never do any harm to humans – under any circumstance. This corresponds to the three laws of robotics by Isaac Asimov and is a indispensable prerequisite for the usefulness of AI. Asimov himself characterises his laws as necessary but not sufficient. Ethical principles for algorithms are problematic because they do not own a personality that can have the experience of birth, happiness, pain, sickness, and death. In case this happens one day nonetheless, this would create a completely new set of challenges.

The tasks that lie ahead of us can only be tackled cooperatively. We would like to extend our support to all who are willing to utilise the potentials of the digital transformation only to such an extent and in such a manner that does not endanger human health, life, or the environment and does not imperil social wellbeing. Below the threshold of existential threats, an intensive societal and political debate is called for, which determines the positive, neutral, or adverse impacts of the digital transformation on common welfare,

We look forward to cooperating with you!

Bibliography

The World We Want to Live In – An Introduction

Barrat, James (2013): Our Final Invention. New York

Becker, Philipp von (2015): The New Faith in Immortality. Transhumanism, biotechnology and digital capitalism. Vienna

Fourest, Carone (2020): Generation Beleidigt.Von der Sprachpolizei zur Gedankenpolizei. Berlin

Hamilton, Isabel Asher (2019): Inside the science behind Elon Musk's crazy plan to put chips in people's brains and create human-AI hybrids. Available at: https://www.businessinsider.de/international/we-spoke-to-2-neuroscientists-about-how-exciting-elon-musks-neuralink-really-is-2019-9/?r=US&IR=T (last updated 12.6.2021)

Kastner, Jens/Susemichel, Lea (2020): Identitätspolitiken: Konzepte und Kritiken in Geschichte und Gegenwart der Linken. Münster

Kelly, Grace (2021): Elon Musk's Neuralink plans to implant chips in humans this year. Available at: https://www.msn.com/de-de/finanzen/top-stories/elon-musks-neuralink-plant-noch-in-diesem-jahr-die-implantation-von-chips-beim-menschen/ar-BB1fK6Dp

Musk, Elon (2021): https://www.youtube.com/watch?v=rsCul1sp4hQ und https://twitter.com/elonmusk/status/1380313600187719682?ref_src=twsrc%5Etfw%7Ctwcamp%5Etweetembed%7Ctwterm%5E1380336699847233537%7Ctwgr%5E%7Ctwcon%5Es2_&ref_url=https%3A%2F%2Fwww.businessinsider.com%2Felon-musk-predicts-neuralink-chip-human-brain-trials-possible-2021-2021-2 (zuletzt 12.6.2021)

Puschmann, Cornelius/Fischer, Sarah (2020): Wie Deutschland über Algorithmen schreibt – Eine Analyse des Mediendiskurs über Algorithmen und Künstliche Intelligenz (2005–2020). Gütersloh

Schwab, Klaus and Mallert, Thiery (2020): The Great Reset. World Economic Forum. Cologne/Geneva

Wagenknecht, Sahra (2021): Die Selbstgerechten – Mein Gegenprogramm – für Gemeinsinn und Zusammenhalt. Frankfurt/Main

Chapter 1: Datafication, Disciplining, Demystification

Bentham, Jeremy. (2013): Panoptikum Oder Das Kontrollhaus. Orig. 1791. Berlin.

Bernoulli, Christoph (1841): Handbuch der Populationistik. Stettin.

Bourne, Charles (1963): Methods of Information Handling. New York.

BVerfG (Federal Constitutional Court) (1983): Population census judgement of 15.12.1983 – 1 BvR 209/83, 1 BvR 269/83, 1 BvR 362/83, 1 BvR 420/83, 1 BvR 440/83, 1 BvR 484/83.

CCC (Chaos Computer Club) (1998): Hackerethik. http://dasalte.ccc.de/hackerethics?language=de. (1.4.2021)

Coy, Wolfgang et al., eds. (1992): Sichtweisen Der Informatik. Braunschweig/Wiesbaden.

Criado-Perez, Caroline (2020): Invisible Women. How a world dominated by data ignores half the population. Munich.

EUCOM (European Commission) (2020a): Proposal for a Regulation of the European Parliament and of the Council on European data governance (Data Governance Act), COM/2020/767 final, 25 November 2020. https://eur-lex.europa.eu/legal-content/EN/ALL/?uri=CELEX:52020PC0767 (1.4.2021).

EUCOM (European Commission) (2020b): Impact Assessment Report Accompanying the Document Proposal for a Regulation of the European Parliament and of the Council on European data governance (Data Governance Act), COM/2020/767 final, 25 November 2020. https://ec.europa.eu/newsroom/dae/document.cfm?doc_id=71225. (1.4.2021)

Foucault, Michel (1975): Surveiller et punir. Naissance de la prison. Paris.

Klein, Jürgen/Giglioni, Guido (2020): "Francis Bacon", In: The Stanford Encyclopedia of Philosophy (Fall 2020 Edition), Edward N. Zalta (ed.). https://plato.stanford.edu/archives/fall2020/entries/francis-bacon/. (1.4.2021)

Knaut, Andrea (2017): Fehler von Fingerabdruckerkennungssystemen Im Kontext. Disssertation at the Humboldt University Berlin. https://edoc.hu-berlin.de/handle/18452/19001 (1.4.2021)

Leibniz, Gottfried Wilhelm (1685): Entwurf gewisser Staatstafeln. In: Politische Schriften I, ed. by Hans Heinz Holz, Frankfurt am Main, 1966, pp. 80–89.

Michie, Donald (1961): Trial and Error. In: Barnett, S. A., & McLaren, A.: Science Survey, Part 2, 129–145. Harmondsworth.

Rosling, Hans (2006): Debunking myths about the "third world". TED Talk, Monterey. https://www.gapminder.org/videos/hans-rosling-ted-2006-debunking-myths-about-the-third-world/

Ullrich, Stefan (2019a): Algorithms, Data and Ethics. In: Bendel, Oliver (ed.), Handbuch Maschinenethik. Wiesbaden. S. 119–144.

Ullrich, Stefan (2019b): Boulevard Digital, Berlin. 2019.

Ullrich, Stefan/Messerschmidt, Reinhard/Hilbig, Romy/Butollo, Florian/Serbanescu, Diana (2019c): Demystifying IT Systems. In: Anja Höfner, Vivian Frick (eds.), What Bits and Trees Connect. Shaping digitalisation sustainably. Munich, p. 62.63.

Warren/Brandeis (1890): The Right to Privacy, 4 HARV. L. REV. 193.

Weizenbaum, Joseph (1978): The Power of Computers and the Powerlessness of Reason. Munich.

Chapter 2: Technical Foundations and Mathematical-Physical Limits

Asimov, Isaac. 1991. *Foundation*. New York: Bantam Spectra.

Assion, Simon. 2014. "What Does Case Law Say About Chilling Effects?" https://www.telemedicus.info/was-sagt-die-rechtsprechung-zu-chilling-effects/.

BBC. n.d. "BMW cars found to contain more than a dozen flaws." https://www.bbc.com/news/technology-44224794.

Bengio, Yoshua, Yann LeCun, and Geoffrey Hinton. 2015. "Deep Learning." *Nature* 521 (7553): 436–44.

Bentham, Jeremy. 1995. "Panopticon, or, the Inspection-House." In *The Panopticon Writings*, edited by Miran Božovič, 31–95. London/New York.

Statista 2020. "Bitcoin Network Average Energy Consumption Per Transaction Compared to VISA Network as of 2020." https://www.statista.com/statistics/881541/bitcoin-energy-consumption-transaction-comparison-visa/. Last: 06.5.21

Everling, Oliver. 2020. *Social Credit Rating: Reputation Und Vertrauen Beurteilen*. Springer Gabler, Wiesbaden.

Gallagher, Ryan, and Ludovica Jona. 2019. "We Tested Europe's New Lie Detector For Travellers – And Immediately Triggered A False Positive." The Intercept. https://theintercept.com/2019/07/26/europe-border-control-ai-lie-detector/.

Geisberger, Eva, and Manfred Broy. 2012. *agendaCPS – Integrated Research Agenda Cyber-Physical Systems*. acatech STUDIE.

Gibson, William. 1984. *neuromancer*. New York.

Hofstadter, Douglas R. 1999. *Gödel, Escher, Bach: An Eternal Golden Braid*. Basic Books, Inc.

Huffman, David A. 1952. "A Method for the Construction of Minimum-Redundancy Codes." In *Proceedings of the i.r.e.*, 1098.

EU COM 2020. "Intelligent Parkinson eaRly detectiOn Guiding NOvel Supportive InterventionS." 2020. https://cordis.europa.eu/article/id/421509-can-mobile-selfies-predict-early-parkinson-s-disease/de. Last: 06.5.21

King, and Hope. 2016. "This Startup Uses Battery Life to Determine Credit Scores." CNN Business News. https://money.cnn.com/2016/08/24/technology/lenddo-smartphone-battery-loan/index.html.

Kratzer, Klaus Peter. 1994. *neural networks: fundamentals and applications*. Munich: Hanser.

Langer, Markus, Cornelius J. König, and Vivien Busch. 2020. "Changing the Means of Managerial Work: Effects of Automated Decision Support Systems on Personnel Selection Tasks." *Journal of Business and Psychology*. https://doi.org/10.1007/s10869-020-09711-6.

Lanier, Jaron. 2018. *Ten Arguments for Deleting Your Social Media Accounts Right Now*. Macmillan, New York.

Leeb, Markus R., and Daniel Steinlechner. 2014. "The Kevin Conspiracy." NEWS.at. https://www.news.at/a/kreditwuerdigkeit-namen-kevin-komplott.

Minsky, Marvin. 1974. "A Framework for Representing Knowledge." 306. Cambridge, MA: MIT-AI Laboratory Memo.

Newman, Matthew, Carla Groom, Lori Handelman, and James Pennebaker. 2008. "Gender Differences in Language Use: An Analysis of 14,000 Text Samples." *Discourse Processes* 45: 211–36.

Parnas, David L. 1985. "Software Aspects of Strategic Defense Systems." *Communications of the ACM* 28 (12).

Shannon, Claude E. 1948. "A Mathematical Theory of Communication." In *The Bell System Technical Journal*. Vol. 27.

Tanenbaum, Andrew S. 2012. *Computer Networks*. Munich: Pearson Studium.

"Hunt, Elle 2016. Tay, Microsoft's AI Chatbot, Gets a Crash Course in Racism from Twitter." https://www.theguardian.com/technology/2016/mar/24/tay-microsofts-ai-chatbot-gets-a-crash-course-in-racism-from-twitter. Last: 06.5.21

Turing, Alan. 1950. "Computing Machinery and Intelligence."

"UN News 2020: Bias, Racism and Lies: Facing up to the Unwanted Consequences of AI." https://news.un.org/en/story/2020/12/1080192.

BMU 2020: Federal Ministry for the Environment, Nature Conservation and Nuclear Safety:. "Video-Streaming: Art Der Datenübertragung Entscheidend Für Klimabilanz." 2020. https://www.bmu.de/pressemitteilung/video-streaming-art-der-datenuebertragung-entscheidend-fuer-klimabilanz/. Last 06.5.2021

Vigen, Tyler. n.d. "Spurious Correlations." https://tylervigen.com/spurious-correlations.

Weizenbaum, Joseph. 1966. "ELIZA a Computer Program for the Study of Natural Language Communication Between Man and Machine." *Communications of the ACM* 9 (6).

Chapter 3: Path Dependence and Lock-in

Arthur, W Brian. (1989). Competing technologies, increasing returns, and lock-in by historical events. The Economic Journal, 99(394), 116–131.

Arthur, W Brian. (1994). Increasing returns and path dependence in the economy. Ann Arbor, MI: University of Michigan Press.

Arthur, W Brian. (2013). Comment on Neil Kay's paper – 'Rerun the tape of history and QWERTY always wins'. Research Policy, 6(42), 1186–1187.

BBC. (2016). EU Referendum – Results in full. Retrieved from https://www.bbc.com/news/politics/eu_referendum

Beyer, Jürgen. (2005). Pfadabhängigkeit ist nicht gleich Pfadabhängigkeit! Wider den impliziten Konservatismus eines gängigen Konzepts/Not All Path Dependence Is Alike – A Critique of the "Implicit Conservatism" of a Common Concept. Zeitschrift für Soziologie, 34(1), 5–21.

Beyer, Jürgen. (2010). The Same or Not the Same – On the variety of mechanisms of path dependence. International Journal of Social Sciences 5(1).

Bluedot. (2020). Anticipate outbreaks. Mitigate risk. Build resilience. Retrieved from http://bluedot.global

Brunnermeier, Markus K. (2009). Deciphering the liquidity and credit crunch 2007–2008. Journal of Economic Perspectives, 23(1), 77–100.

Clarke, Harold D, Goodwin, Matthew, & Whiteley, Paul. (2017). Why Britain voted for Brexit: an individual-level analysis of the 2016 referendum vote. Parliamentary Affairs, 70(3), 439–464.

Collier, Ruth Berins, & Collier, David. (1991). Shaping the political arena: Critical junctures, the labor movement, and regime dynamics in Latin America. Princeton, MA: Princeton University Press.

David, Paul A. (1985). Clio and the Economics of QWERTY. The American economic review, 75(2), 332–337.

David, Paul A. (1997). Path dependence and the quest for historical economics: one more chorus of the ballad of QWERTY (Vol. 20): Nuffield College Oxford.

David, Paul A. (2001). Path dependence, its critics and the quest for 'historical economics'. Evolution and path dependence in economic ideas: Past and present, 15, 40.

David, Paul A. (2007). Path dependence: a foundational concept for historical social science. Cliometrica, 1(2), 91–114.

Garud, Raghu, & Karnøe, Peter. (2001). Path creation as a process of mindful deviation. Path dependence and creation, 138.

Giddens, Anthony. (1984). The constitution of society: Outline of the theory of structuration: Univ. of California Press.

Hänska Ahy, Maximillian. (2016). Networked communication and the Arab Spring: Linking broadcast and social media. New Media & Society, 18(1), 99–116.

Kominek, Jasmin. (2012). Global Climate Policy Reinforces Local Social Path-Dependent Structures: More Conflict in the World? In Climate change, human security and violent conflict (pp. 133–147): Springer.

Kominek, Jasmin, & Scheffran, Jürgen. (2012). Cascading processes and path dependency in social networks. In Transnationale Vergesellschaftungen (pp. 1288 auf CD-ROM): Springer VS.

Liebowitz, Stan, & Margolis, Stephen E. (2014). Path Dependence and Lock-In: Edward Elgar Publishing.

Link, Jasmin S. A. (2018). Induced Social Behavior: Can Path Dependence or Climate Change induce Conflict? (Doktorarbeit, Ph. D. Thesis). Universität Hamburg, Hamburg. Retrieved from https://ediss.sub.uni-hamburg.de/handle/ediss/6264

Mahoney, James. (2000). Path dependence in historical sociology. Theory and society, 29(4), 507–548.

Niiler, Eric. (2020). An AI Epidemiologist Sent the First Warnings of the Wuhan Virus. Retrieved from https://www.wired.com/story/ai-epidemiologist-wuhan-public-health-warnings/

North, Douglass C. (1990). Institutions, institutional change and economic performance: Cambridge university press.

Pierson, Paul. (2000). Increasing returns, path dependence, and the study of politics. American Political Science Review, 94(02), 251–267.

Stein, Mark. (2011). A culture of mania: A psychoanalytic view of the incubation of the 2008 credit crisis. Organization, 18(2), 173–186.

Sydow, Jörg, Schreyögg, Georg, & Koch, Jochen. (2005). Organizational paths: Path dependency and beyond.

Sydow, Jörg, Schreyögg, Georg, & Koch, Jochen. (2009). Organizational path dependence: Opening the black box. Academy of management review, 34(4), 689–709.

Tagesschau. (2020). Sorgten TikTok-User für die leeren Ränge? Retrieved from www.tagesschau.de/ausland/usa-wahlkampf-trump-tulsa-tiktok-101.html

Torevell, Terri. (2020). Anxiety UK study finds technology can increase anxiety. Retrieved from https://www.anxietyuk.org.uk/for-some-with-anxiety-technology-can-increase-anxiety/

Tufekci, Zeynep. (2018). How social media took us from Tahrir Square to Donald Trump. MIT Technology Review, 14, 18.

Usherwood, Bob. (2017). The BBC, Brexit and the Bias against Understanding. Retrieved from https://www.vlv.org.uk/issues-policies/blogs/the-bbc-brexit-and-the-bias-against-understanding/

Vergne, Jean-Philippe, & Durand, Rodolphe. (2010). The missing link between the theory and empirics of path dependence: conceptual clarification, testability issue, and methodological implications. Journal of Management Studies, 47(4), 736–759.

Wikipedia. (2020). Landtagswahl in Baden-Württemberg 2011. Retrieved from https://de.wikipedia.org/wiki/Landtagswahl_in_Baden-W%C3%BCrttemberg_2011

Zuckerberg, Mark. (2017, February 16, 2017). Building Global Community. Retrieved from https://www.facebook.com/notes/mark-zuckerberg/building-global-community/10154544292806634

Chapter 4: Questions in the Philosophy of Technology

Anders, Günther (1957/1980): Die Antiquiertheit des Menschen, Volumes I and II. Munich

Floridi, Luciano (2015): Die 4. Revolution – Wie die Infosphäre unser Leben verändert. Berlin

Hacking, Ian (1996): Einführung in die Philosophie der Naturwissenschaften. Stuttgart

Janich, Peter (1997): Kleine Philosophie der Naturwissenschaften. Munich

Kapp, Ernst (1877): Grundlinien einer Philosophie der Technik. Braunschweig

Mitcham, Carl (1994): Thinking through Techology. Chicago

Nordmann, Alfred (2008): Technikphilosophie. Hamburg

Weizenbaum, Joseph (1977): The Power of Computers and the Powerlessness of Reason. Frankfurt

Wiegerling, Klaus (2011): Philosophie intelligenter Welten. Paderborn

Chapter 5: Digital Extensions, Transhumanism, and Posthumanism

Anders, Günther (1956/1980): Die Antiquiertheit des Menschen. On the Soul in the Age of the Second Industrial Revolution. 5th edition. Munich

Barrat, James (2013): Our Final Invention. New York

Bauberger, Stefan/Schmiedchen, Frank (2019): Menschenbild und künstliche Intelligenz. Input paper of the VDW Study Group on Technology Assessment of Digitalisation for the VDW Annual Conference 2019. https://vdw-ev.de/wp-content/uploads/2019/10/Inputpapier-VDW-Jahrestagung-2019-Menschenbild-und-KI.pdf. (last 12.6.2021)

Becker, Philipp von (2015): The New Faith in Immortality. Transhumanism, biotechnology and digital capitalism. Vienna

Birnbacher, Dieter (2013): Utilitarianism. In: Grunwald (2013), pp. 153–158

Blumentritt, Siegmar/Milde, Lothar (2008): Exoprosthetics. In: Wintermantel, Erich/Suk-Woo Ha: Medizintechnik, 4th edition, Luxembourg, pp. 1753–1805.

Boström, Nick (2002): Existential Risks – Analyzing Human Extinction Scenarios and Related Hazards. In: Journal of Evolution and Technology, Vol. 9, No. 1 (2002). https://www.nickbostrom.com/existential/risks.html. (last 12.6.2021)

Boström, Nick (2003): Human Genetic Enhancements: A Transhumanist Perspective. In: Journal of Value Inquiry, Vol. 37, No. 4, pp. 493–506. https://www.nickbostrom.com/ethics/genetic.html. (02.3.2021)

Boström, Nick (2008): Why I Want To Be a Posthuman When I Grow Up. In: Gordijn/Chadwick, p. 107–137

Braidotti, Rosi (2014): Posthumanismus – Leben jenseits des Menschen. Frankfurt/New York

Case, Anne/ Deaton, Angus (2017): Mortality and Morbidity in the 21st Century. https://www.brookings.edu/wp-content/uploads/2017/08/casetextsp17bpea.pdf. (02.3.2021)

Clausen, Jens (2006): Ethical Aspects of Brain-Computer Interfaces in Motor Neuroprostheses. In: International Review of Information Ethics, 09/2006 (Vol. 5), pp. 25–32. https://bmcmedethics.biomedcentral.com/articles/10.1186/s12910-017-0220-y. (02.3.2021)

Clausen, Jens (2011): Technology in the Brain. Ethical, theoretical and historical aspects of modern neurotechnology. Medizin-Ethik 23 – Jahrbuch des Arbeitskreises Medizinischer Ethik-Kommissionen in der Bundesrepublik Deutschland. Cologne

Clausen, Jens/Levy, Neil (eds.) (2015): Handbook of Neuroethics. Dordrecht, Heidelberg, New York, London

Coekelbergh, Mark (2013): Human Beings @ Risk – Enhancement, Technology, and the Evaluation of Vulnerability Transformations. Dordrecht

Coeckelbergh, Mark (2018): Transcendence Machines: Der Transhumanismus und seine(technisch-) religiöse Quellen.In: Göcke, Benedikt P./Meier-Hamidi, Frank (eds.) (2018): Designobjekt Mensch: Die Agenda des Transhumanismus auf dem Prüfstand. Freiburg, pp. 81–93.

Erdmann, Martin (2005): Extropianismus – Die Utopie der technologischen Freiheit. https://www.bucer.de/fileadmin/_migrated/tx_org/mbstexte035.pdf. (03.01.2021)

Fenner, Dagmar (2019): Selbstoptimierung und Enhancement. An ethical outline.Tübingen

Flessner, Bernd (2018): Die Rückkehr der Magier. In: Spreen, dierk u.a. (2018), p. 63–106

Friesinger, Günther/Schossböck, Judith (eds.) (2014): The next Cyborg. Vienna

Fuller, Steve (2020): Nietzschean Meditations. Untimely Thoughts at The Dawn of The Transhuman Era. Basel

Gordijn, Bert, Chadwick, Ruth (eds.) (2008): Medical Enhancement and Posthumanity. Dordrecht, Heidelberg, New York, London

Grunwald, Armin (ed.) (2013): Handbuch Technikethik. Stuttgart/Weimar

Gunz, Philipp (2017): Homo sapiens is older than thought. https://www.mpg.de/11322546/homo-sapiens-ist-aelter-als-gedacht. (02.3.2021)

Hansmann, Otto (2015): Transhumanism – Vision und Wirklichkeit. Berlin

Hansmann, Otto (2016): Zwischen Kontrolle und Freiheit: Die vierte industrielle Revolution und ihre Gesellschaft. Berlin

Harari, Yuval Noah (2017): Homo Deus. A History of Tomorrow. Munich

Hayes, Shaun (2014): Transhumanism. In: https://www.britannica.com/topic/transhumanism. (01.3.2021)

Heinrichs, Jan-Hendrik (2019): Neuroethik. An introduction. Berlin

Herder, Johann Gottfried (1769): Abhandlung über den Ursprung der Sprache. https://www.projekt-gutenberg.org/herder/sprache/sprach01.html. (02.02.2021)

Jansen, Markus (2015): Digitale Herrschaft. Stuttgart

Kai-Fu, Lee (2019): AI Super Powers. Frankfurt/New York

Koops, Bert-Jaap et al (eds.) (2013): Engineering the Human – Human Enhancement Between Fiction and Fascination. Heidelberg/New York/Dordrecht/London

Krüger, Oliver (2019): Virtuality and Immortality. Freiburg/Breisgau/Vienna/Berlin

Kurthen, Martin (2009): White and Black Posthumanism – After Consciousness and the Unconscious. Vienna/New York

Kurzweil, Raymond (2014): Menschheit 2.0. Berlin.

Loh, Janina (2018): Trans- und Posthumanismus zur Einführung. Hamburg

Lüthy, Christoph (2013): Historical and Philosophical Reflections on Natural, Enhanced and Artificial Men and women. In: Koops et al., pp. 11–28

Moravec, Hans (1990): Mind Children. The Race between Human and Artificial Intelligence. Hamburg

More, Max (2013): The Philosophy of Transhumanism. At: https://media.johnwiley.com.au/product_data/excerpt/10/11183343/1118334310-109.pdf. (02.3.2021)

Mulder, Theo (2013): Changing the Body Through the Centuries. In Koops et al., pp. 29–44

Müller, Oliver/Clausen, Jens/Maio, Giovanni (2009): Der technische Zugriff auf das menschliche Gehirn. In: The Technicised Brain – Neurotechnologies as a Challenge for Ethics and Anthropology. Paderborn

Nida-Rümelin, Julian/Weidenfeld, Nathalie (2018): Digitaler Humanismus- Eine Ethik für das Zeitalter der Künstlichen Intelligenz. Munich

Ohly, Lukas (2019): Ethics of Robotics and Artificial Intelligence. Berlin

Rössl, Philipp (2014): Ethical dimensions of neuroprosthetics, p. 15. In: Friesinger/ Schossböck (2014), pp. 15–35.

Samuelson, Norbert/Tirosh-Samuelson, Hava (2012): Jewish Perspectives on Transhumanism. in: Samuelson/Mossmann, S. 105–132

Samuelson, Hava Tirosh/Mossmann, Kenneth L. (eds.) (2012) Building Better Humans? Refocussing the Debate on Transhumanism. Frankfurt/M.

Schmiedchen, Frank et al (2018): Statement on the Asilomar Principles on Artificial Intelligence. Statement of the VDW Study Group on Technology Assessment of Digitalisation. Berlin. https://vdw-ev.de/wp-content/uploads/2018/05/Stellungnahme-SG-TA-Digitalisierung-der-VDW_April-2018.pdf. (01.03.2021)

Schwab, Klaus (2019): The Future of the Fourth Industrial Revolution. Munich

Sorgner, Stefan Lorenz (2018): Beautiful New Man. Berlin

Spreen, Dierk et al. (2018): Kritik des Transhumanismus. On an ideology of the optimisation society. Bielefeld

Wagner, Thomas (2015): Robocracy. Google, Silicon Valley and the human as a discontinued model. Cologne.

Weizenbaum Joseph/Pörksen, Bernhard (2000): Das Menschenbild der Künstlichen Intelligenz. Ein Gespräch. https://www.nomos-elibrary.de/10.5771/0010-3497-2000-1-4.pdf?download_full_pdf=1. (01.3.2021)

Wiegerling, Klaus (2014): Die Veränderung des Gesundheitsverständnisses in Zeiten der technischen Aufrüstung des menschlichen Körpers und seine Auswirkungen auf das leibliche Selbstverständnis. In: Friesinger/Schossböck, p. 35–51

WBGU (German Advisory Council on Global Change) (2019): Main Report 2019 – Our Common Digital Future. Berlin

Chapter 6: Machine Rights

Bauberger, Stefan (2020): Which AI? Munich.

Bayern, Shawn J. (2015): The Implications of Modern Business-Entity Law for the Regulation of Autonomous Systems (Octobert 31, 2015). 19 Stanford Technology Law Review 93 (2015); FSU College of Law, Public Law Research Paper No. 797; FSU College of Law, Law, Business & Economics Paper No. 797. Available at SSRN: https://ssrn.com/abstract=2758222 (last 13.5.2021).

Bayern, Shawn J. and Burri, Thomas and Grant, Thomas D. and Häusermann, Daniel Markus and Möslein, Florian and Williams, Richard (2017): Company Law and Autonomous Systems: A Blueprint for Lawyers, Entrepreneurs, and Regulators (October 10, 2016). 9 Hastings Science and Technology Law Journal 2 (Summer 2017) 135–162. Available at SSRN: https://ssrn.com/abstract=2850514 (last accessed 13.5.2021).

Burri, Thomas (2018): The EU is right to refuse legal personality for Artificial Intelligence. Available at https://www.euractiv.com/section/digital/opinion/the-eu-is-right-to-refuse-legal-personality-for-artificial-intelligence/ (last 13.5.2021).

Bostrom, Nick (2016): Superintelligence. Berlin.

Collins, Harry (2018): Artificial Intelligence. Cambridge.

Darling, Kate (2016): Extending legal protection to social robots: The effects of anthropomorphism, empathy, and violent behaviour towards robotic objects. In Robot Law, ed. Ryan Calo, A. Michael Froomkin, and Ian Kerr, pp. 213–231. Northampton, MA: Edward Elgar.

European Parliament (2017): European Parliament resolution of 16 February 2017 with recommendations to the Commission on Civil Law Rules on Robotics (2015/2103(INL)). Available at: https://www.europarl.europa.eu/doceo/document/TA-8-2017-0051_EN.html (last 13.5.2021).

Görz, Günther and Schmid, Ute and Braun, Tanya (eds.) (2021): Handbuch der künstlichen Intelligenz. Oldenburg, 6th edition.

Gunkel, David (2018): Robot Rights. Cambridge 2018.

Janich, Peter (2006): What is Information? Frankfurt.

Kant, Immanuel (2014): Grundlegung zur Metaphysik der Sitten, ed. v. T. Valentiner u. eingel. v. H. Ebeling.

Kant, Immanuel (2012): Grundlegung zur Metaphysik der Sitten. Hamburg.

Kaplan, Jerry (2017): Artificial Intelligence. An introduction. Frechen.

Kurzweil, Raymond (2015): Menschheit 2.0. Berlin.

Moravec, Hans (1999): Computer übernehmen die Macht: Vom Siegeszug der künstlichen Intelligenz. Hamburg 1999.

Nagel, Thomas (1974): What Is it Like to Be a Bat? Philosophical Review. LXXXIII (4): 435–450. October 1974.

Nevejans, Nathalie et. al. (2018): OPEN LETTER TO THE EUROPEAN COMMISSIONARTIFICIAL INTELLIGENCE AND ROBOTICS, 05/04/2018, Available at: https://g8fip1kplyr-33r3krz5b97d1-wpengine.netdna-ssl.com/wp-content/uploads/2018/04/RoboticsOpenLetter.pdf (last 13.5.2021).

Tegmark, Max (2017): Life 3.0. Berlin.

Chapter 7: Liability Issues

Federal Ministry for Economic Affairs and Energy (BMWi), April 2019: "Artificial Intelligence and Law in the Context of Industry 4.0." Available at https://www.plattform-i40.de/PI40/Redaktion/DE/Downloads/Publikation/kuenstliche-intelligenz-und-recht.pdf?__blob=publicationFile&v=6 (last 18.4.2021)

European Commission, Report from the Commission to the European Parliament, the Council and the European Economic and Social Committee (COM (2020) 64 final), 19.2.2020: 'Report on the security and liability implications of artificial intelligence, the internet of things and robotics'. Available at https://ec.europa.eu/info/sites/info/files/report-safety-liability-artificial-intelligence-feb2020_de.pdf (last 18.4.2021).

European Commission, Directorate-General for Justice and Consumers, (2019): "Liability for Artificial Intelligence and other emerging digital technologies." *Report from the Expert Group on Liability and New Technologies – New Technologies Formation.* Available at https://op.europa.eu/en/publication-detail/-/publication/1c5e30be-1197-11ea-8c1f-01aa75ed71a1/language-en/format-PDF (last accessed 18.4.2021).

Chapter 8: Norms and Standards

8 A: Norms and Standards for Digitalisation

Blind, K. (n.d.), Jungmittag, A., Mangelsdorf, A.: Der gesamtwirtschaftliche Nutzen der Normung. An update of the DIN study from 2000, DIN-Berlin.

Hawkins, R. (1995), Mansell, R., Skea, J. (eds.): Standards, Innovations and Competitiveness. The Politics and Economics of Standards in Natural and Technical Environments, Aldershot, UK

Hesser, W. (2006) Feilzer, A., de Vries, H. (eds.): Standardisation in Companies and Markets, Helmut Schmidt University Germany, Erasmus University of Rotterdam Netherlands

Czaya, A. (2008): Das Europäische Nomungssystem aus der Perspektive der Neuen Institutionenökonomik, Diss. Hamburger Universität der Bundeswehr, Schriften zur Wurtschaftstheorie und Wirtschaftspolitik, Frankfurt/M. u.a.

DIN, 2018: Digitalisation only succeeds with norms and standards. 18-02-din-position-paper-digitisation-succeeds-only-with-norms-and-standards-data.pdf (last 15.6.2021)

DIN/DKE (2020): Deutsche Normungsroadmap Industrie 4.0, Berlin Frankfurt/M.

Sabautzki, W. (2020): Wer beherrscht die "Nationale Plattform Zukunft der Mobilität"?, in: ISW sozial-ökologische Wirtschaftsforschung e.V., München 10.1.2020, https://www.isw-muenchen.de/2020/01/wer-beherrscht-die-nationale-plattform-zukunft-der-mobilitaet/

DIN/DKE/VDE (n.d.) : Whitepaper 'Ethics and Artificial Intelligence: What can technical standards and norms achieve?, Berlin/Frankfurt/M., funded by the BMWi on the basis of a resolution of the German Bundestag.

ISO-survey (2019): Certification & Conformity (https://www.iso.org/the-iso-survey.html)

Gayko, J. (2021, 29.4.): Overview Normung und Standardisierung zur Industrie 4.0, Standardisation Council Industry 4.0, Frankfurt, Germany (www.sci40.com),

Deussen, P. (2020, 7.12.): Themenschwerpunkt: Grundlagen, Welche Grundlagen müssen werden geschaffen, um KI-basierte Systeme in den Markt zu bringen, DIN/DKE/VDE -National Standards Officer Germany, Berlin/Frankfurt/M.

European Commission (2021, 21.4.):) Proposal for a regulation of the European parliament and of the council establishing harmonised rules for artificial intelligence (Artificial Intelligence Law) and amending certain legislation of the union (2021/0106(COD) (https://eur-lex.europa.eu/legal-content/de/txt/?uri=celex:52021pc0206)

DIN/FDKE/VDE (2021, 9.6.): Position paper on the Artificial Intelligence Act: Standards as a central building block of European AI regulation, Berlin/Frankfurt/M. (https://www.dke.de/de/arbeitsfelder/core-safety/news/standards-als-baustein-der-europaeischen-ki-regulierung)

Wahlster, W. (2020): Artificial intelligence: It won't work without norms and standards (https://www.din.de/de/forschung-und-innovation/themen/kuenstliche-intelligenz)

Heinrich Böll Foundation/Brussels (2020) : Technical standardisation, China and the future international order – A European perspective, by Dr Tim N. Rühlig/Stockholm, Brussels (E-books: http://eu.boell.org/, Brussels)

Deutsche Bundestag (2021): Öffentliche Anhörung des Auswärtigen Ausschusses, am 7.6.2021 – 'Innovative Technologien und Standardisierungen in geopolitischer Perspektive' (Stellungnahme von: D. Voelsen/SWP-Berlin, T. N. Rühlig/SIIA, A.Müller-Maguhn/Jornalist, Sibylle Gabler/DIN) https://www.bundestag.de/auswaertiges#url=L2F1c3NjaHVlc3NlL2EwMy9BbmhvZXJJbmdl-bi84NDM2MjgtODQzNjI4&mod=mod538410

8 B: Standardisation as a Geopolitical-Technological Instrument of Power

Arcesati, Rebecca (2019): Mercator Institute for China Studies: Chinese Tech standards put the screw on European companies, Berlin 2019, available online at: https://merics.org/de/analyse/chinese-tech-standards-put-screws-european-companies

Bartsch, Bernhard (2016): Bertelsmann-Stiftung – China 2030, Szenarien und Strategien für Deutschland, Gütersloh, available online at: https://www.bertelsmann-stiftung.de/fileadmin/files/BSt/Publikationen/GrauePublikationen/Studie_DA_China_2030_Szenarien_und_Strategien_fuer_Deutschland.pdf

Bundestag (2021): Public Hearing of the Foreign Affairs Committee on 07.06.2021: Innovative Technologies and Standardisation in a Geopolitical Perspective, agenda and contributions available online: https://www.bundestag.de/ausschuesse/a03/Anhoerungen#url=L2F1c3NjaaHVlc3NlL2EwMy9BbmhvZXJ1bmdlbi84NDM2MjgtODQzNjI4&mod=mod685898

Bundesverband der deutschen Industrie (ed.) (2019): Grundsatzpapier China, Partner und systemischer Wettbewerber – Wie geht wir mit Chinas staatlich gelenkter Volkswirtschaft um, Berlin 2019, available online at: https://www.politico.eu/wp-content/uploads/2019/01/BDI-Grundsatzpapier_China.pdf

De la Bruyère, Emily/Picarsic, Nathan (2020): China Standards Series, China Standards 2035 – Beijing's Platform Geopolitics and "Standardisation Work in 2020", Washington DC/ New York 2020. Available online at: https://issuu.com/horizonadvisory/docs/horizon_advisory_china_standards_series_-_standard; here also the translation of the strategy "China Standards 2035".

Ding, Jeffrey (2020): Balancing Standards: U.S. and Chinese Strategies for Developing Technicsal Standards in AI, Oxford 2020, available online at: https://www.nbr.org/publication/balancing-standards-u-s-and-chinese-strategies-for-developing-technical-standards-in-ai/

EU COM (2021): Rolling Plan for ICT Standardisation 2021 of the European Commission, Brussels, available online at: https://ec.europa.eu/docsroom/documents/44998

Lamade, Johannes (2020): Wirtschaftspolitische Ziele und Diskurse, in: Darimont, Barbara (ed.): Wirtschaftspolitik der Volksrepublik China, pp. 51–68; Wiesbaden 2020.

Rühlig, Tim Nicholas (2020): Technical Standardisation, China and the future international order, Heinrich Böll Foundation, Brussels 2020, available online at: https://eu.boell.org/sites/default/files/2020-03/HBS-Techn%20Stand-A4%20web-030320-1.pdf?dimension1=anna2020

Rühlig, Tim Nicholas (2021): Background paper, Berlin 2021, available online at: https://www.bundestag.de/resource/blob/845192/722efed99b71b971bc62cdd43579dd5b/Stellungnahme-Dr-Tim-Nicholas-Ruehlig-data.pdf

Schallbruch, Martin (2020): Director Digital Society Institue ESMT, in Wirtschaftswoche, December 2020, available online at: https://www.wiwo.de/politik/deutschland/streit-um-huawei-das-ist-eine-technologiepolitische-katastrophe/26687180.html

Semerijan, Hratch G (2016/2019): China, Europe, and the use of standards as trade barriers. How should the U.S. respond; Gaithersburg 2016, updated 2019, available online at: https://www.nist.gov/speech-testimony/china-europe-and-use-standards-trade-barriers-how-should-us-respond

Steiger, Gerhard (2020): Background article vdma – New standardisation strategy "China Standards 2035", Frankfurt 2020, available online at: http://normung.vdma.org/viewer/-/v2article/render/50001829

Chapter 9: Intellectual Property Rights

Association Internationale pour la Protection de la Propriété Intellectuelle (AIPPI), 14 February 2020: "Written Comments on the WIPO Draft Issues Paper on Intellectual Property Policy and Artificial Intelligence. WIPO Conversation on Intellectual Property (IP) and Artificial Intelligence

(AI)". See https://www.wipo.int/export/sites/www/about-ip/en/artificial_intelligence/call_for_comments/pdf/org_aippi.pdf (last: 18.4.2021).

Brazil and Argentina, *Joint Statement on Electronic Commerce. Electronic Commerce and Copyright.* WTO JOB/GC/200/Rev. 1 of 24 September 2018.

Federal Ministry for Economic Affairs and Energy (BMWi), April 2019: "Artificial Intelligence and Law in the Context of Industry 4.0." Available at https://www.plattform-i40.de/PI40/Redaktion/DE/Downloads/Publikation/kuenstliche-intelligenz-und-recht.pdf?__blob=publicationFile&v=6 (last: 18.4.2021)

Czernik, A. (2016): "What is an Algorithm – Definition and Examples". Available at https://www.datenschutzbeauftragter-info.de/was-ist-ein-algorithmus-definition-und-beispiele/ (last: 18.4.2021)

European Commission, Communication *COM(2020) 66 final of 19.02.2020, A European Data Strategy.* See https://ec.europa.eu/info/sites/info/files/communication-european-strategy-data-19feb 2020_de.pdf (last: 18.4.2021)

Free, R./Jane Hollywood, J.: "Diagnosis AI", in *Intellectual Property Magazine* December 2019/January 2020, p. 32.

Herfurth, U. (2019): "Künstliche Intelligenz und Recht", available at https://www.herfurth.de/wp-content/uploads/2019/01/hp-compact-2019-01-knstliche-intelligenz.pdf (last accessed 18.4.2021)

Hugenholtz, B. (2019): 'The New Copyright Directive: Text and Data Mining (Articles 3 and 4)', Kluwer Copyright Blog, 24.7.2019. Accessed at http://copyrightblog.kluweriplaw.com/2019/07/24/the-new-copyright-directive-text-and-data-mining-articles-3-and-4/ (last accessed 18.4.2021).

Levy, K./Fussell, J/Streff Bonner, A.: "Digital disturbia", in *Intellectual Property Magazine* December 2019/January 2020, pp. 30/31.

Okediji, R. (2018): 'Creative Markets and Copyright in the Fourth Industrial Era: Reconfiguring the Public Benefit for a Digital Trade Economy', ICTSD Issue Paper No. 43; https://pdffox.com/creative-markets-and-copyright-in-the-fourth-industrial-era-pdf-free.html (last accessed 18.4.2021).

Samuelson, P./Scotchmer, S. (2001): "The Law and Economics of Reverse Engineering", 111 Yale L. J. 1575 (2001). Available at https://scholarship.law.berkeley.edu/facpubs/62/ (last 18.4.2021).

Schmiedchen, F. et al (2018): "Statement on the Asilomar Principles on Artificial Intelligence." Digitalisation Technology Assessment Study Group. Berlin https://vdw-ev.de/wp-content/uploads/2018/05/Stellungnahme-SG-TA-Digitalisierung-der-VDW_April-2018.pdf. (last 18.4.2021)

Schönenberger, D. (2017): 'Deep Copyright: Up- and Downstream Questions Related to Artificial Intelligence (AI) and Machine Learning (ML)', in *Droit d'Auteur 4.0 / Copyright 4.0. PI-IP – Series on Intellectual Property Law,* University of Geneva, Faculty of Law, pp. 145–172. Available at https://www.unige.ch/droit/pi/publications/publications/vol10/ (last accessed 18.4.2021).

Schürmann/Rosental/Dreyer (2019): "Artificial Creativity"? – Protection of AI and its works by copyright. Available at https://www.srd-rechtsanwaelte.de/blog/kuenstliche-intelligenz-urheberrecht/ (last accessed 18.4.2021).

Swiss Federal Administration, *Strategy for Open Administrative Data in Switzerland 2019–2023.* Available at https://www.admin.ch/opc/de/federal-gazette/2019/879.pdf (last accessed 18.4.2021).

Skinner, B. (2020): "Eine Strategie des Bundes für den Papierkorb". *Neue Zürcher Zeitung of* 8 May 2020, p. 9.

United Nations Conference on Trade and Development (UNCTAD) *e-commerce week 2020,* webinar on "Who owns our data? What is the role of intellectual property? Summary available at https://unctad.org/system/files/official-document/DTL_STICT_2020.11.05_eWeek2020finalsummaryreport_en.pdf (last 18.4.2021)

Van Asbroeck, B./Debussche, J./César, J. (2017): "Building the European Data Economy – Data Ownership". White Paper, Bird & Bird.

World Intellectual Property Organisation (WIPO), "The WIPO Conversation on Intellectual Property and Artificial Intelligence", 2019, available at https://www.wipo.int/about-ip/en/artificial_intelligence/conversation.html (last accessed 18.4.2021).

Chapter 10: Lethal Autonomous Weapons Systems – New Threat and New Arms Race?

Alwardt, Christian/Neuneck, Götz/Polle, Johanna et al. (2017): Sicherheitspolitische Implikationen und Möglichkeiten der Rüstungskontrolle autonomer Waffensysteme, Institute for Peace Research and Security Policy at the University of Hamburg (IFSH). Unpublished expert report for the Office of Technology Assessment at the German Bundestag.

Arraf, Jane/Schmitt, Eric (2021): Iran's Proxies in Iraq Threaten U.S. With More Sophisticated Weapons, The New York Times, 4 June 2021.

BICC (2013): *The UN Arms Convention*. Arms control thematic module. https://sicherheitspolitik.bpb.de/m7/articles/m7-07 (last 13.6.2021)

Boulanin, Vincent (2016): Mapping the development of autonomy in weapon systems. A primer on autonomy, SIPRI 2016: https://www.sipri.org/sites/default/files/Mapping-development-autonomy-in-weapon-systems.pdf.

Boulanin, Vincent/Verbruggen, M. (2017): Mapping the Development of Autonomy in Weapon Systems, https://www.sipri.org/sites/default/files/2017-11/siprireport_mapping_the_development_of_autonomy_in_weapon_systems_1117_0.pdf. (5.8.2020).

Cramer, Maria (2021): A.I. Drone May Have Acted on Its Own in Attacking Fighters, U.N. Says, The New York Times, 3. June 2021

Department of Defense (2018) (ed.): Summary of the National Defense Strategy; Washington D.C. p. 3 https://dod.defense.gov/Portals/1/Documents/pubs/2018-National-Defense-Strategy-Summary.pdf (last 13.6.2021)

DSB (Defense Science Board) (2012): The Role of Autonomy in DoD Systems. https://fas.org/irp/agency/dod/dsb/autonomy.pdf. (5.8.2020)

DSB (2016): Summer Study on Autonomy. https://fas.org/irp/agency/dod/dsb/autonomy-ss.pdf. (5.8.2020) Dutch Government (2016)

Erhart, Hans-Georg (2021): Meeting and Killing in: Der Freitag, No. 2 (2021) 14 January 2021.

Forstner, Christian/Neuneck, Götz (eds.) (2018): Physik, Militär und Frieden. Physicists between arms research and the peace movement, Springer Spektrum, Wiesbaden

Geiss, Robin (2015): Die völkerrechtliche Dimension autonomer Waffenysteme, Friedrich-Ebert-Stiftung, June 2015.

Goose, Stephan D./Wareham, Mary (2016): The Growing International Movement Against Killer Robots, in: Harvard International Review, Vol. 37, N°. 4, pp. 28–33.

Grünwald, Reinhard/Kehl, Christoph (2020): Autonome Waffensysteme. Final Report on the TA Project, Working Report No. 187. Office of Technology Assessment at the German Bundestag.

Hofman, F. G. (2017/2018): Will War's Nature Change in the Seventh Military Revolution? Parameters 47, no. 4

Horowitz, Michael/Schwartz, Joshua A./Fuhrmann, Matthew (2020): China has made Drone Warfare Global, Foreign Affairs. https://www.foreignaffairs.com/articles/china/2020-11-20/china-has-made-drone-warfare-global (last accessed 13.6.2021)

ICRC (2019): Autonomy, artificial intelligence and robotics: Technical aspects of human control, Geneva

Kallenborn, Zachary (2021): If a Killer Robot were Used, would we know? Bulletin of the Atomic Scientists, June 4, 2021.

Keegan, John (1995): The Culture of War, Rowohlt

Neuneck, Götz (2009): Atomares Wettrüsten der Großmächte – kein geschlossenes Kapitel. In the conference proceedings: "Kampf dem Atomtod", Dölling und Galitz Verlag, Hamburg, pp. 91–119.

Neuneck, Götz (2011): Revolution oder Evolution. On the development of the RMA concept. Science and Peace. 29(2011): S. 6–13

Neuneck, Götz et al. (2017): Sicherheitspolitische Implikationen und Möglichkeiten der Rüstungskontrolle autonomer Waffensysteme, IFSH an der Universität Hamburg. Unpubl. Report for the Office of Technology Assessment at the German Bundestag.

Payne, Kenneth (2018): Artificial Intelligence: A Revolution in Strategic Affairs? In: Survival Vol. 60(5), October-November 2018, pp. 7–32.

Chapter 11: Education and Digitalisation – Technology Assessment and the Demystification of "Digital Education" in Theory and Practice

Allert, Heidrun: Algorithms and Inequality. In: merz. Zeitschrift für Medienpädagogik, 03 (2020). Available at: https://www.merz-zeitschrift.de/alle-ausgaben/pdf/algorithmen-und-ungleichheit/ [14.02.2021].

Altenrath, Maike; Helbig, Christian; Hofhues, Sandra: Deutungshoheiten: Digitalisation and Education in German and EU Programmes and Funding Guidelines. Journal MedienPädagogik 17. Jahrbuch Medienpädagogik (2020). S. 565–594. Available at: https://www.medienpaed.com/issue/view/83 [14.02.2021].

Anderson, Susan; Maninger, Robert (2007): Preservice Teachers' Abilities, Beliefs, and Intentions regarding Technology Integration. Journal of Educational Computing Research. Verfügbar unter: https://journals.sagepub.com/doi/10.2190/H1M8-562W-18J1-634P [14.02.2021].

Antonowsky, Aaron: Salutogenesis. On Demystifying Health. German expanded edition by Alexa Franke. Tübingen: dgvt-Verlag, 1997.

Balslev, Jesper (2020): Evidence of a potential. Dissertation. Available at: https://jesperbalslev.dk/evidence-of-a-potential-ph-d-thesis/ [14.02.2021].

Barr, Rachel et al. (2020): Beyond Screen Time: A Synergistic Approach to a More Comprehensive Assessment of Family Media Exposure During Early Childhood. Frontiers in Psychology. 11. Available at: https://pubmed.ncbi.nlm.nih.gov/32754078/ [08.03.2021].

Beißwenger, Michael; Bulizek, Björn; Gryl, Inga; Schacht, Florian (eds.): Digitale Innovationen und Kompetenzen in der Lehramtsausbildung. Duisburg: Universitätsverlag Rhein-Ruhr, 2020.

Best, Alexander et al. (2019). Kompetenzen für informatische Bildung im Primarbereich. Bonn: Gesellschaft für Informatik e. V. Available at: https://dl.gi.de/bitstream/handle/20.500.12116/20121/61-GI-Empfehlung_Kompetenzen_informatische_Bildung_Primarbereich.pdf?sequence=1&isAllowed=y [14.02.2021].

Bitzer, Eva Maria; Bleckmann, Paula; Mößle, Thomas: Prävention problematischer und suchtartiger Bildschirmmediennutzung. A Germany-wide survey of practical facilities and experts. KFN Research Report 125. Hannover: Criminological Research Institute, 2014.

Bleckmann, Paula: Medienmündig. How our children learn to use screens in a self-determined way. 1st edition. Stuttgart: Klett-Cotta, 2012.

Bleckmann, Paula; Mößle, Thomas: Position on problem dimensions and prevention strategies of screen use. In: SUCHT 2014/60 (2014), pp. 235–247.

Bleckmann, Paula; Mößle, Thomas; Siebeneich, Anke: ECHT DABEI – gesund groß werden im digitalen Zeitalter. Manual for primary school teachers. Berlin: BKK Dachverband 2018, pp. 56–59.

Bleckmann, Paula: Experiential Education in the Digital Age. Between appropriation, compensation and active media work. e & l International Journal for Action-Oriented Learning, 3&4 (2019), pp. 26–31.

Bleckmann, Paula; Lankau, Ralf (eds.): Digitale Medien und Unterricht. A controversy. Weinheim: Julius Beltz, 2019.

Bleckmann, Paula; Zimmer, Jasmin (2020): "Small-scale technology assessment" for media literacy in teacher training: Weighing opportunities and risks of analogue and digital learning scenarios on two levels. In: Beißwenger, Michael; Bulizek, Björn; Gryl, Inga; Schacht, Florian (eds.): Digitale Innovationen und Kompetenzen in der Lehramtsausbildung. Duisburg: Universitätsverlag Rhein-Ruhr 2020, pp. 303–329.

Bleckmann, Paula (2020): Keynote on "Development-oriented media education: (Further) teacher training requirements", Online Workshop on "Development-oriented and age-appropriate media education" as part of LifeLongLearningWeek 2020, 02.12.2020. Available at: https://www.youtube.com/watch?v=NH_MgV7APUw&list=PL-alwtxqu7HEKj5U6rmi5pY-vpV7R3i4E&index=4&t=12s [14.02.2021].

Bleckmann, Paula et al. (2020): What are possible health consequences (psycho-social)? Available at: https://unblackthebox.org/wp-content/uploads/2021/02/UBTB_Onepager_Psychosoziale_Folgen.pdf [08.03.2021].

Bleckmann, Paula; Pemberger, Brigitte; Stalter, Stephanie; Siebeneich, Anke: ECHT DABEI – Manual für Kita-Fachkräfte. 5th edition. Prevention programme ECHT DABEI – Gesund großwerden im digitalen Zeitalter (ed.). Freiburg i. Br., 2021.

Brunnengräber, Achim; Zimmer, Fabian: Digital in den Stau? Why digitalisation and electrification do not necessarily accelerate the sustainable mobility transition. In: Göpel, Maja; Leitschuh, Heike; Brunnengräber, Achim et al. (eds.): Yearbook Ecology 2019/20. The Ecology of the Digital Society. Stuttgart: Hirzel 2020, pp. 83–98.

Curzon, Paul; Mc Owan, Peter William: The Power of Computational Thinking. Games, magic and puzzles to help you become a computational thinker. London: World Scientific, 2018.

Van Deursen, Alexander; Helsper, Ellen (2015). The Third-Level Digital Divide: Who Benefits Most from Being Online? Communication and Information Technologies Annual. Available at: https://www.researchgate.net/publication/287277656_The_Third-Level_Digital_Divide_Who_Benefits_Most_from_Being_Online [14.02.2021].

Dittler, Ullrich; Hoyer, Michael: Growing up in virtual media worlds. Chances and dangers of digital media from a media-psychological and media-pedagogical perspective. Munich: Kopaed, 2008.

Dräger, Jörg; Müller-Eiselt, Ralph: The digital education revolution. The radical transformation of learning and how we can shape it. Munich: Deutsche Verlags-Anstalt, 2015.

Felschen, Christina (2020): Covid-19. Jugendliche verbringen deutlich mehr Zeit mit Computerspielen. ZEIT ONLINE 29.07.2020. Available at: https://www.zeit.de/wissen/gesundheit/2020-07/covid-19-jugendliche-computerspiele-studie-mediensucht?utm_referrer=https://www.google.com/ [08.03.2021].

Förschler, Annina (2018): Das ‚Who is who?' der deutschen Bildungs-Digitalisierungsagenda – eine kritische Politiknetzwerk-Analyse. In: Pädagogische Korrespondenz, 58 (2). Pp. 31–52. Budrich Academic Press. Available at: https://www.researchgate.net/publication/330280542_Das_Who_is_who_der_deutschen_Bildungs-Digitalisierungsagenda_-_eine_kritische_Politiknetzwerk-Analyse_In_Pädagogische_Korrespondenz_582_31-52 [12.02.2021].

Gotsch, Matthias: Impacts of Digital Transformation on Environment and Climate Protection. Development of an analytical assessment scheme. In: Göpel, Maja; Leitschuh, Heike; Brunnengräber, Achim et al. (eds.): Yearbook Ecology 2019/20. The Ecology of Digital Society. Stuttgart: Hirzel 2020, pp. 99–109.

Grunwald, Armin (2020a): Künstliche Intelligenz – Gretchenfrage 4.0. Süddeutsche Zeitung, Kultur 03.01.2020. Available at: www.sueddeutsche.de/kultur/kuenstliche-intelligenz-gretchenfrage-4-0-1.4736017 [14.02.2021].

Grunwald, Armin: Digital disasters. Experiences from Technology Assessment. In: Göpel, Maja; Leitschuh, Heike; Brunnengräber, Achim et al. (eds.): Jahrbuch Ökologie 2019/20. Die Ökologie der digitalen Gesellschaft. Stuttgart: Hirzel 2020b, pp. 65–71.

Hanses, Andreas: Gesundheit und Biographie – eine Gradwanderung zwischen Selbstoptimierung und Selbstsorge als gesellschaftliche Kritik. In: Paul, Bettina; Schmidt-Semisch, Henning (eds.): Risiko Gesundheit. On risks and side effects of the health society. Wiesbaden: VS Verlag 2010, pp. 89–104.

Hartong, Sigrid: Between assessments, digital technologies and big data: The growing influence of 'hidden' data mediators in education. European Educational Research Journal, 15/5 (2016), pp. 523–536.

Hartong, Sigrid; Förschler, Annina; Dabisch, Vito (2019): Translating Educational Inequality into Data Infrastructures: The Example of Digital School Monitoring Systems in German and US State Education Agencies. Available at: https://www.researchgate.net/publication/335836260_Translating_Educational_Inequality_into_Data_Infrastructures_The_Example_of_Digital_School_Monitoring_Systems_in_German_and_US_State_Education_Agencies [14.02.2021].

Hartong, Sigrid u.a. (2020): Wissenschaftliche Netzwerkinitiative "unblackthebox" – Für einen (selbst) bewussten Umgang mit digitalen Datentechnologien in Bildungseinrichtungen. Available at: https://unblackthebox.org/ [16.01.2021].

Hattie, John: Making Learning Visible. 3rd, expanded edition with index and glossary. Beywl, Wolfgang/Zierer, Klaus (eds.). Baltmannsweiler: Schneider Verlag Hohengehren, 2015.

Hauser, Kornelia: Critical Sociology of Education. Domination and (Self)Liberation. In: Sandoval, Marisol (ed.): Bildung MACHT Gesellschaft. Münster: Verlag Westfälisches Dampfboot 2011, pp. 26–38.

Hauser, Urs; Hromkovič, Juraj; Klingenstein, Petra; Lacher, Regula; Lütscher, Pascal; Staub, Jacqueline: Einfach Informatik Zyklus 1. Baar/CH: Klett und Balmer, 2020.

Held, Martin; Schindler, Jörg: Metals. The material prerequisite of the digital transformation. In: Göpel, Maja; Leitschuh, Heike; Brunnengräber, Achim et al. (eds.): Yearbook Ecology 2019/20. The Ecology of the Digital Society. Stuttgart: Hirzel 2020, pp. 125–137.

Hromkovič, Juraj; Lacher, Regula: Einfach Informatik 5/6, Primarstufe. Baar/CH: Klett und Balmer, 2019.

Hübner, Edwin: Anthropological Media Education. Frankfurt am Main: Peter Lang, 2005.

Humbert, Ludger; Magenheim, Johannes; Schroeder, Ulrik; Fricke, Martin; Bergner, Nadine (2019): Informatik an Grundschulen (IaG) – Einführung – Grundlagen. Teacher's guide. Available at: https://www.schulministerium.nrw/sites/default/files/documents/Handreichung-fuer-Lehrkraefte.pdf [07.05.21]

Illich, Ivan: Self-Limitation. A Political Critique of Technology. Rowohlt, Reinbek 1975. German by N. T. Lindquist. Original title: Tools for Conviviality. New York: Harper and Row, 1973.

Illich, Ivan: Genus. Towards a historical critique of equality. Munich: Beck, 1982.

Yearbook Ecology 2019/20. The Ecology of the Digital Society. Göpel, Maja; Leitschuh, Heike; Brunnengräber, Achim; Ibisch, Pierre; Loske, Reinhard; Müller, Michael; Sommer, Jörg; von Weizsäcker, Ernst Ulrich (eds.). Stuttgart: Hirzel, 2020.

Kassel, Laura; Fröhlich-Gildhoff, Klaus; Rauh, Katharina: Bestands- und Bedarfserhebung 2015/16 Ergebnisse. Böttinger, Ullrich; Fröhlich-Gildhoff, Klaus (eds.). Prevention Network Ortenaukreis. Offenburg: Landratsamt Ortenaukreis and Freiburg: Zentrum für Kinder- und Jugendforschung, 2017.

Kernbach, Julia; Tetzlaff, Frederik; Bleckmann, Paula; Pemberger, Brigitte: Einstellungen und Bewertungen von Eltern an reformpädagogischen Schulen zur medienerzieherischen Praxis. Results based on innovative questioning dimensions of the quantitative-explorative MünDig study. Themenheft Eltern, Zeitschrift MedienPädagogik, 2021 submitted.

Köhler, Katja; Schmid, Ute; Weiß, Lorenz; Weitz; Katharina (n.d.): Pixel & Co. – Informatik in der Grundschule. Commentary for teachers. Braunschweig: Westermann, 2021.

Krapp, Andreas; Weidenmann, Bernd (eds.): Pädagogische Psychologie. A Textbook. 4th, fully revised ed. Weinheim: Beltz, 2001.

Langmeyer, Alexandra; Guglhör-Rudan, Angelika; Naab, Thorsten; Urlen, Marc; Winklhofer, Ursula (2020): Kind sein in Zeiten von Corona. Result report on the situation of children during the lockdown in spring 2020. Available at: https://www.dji.de/themen/familie/kindsein-in-zeiten-von-corona-studienergebnisse.html [08.03.2021].

Lankau, Ralf; Burchardt, Matthias (2020): Humane Bildung statt Metrik und Technik. Available at: http://futur-iii.de/wp-content/uploads/sites/6/2020/07/aufruf_zur_besinnung_pub.pdf [14.02.2021].

Lankau, Ralf: No one learns digitally. On the sensible use of new media in the classroom. Weinheim: Beltz, 2017.

Linn, Susan: Consuming Kids. Protecting Our Children from the Onslaught of Marketing and Advertising. New York: Random House, 2005.

McDaniel, Brandon T.; Radesky, Jenny S.: Technoference: Parent Distraction With Technology and Associations With Child Behavior Problems. Child Development, Volume 89, Issue1, January/February (2018), pp. 100–110.

Medienberatung NRW (2019): Medienkompetenzrahmen NRW. 2nd edition. Available at: https://medienkompetenzrahmen.nrw/fileadmin/pdf/LVR_ZMB_MKR_Broschuere.pdf [14.02.2021].

Möller, Christoph (ed.): Internet- und Computersucht. A practical handbook for therapists, educators and parents. Stuttgart: Kohlhammer, 2012.

Mößle, Thomas: Dick, dumm, abhängig, gewalttätig? Problematic media use patterns and their consequences in childhood. Results of the Berlin Longitudinal Media Study. Baden Baden: Nomos Verlag, 2012.

Mößle, Thomas; Bleckmann, Paula; Rehbein, Florian; Pfeiffer, Christian: The Influence of Media on School Performance. In: Möller, Christoph (ed.): Internet and Computer Addiction. A practical handbook for therapists, educators and parents. Stuttgart: Kohlhammer 2012, pp. 67–76.

Muuß-Merholz, Jöran (2019). The great amplifier. Is digitalisation dividing the education world? – Essay. Federal Agency for Civic Education (Aus Politik und Zeitgeschichte)(ed.). Available at: http://www.bpb.de/apuz/293120/der-grosse-verstaerker-spaltet-die-digitalisierung-die-bildungswelt?p=all [14.02.2021].

Nistor, Nicolae; Lerche, Thomas; Weinberger, Armin; Ceobanu, Ciprian; Heymann, Jan Oliver: Towards the integration of culture in the Unified Theory of Acceptance and Use of Technology. British Journal of Educational Technology, 45/1 (2014), pp. 36–55.

Paul, Bettina; Schmidt-Semisch, Henning (eds.): Risiko Gesundheit. On risks and side effects of the health society. Wiesbaden: VS Verlag, 2010.

Pausder, Verena (2020): Interview with Verena Pausder. "Corona was an ideal tutor for the schools". Available at: https://www.rbb24.de/panorama/thema/2020/coronavirus/beitraege_neu/2020/09/gruenderszene.html [14.02.2021].

Pemberger, Brigitte; Bleckmann, Paula (2021): "Analog-Digidaktik – Wie Kinder ohne Bildschirm fit fürs digitale Zeitalter werden". Available at: https://www.alanus.edu/de/forschung-kunst/wissenschaftliche-kuenstlerische-projekte/detail/analog-digidaktik-wie-kinder-ohne-bildschirm-fit-fuers-digitale-zeitalter-werden [26.06.2021].

Pfeiffer, Christian; Mößle, Thomas; Kleimann, Matthias; Rehbein, Florian: The PISA losers and their media consumption. An analysis based on various empirical studies. In: Dittler, Ullrich; Hoyer, Michael: Growing up in virtual media worlds. Chances and dangers of digital media from a media-psychological and media-pedagogical perspective. Munich: Kopaed 2008, pp. 275–306.

Pop-Eleches, Cristian; Malamud, Ofer (2010): Home Computer Use and the Development of Human Capital. University of Chicago and Columbia University. Available at: http://www.columbia.edu/~cp2124/papers/popeleches_QJE.pdf [14.02.2021].

Puentedura, Ruben R. (2006): Transformation, Technology, and Education. Available at: http://www.hippasus.com/resources/tte/ [14.02.2021].

Radesky, Jenny; Miller, Alison L.; Rosenblum, Katherine L.; Appugliese, Danielle; Kaciroti, Niko; Lumeng, Julie C.: Maternal mobile device use during a structured parent-child interaction task. Academic Pediatrics 15/2 (2014), pp. 238–244.

Reinhardt, Ulrich (2018): Freizeit Monitor 2018. Stiftung für Zukunftsfragen. Available at: http://www.freizeitmonitor.de/fileadmin/user_upload/freizeitmonitor/2018/Stiftung-fuer-Zukunftsfragen_Freizeit-Monitor-2018.pdf [14.02.2021].

Reinmann-Rothmeier, Gabi; Mandl, Heinz: Unterrichten und Lernumgebungen gestalten. In: Krapp, Andreas; Weidenmann, Bernd (eds.): Pädagogische Psychologie. A Textbook. 4th, fully revised ed. Weinheim: Beltz 2001, pp. 601–646.

Rekus, Jürgen; Mikhail, Thomas: Neues schulpädagogisches Wörterbuch. 4th, revised edition, new edition. Weinheim: Beltz Juventa, 2013.

Rosa, Lisa (2017): Lernen im digitalen Zeitalter. Prezi presentation at eEduca 2017 in Salzburg. Available at: https://shiftingschool.wordpress.com/2017/11/28/lernen-im-digitalen-zeitalter/ [14.02.2021].

Sandoval, Marisol (ed.): Bildung MACHT Gesellschaft. Münster: Verlag Westfälisches Dampfboot, 2011.

Schmidt, Robin (2020): ICT Professionalisation and ICT Beliefs. Professionalisation of prospective teachers in the digital transformation and their professional beliefs about digital information and communication technologies (ICT). Available at: https://edoc.unibas.ch/76795/1/Schmidt_Robin_ICT-Beliefs_Professionalisierung-Dissertation.pdf [14.02.2021].

Selwyn, Neil (2014): Data entry: towards the critical study of digital data and education. Learning, Media and Technology, pp. 64–82. Available at: https://www.tandfonline.com/action/showCitFormats?doi=10.1080%2F17439884.2014.921628 [14.02.2021].

Sigman, Aric (2017): Screen Dependency Disorders: a new challenge for child neurology. Journal of the International Child Neurology Association. Available at: https://www.researchgate.net/publication/317045692_Screen_Dependency_Disorders_a_new_challenge_for_child_neurology [14.02.2021].

Sommer, Jörg; Ibisch, Pierre L.; Göpel, Maja: The Ecology of the Digital Society. Towards a meaningful use of technology for a socio-ecological transformation. In: Göpel, Maja; Leitschuh, Heike; Brunnengräber, Achim et al. (eds.): Yearbook Ecology 2019/20. The Ecology of Digital Society. Stuttgart: Hirzel 2020, pp. 232–246.

Spitzer, Manfred: Digital Dementia. How we drive ourselves and our children crazy. Munich: Droemer, 2012.

Stiftung "Haus der kleinen Forscher": Frühe informatische Bildung – Ziele und Gelingensbedingungen für den Elementar- und Primarschulbereich. Scientific studies on the work of the "Haus der kleinen Forscher" foundation. Volume 9. Opladen, Berlin, Toronto: Verlag Barbara Budrich, 2018.

Sümmchen, Corinna: Analogue Social Network or Social Media Unplugged. Action-oriented prevention of cyber risks. In: e & l International Journal for Action-Oriented Learning, 3&4 (2019), pp. 40–43.

Süss, Daniel; Lampert, Claudia; Trueltzsch-Wijnen, Christine: Medienpädagogik. An introductory study book. 2nd, revised and updated ed. Wiesbaden: Springer VS, 2013.

Turkle, Sherry: Alone together – Why We Expect More from Technology and Less from Each Other. New York: Basic Books, 2011.

Twenge, Jean Marie: iGen: Why Today's Super-Connected Kids Are Growing Up Less Rebellious, More Tolerant, Less Happy--and Completely Unprepared for Adulthood--and What That Means for the Rest of Us. New York: Atria Books, 2017.

Federation of German Scientists (2019): VDW Position Paper for the Annual Conference 2019 – The Ambivalences of the Digital. Humans and technology between new spaces of possibility and (un) noticeable losses. Available at: https://vdw-ev.de/wp-content/uploads/2019/09/VDW-Positions-papier-Digitalisierung-Jahrestagung-2019.pdf [14.02.2021].

Wagenschein, Martin: Teaching Understanding. With an introduction by Hartmut von Hentig. 4th ed. Weinheim/Basel: Beltz, 2008.

Te Wildt, Bert: Digital Junkies – Internet addiction and its consequences for us and our children. Munich: Droemer, 2015.

Williamson, Ben; Hogan, Anna (2020): Commercialisation and privatisation in/of education in the context of Covid-19. Education International. Available at: https://issuu.com/educationinterna-tional/docs/2020_eiresearch_gr_commercialisation_privatisation [14.02.2021].

Zierer, Klaus: Hattie for Stressed Teachers. Key messages and recommendations for action from John Hattie's "Visible Learning" and "Visible Learning for Teachers". Corrected reprint. Baltmanns-weiler: Schneider Verlag Hohengehren GmbH, 2015.

Zierer, Klaus: Lernen 4.0. Pädagogik vor Technik: Möglichkeiten und Grenzen einer Digitalisierung im Bildungsbereich. 2nd, ed. Baltmannsweiler: Schneider Verlag Hohengehren GmbH, 2018.

Zimmer, Jasmin; Bleckmann, Paula; Pemberger, Brigitte: Technikfolgenabschätzung bei "Digitaler Bildung". In: Bleckmann, Paula; Lankau, Ralf (eds.): Digitale Medien und Unterricht. A controversy. Weinheim: Julius Beltz 2019, pp. 13–25.

Chapter 12: When Does 'Digitalisation' Contribute to UN Sustainable Development Goal 3 'Health and Well-Being'?

Adorno, T. W. (1982): Negative Dialectic, Frankfurt a. M.

Altmeyer, M. (2016): In Search of Resonance. How the life of the soul is changing in digital modernity. Göttingen

Arendt, H. (2002) : Vita Activa oder vom tätigen Leben. Munich: Pieper. (US-American original edition 1958: The Human Condition. Chicago

Baranski, B.; Behrens, J. & Westerholm P. (eds.) (1997): Occupational Healt Policy, Practise and Evaluation. Copenhagen/Geneva

Behrens, Johann (2020): 'Digitalisation' or 'Health as Self-Determined Participation'? In: Korczak, Dieter in Vereinigung Deutscher Wissenschaftler (ed.) Digitale Heilsversprechen – Zur Ambivalenz von Gesundheit, Algorithmen und Big Data. Frankfurt a. M.

Behrens, J. (2019): Theory of Nursing and Therapy. Bern, Göttingen, Toronto: Hogrefe

Behrens J, (1982): The differentiation of work, in: K. O. Hondrich (ed.), Soziale Differenzierung, pp. 129–209. Frankfurt/New York.

Behrens, J., Zimmermann, M. (2017): Socially unequally treated? A. Sens and P. Bourdieu's theories and social inequality in health care – using preventive rehabilitation as a case study. Bern, Göttingen, Toronto

Behrens, J., Langer, G. (2016): Methods and Ethics of Nursing Practice and Health Services Research – Confidence-Building Demystification of 'Science', Bern, Göttingen, Toronto

Behrens J, Weber A, Schubert M (eds.) (2012): Von der fürsorglichen Bevormundung/ über die organisierte Unverantwortlichkeit/ zur professionsgestützten selbstbestimmten Teilhabe? Health and social policy after 1989. Opladen/Toronto

Behrens J, Morschhäuser, M, Viebrock, H, Zimmermann, E. 1999 Länger erwerbstätig, aber wie? Wiesbaden

Behrens J, Leibfried, S. Social policy research. A survey of university and university-related work, in: Zeitschrift für Sozialreform, Jg. 33, Heft 1/1987, pp. 1–19.

Behrens J, v. Carnap, S.; Zerb, P Grundrente, in: Fetscher, I. Grundbegriffe des Marxismus, Hamburg 1975, pp. 252–266.

Etymological Dictionary of German (2015). Berlin

Ferber, L. v. & Behrens, J. (eds. (1997): Public Health Research with Health and Social Data – Status and Perspectives. Sankt Augustin

Grount, J. (1662): Observations. London

Habermas, J. (2019): Also a history of philosophy. Berlin

Habermas, J. (2015): Zwischen Naturalismus und Religion. Philosophical Essays. Frankfurt a.M.

Heyen, N. B. (2016): Digitale Selbstvermessung und Quantified Self. Potentials, risks and options for action. Karlsruhe

Husserl, E. (1992): Die Krisis der europäischen Wissenschaften und die transzendentale Phänomenologie. Collected Works (Husserliana) Volume VI. Hamburg

Jonas, H. (2003): Prinzip Verantwortung. Frankfurt a.M.

Karner, S., Stenner, H., Spate, M., Behrens, J., & Krakow, K. (2019). Effects of a robotic intervention on visuospatial hemineglect in postacute stroke patients: A randomized controlled trial. Clinical rehabilitation, 33(12), 1940–1948.

Kluge, Friedrich (1999): Etymologisches Wörterbuch der deutschen Sprache. 23rd extended new edition. Edited by Elmar Seebold. Berlin

Korczak, D. in Vereinigung Deutscher Wissenschaftler (ed.) (2020): Digitale Heilsversprechen – Zur Ambivalenz von Gesundheit, Algorithmen und Big Data. Frankfurt a.M.

Korczak, D. and Wilken, M. (2008): Scoring im Praxistest: Informative value and application of scoring procedures in lending and conclusions. Study. Munich

Müller, Klaus-Robert, Lecture Leibniz Day of the Berlin-Brandenburg Academy of Sciences and Humanities on 30.6.2018

Nassehi, A. (2019): Theorie der digitalen Gesellschaft, Munich

Pearl, J. (2009): Causality: Models, Reasoning and Inference. 2nd ed. Cambridge

Peirce, C. S. (2009): The Logic of Interdisciplinarity. Berlin

Quetelet, A. (1869): Physique sociale ou essai sur le développement des facultés de l'homme, Brussels

Quetelet, A. (1870): Anthropométrie ou mesure des différentes facultés de l'homme, Brussels

Uexküll, Jakob v. (1973): Theoretische Biologie. Frankfurt a.M.

Uexküll, Thure v. (1990): Lehrbuch der Psychosomatischen Medizin. Munich

Watzlawik, Paul. (2020): 100 years of Paul Watzlawik. From 'Human Communication'. Bern

WBGU German Advisory Council on Global Change (2019): Our Common Digital Future. Berlin

Windeler J, Antes G, Behrens J, Donner-Banzoff N, Lelgemann M (2008): Randomised controlled trials: Critical evaluation is an essential feature of medical practice, in: Dtsch. Ärztebl. 2008; 105(11): A-565

Chapter 13: Reductionist Temptations: Artificial Intelligence and Sustainability

Abshagen, M. L. and Grotefendt, N., 2020: A Question of Power – Sustainability and Digitalisation. In: FIfF Communication 1/20, pp. 33–36. Available online at: https://www.fiff.de/publikationen/fiff-kommunikation/fk-2020/fk-2020-1/fk-2020-1-content/fk-1-20-p33.pdf [as of: 14.06.2021].

acatech, 2020: European Public Sphere. Shaping Europe's Digital Sovereignty. Available online at: https://www.acatech.de/publikation/european-public-sphere/ [as of: 14.06.2021].

Betz, S., 2020: Impact of Software Systems on Sustainability and the Shared Responsibility of Software Developers. In: FIfF Communication, 3/20: pp. 28–31. https://www.fiff.de/publikationen/fiff-kommunikation/fk-2020/fk-2020-3/fk-2020-3-content/fk-3-20-p28.pdf [as of: 14.06.2021].

Bietti, E. and Vatanparast, R., 2020: Data Waste. In: Harvard Inernational Law Journal Frontiers, Vol. 61/2020. Available online at: https://harvardilj.org/wp-content/uploads/sites/15/Bietti-and-Vatanparast-PDF-format.pdf [Accessed: 14.06.2021].

BMBF – Federal Ministry of Education and Research, 2019: Natürlich. Digital. Sustainable. An action plan of the BMBF. Available online at: https://www.bmbf.de/upload_filestore/pub/Natuerlich_Digital_Nachhaltig.pdf [As of: 14.06.2021].

BMEL – Bundesministerium für Ernährung und Landwirtschaft, 2020: Mehr als 50 Millionen Euro für digitale Experimentierfelder in der Landwirtschaft. Available online at: https://www.bmel.de/DE/themen/digitalisierung/digitale-experimentierfelder.html [As of: 14.06.2021].

BMU – Federal Ministry for the Environment, Nature Conservation and Nuclear Safety, 2020: Umweltpolitische Digitalagenda. Available online at: https://www.bmu.de/fileadmin/Daten_BMU/Pools/Broschueren/broschuere_digitalagenda_bf.pdf [Status: 14.06.2021].

Boedicker, D. 2020: Technology for or against ecology? Sustainability goals for IT. In: FIfF Communication, 3/20: pp. 65–68. Available online at: https://www.fiff.de/publikationen/fiff-kommunikation/fk-2020/fk-2020-3/fk-2020-3-content/fk-3-20-p65.pdf [as of: 14.06.2021].

Daum, T., 2019: Missing Link: Bye-bye car, hello robo-taxi! Transport à la Silicon Valley. Available online at: https://www.heise.de/newsticker/meldung/Missing-Link-Tschuess-Auto-hallo-Robo-Taxi-Verkehr-a-la-Silicon-Valley-4554167.html [as of: 14.06.2021].

Dhar, S., Guo, J., Liu, J., Tripathi, S., Kurup, U. and Shah, M., 2020: On-Device Machine Learning: An Algorithms and Learning Theory Perspective. Available online at: https://arxiv.org/pdf/1911.00623.pdf [Accessed: 14.06.2021].

Gröger, J. 2020: Digital Ecological Footprint. How sustainable is our digital lifestyle? In: FIfF Communication, 3/20: pp. 22–24. Available online at: https://www.fiff.de/publikationen/fiff-kommunikation/fk-2020/fk-2020-3/fk-2020-3-content/fk-3-20-p22.pdf [as of: 14.06.2021].

Henderson, P., Hu, J., Romoff, J., Brunskill, E., Jurafsky, D. and Pineau, J., 2020: Towards the systematic reporting of the energy and carbon footprints of machine learning. A working paper. Available online at: http://stanford.edu/~phend/papers/ClimateML.pdf [Accessed: 14.06.2021].

Hilty, L. M., 2020: Software and Sustainability: How Extraneous Software Devalues Material Resources. In: FIfF Communication, 3/20: pp. 31–35. Available online at: https://www.fiff.de/publikationen/fiff-kommunikation/fk-2020/fk-2020-3/fk-2020-3-content/fk-3-20-p31.pdf [as of: 14.06.2021].

Höfner, A. and Frick, V., 2019: What Bits and Trees Have in Common. Shaping digitalisation sustainably. Munich: Oekom. Available online at: https://www.oekom.de/buch/was-bits-und-baeume-verbindet-9783962381493 [As of: 14.06.2021].

Ito, J., 2018: The practice of change. Doctoral Dissertation. Academic Year 2018. Keio University. Graduate School of Media & Governance. Available online at: https://dam-prod.media.mit.edu/x/2018/11/01/ito_phd_diss_v1.11.pdf [Accessed: 14.06.2021].

Ito, J., 2018a: Resisting Reduction: A Manifesto. Designing our Complex Future with Machines. Available online at: https://jods.mitpress.mit.edu/pub/resisting-reduction/release/17 [Accessed: 14.06.2021].

Jonas, H., 1979: The Principle of Responsibility. An attempt at ethics for technological civilisation. Frankfurt a. M.: Suhrkamp.

Köhn, M., 2020: Does software have an environmental impact? In: FIfF Communication, 3/20: pp. 36–38. Available online at: https://www.fiff.de/publikationen/fiff-kommunikation/fk-2020/fk-2020-3/fk-2020-3-content/fk-3-20-p36.pdf [as of: 14.06.2021].

Lange, S. and Santarius, T., 2018: Smart Green World? Digitalisation between surveillance, consumption and sustainability. Available online at: http://www.santarius.de/wp-content/uploads/2018/11/Smarte-gru%CC%88ne-Welt-Offizielles-E-Book.pdf [As of: 14.06.2021].

Lapuschkin, S., Wäldchen, S., Binder, A., Montavon, G., Samek, W. and Müller, K.-R., 2019: Unmasking Clever Hans predictors and assessing what machines really learn. In: Nature Communications, 10 (1), p. 60. Available online at: https://www.nature.com/articles/s41467-019-08987-4 [Accessed: 14.06.2021].

Lessig, L., 2009: Code: And other laws of cyberspace. New York: Basic Books.

Marcus, G., 2020: The Next Decade in AI: Four Steps Towards Robust Artificial Intelligence. Available online at: https://arxiv.org/abs/2002.06177v2 [as of: 14.06.2021].

Marcus, G. and Davies, E., 2019: Rebooting AI. Building Artificial Intelligence We Can Trust. New York: Penguin Books.

Marcus, G., 2018: Deep Learning: A Critical Appraisal. Available online at: https://arxiv.org/abs/1801.00631v1 [Accessed: 14.06.2021].

Meadows, D., 1999: Leverage Points: Places to Intervene in a System. Available online at: http://www.donellameadows.org/wp-content/userfiles/Leverage_Points.pdf [As of: 14.06.2021].

Meinel, C., 2020: Only sustainable digitalisation can save the climate. Guest article. Available online at: https://www.spiegel.de/netzwelt/netzpolitik/green-it-nur-nachhaltige-digitalisierung-kann-das-klima-retten-a-540d6972-bf67-4571-841d-9c1479df37e1 [As of: 14.06.2021].

Messerschmidt, R. and Ullrich, S., 2020: A European way towards sustainable AI. In: Social Europe. Available online at: https://socialeurope.eu/a-european-way-towards-sustainable-ai [as of: 14.06.2021].

Messerschmidt, R., 2020: Discourse analysis of the recommendation landscape on digitalisation and sustainability 2017–2018. Available online at: https://www.wbgu.de/fileadmin/user_upload/wbgu/publikationen/hauptgutachten/hg2019/pdf/Expertise_Messerschmidt_HGD.pdf [as of: 14.06.2021].

Nemitz, P. and Pfeffer, M., 2020: Prinzip Mensch. Power, Freedom and Democracy in the Age of Artificial Intelligence. Bonn: Dietz.

Northcutt, C. G., Athalye, A., Mueller, J., 2021: Pervasive Label Errors in Test Sets Destabilize Machine Learning Benchmarks. Available online at: https://arxiv.org/pdf/2103.14749.pdf [Accessed: 14.06.2021].

ÖFIT – Kompetenzzentrum Öffentliche IT, 2021: Gesellschaftliche Technikgestaltung – Orientierung durch eine Metaperspektive auf Schlüsselelemente. Available online at: https://www.oeffentliche-it.de/publikationen?doc=162868&title=Gesellschaftliche%20Technikgestaltung%20-%20Orientierung%20durch%20eine%20Metaperspektive%20auf%20Schl%C3%BCsselelemente [Status: 14.06.2021].

ÖFIT – Kompetenzzentrum Öffentliche IT, 2020: Der Staat auf dem Weg zur Plattform. Available online at: https://www.oeffentliche-it.de/publikationen?doc=113399&title=Der%20Staat%20auf%20dem%20Weg%20zur%20Plattform [as of: 14.06.2021].

O'Neill, C., 2017: Weapons of math destruction. How big data increases inequality and threatens democracy. New York: Crown Publishing Group.

Schischke, K., 2020: The Design Determines the Life Cycle Assessment: Mobile Devices in Environmental Focus. In: FIfF Communication, 3/20: pp. 25–27. Available online at: https://www.fiff.de/publikationen/fiff-kommunikation/fk-2020/fk-2020-3/fk-2020-3-content/fk-3-20-p25.pdf [as of: 14.06.2021].

Schneidewind, U., 2018: The Great Transformation. An introduction to the art of social change. Frankfurt a. M.: S. Fischer.

Spiekermann, S., 2019: Digital Ethics. A value system for the 21st century. Munich: Droemer HC.

Spiekermann, S., 2016: Ethical IT Innovation. A Value-Based System Design Approach. London: CRC Press.

SRU – German Advisory Council on the Environment, 2019: Demokratisch regieren in ökologischen Grenzen – Zur Legitimation von Umweltpolitik. Special Report. Available online at: https://www.

umweltrat.de/SharedDocs/Downloads/DE/02_Sondergutachten/2016_2020/2019_06_SG_Le-gitimation_von_Umweltpolitik.pdf?__blob=publicationFile&v=15 [As of: 14.06.2021].

Strubell, E., Ganesh, A. and McCallum, A., 2019: Energy and Policy Considerations for Deep Learning in NLP. Available online at: https://arxiv.org/abs/1906.02243v1 [Accessed: 14.06.2021].

THE SHIFT PROJECT, 2019: Lean ICT. Towards digital sobriety. Available online at: https://theshift-project.org/wp-content/uploads/2019/03/Lean-ICT-Report_The-Shift-Project_2019.pdf [As of: 14.06.2021].

Wachter, S. and Mittelstadt, B., 2018: A Right to Reasonable Inferences: Re-Thinking Data Protection Law in the Age of Big Data and AI. In: Columbia Business Law Review, 2, pp. 443–493. Available online at: https://doi.org/10.7916/d8-g10s-ka92 [Accessed: 14.06.2021].

WBGU – German Advisory Council on Global Change, 2020: Land Transition in the Anthropocene: From Competition to Integration. Available online at: https://www.wbgu.de/de/publikationen/publikation/landwende [As of: 14.06.2021].

WBGU – German Advisory Council on Global Change, 2019: Our Common Digital Future. Main Report. Available online at: https://www.wbgu.de/de/publikationen/publikation/unsere-gemein-same-digitale-zukunft [As of: 14.06.2021].

WBGU – German Advisory Council on Global Change, 2019a: PPEU

WBGU – German Advisory Council on Global Change, 2019b: PPUN

WBGU – German Advisory Council on Global Change, 2019c: #SustainableDigitalAge – Illustrated Fact Sheet. Available online at: https://www.wbgu.de/de/publikationen/publikation/transforma-tion-unserer-welt-im-digitalen-zeitalter [As of: 14.06.2021].

WBGU – German Advisory Council on Global Change, 2018: Digitalisation: What we need to talk about now. Impulse paper. Available online at: https://www.wbgu.de/de/publikationen/publika-tion/digitalisierung-worueber-wir-jetzt-reden-muessen [Status: 14.06.2021].

WBGU – German Advisory Council on Global Change, 2011: World in Transition: Social Contract for a Great Transformation. Main Report. Available online at: https://www.wbgu.de/de/publika-tionen/publikation/welt-im-wandel-gesellschaftsvertrag-fuer-eine-grosse-transformation [Sta-tus: 14.06.2021].

Weizenbaum, J., 1978: Die Macht der Computer und die Ohnmacht der Vernunft. Frankfurt a.M.: Suhrkamp.

Zuboff, S., 2018: The Age of Surveillance Capitalism. Frankfurt a.M.: Campus.

Chapter 14: Production and Trade in the Age of Digitalisation, Networking, and Artificial Intelligence

Agrawal, Nayank, Dutta, Sumit, Kelley, Richard, and Millán, Ingrid (2021): COVID-19: An inflection point for Industry 4.0. McKinsey.

Altenburg, Tilman, Brahima Coulibaly (2018): Jobs for Africa: Opportunities in a global economy in transition. DIE Bonn

Banga, Karishma, Dirk Willem te Velde (2018): Digitalisation and the future of manufacturing in Afri-ca. Supporting economic transformation, ODI and DFID

Ben Youssef, A. (2020), "How Industry 4.0 can contribute to combating Climate Change?", French Industrial Economics Review (Revue d'Economie Industrielle), 1st Trimester 2020, 161–193, vol. 169

Brynjolfsson, E. and A. McAfee (2014): The second machine age: Work, progress, and prosperity in a time of brilliant technologies. ISBN 978-0-393-23935-5

BTI (2019): Digitisation: Curse or Blessing for Developing Countries? https://blog.bti-project.de/2019/09/18/digitalisierung-fluch-oder-segen-fuer-entwicklungslaender/ [2.2.2021]

Butterfill, James et al. (2017): Disruptive themes drive future commodity demand. In: ETFS Outlook, December 2017, pp. 19–21. London

Chuhan-Pole, Punam et al. (2017): Mining in Africa – Are Local Communities Better Off? A copublication of the Agence Francaise de Développement and the World Bank. Washington D.C.

Cigna, Simone, Lucia Quaglietti (2020): The great trade collapse of 2020 and the amplification role of global value chains. ECB Economic Bulletin, Issue 5/2020, European Central Bank Frankfurt

Dahlman, C., S. Mealy and M. Wermelinger (2016), "Harnessing the digital economy for developing countries", OECD Development Centre Working Papers, No. 334, OECD Publishing, Paris

De Backer, K., T. DeStefano, C. Menon and J. R. Suh (2018): Industrial Robotics and the Global Organisation of Production. OECD Science, Technology and Industry Working Paper no. 2018/03. OECD, Paris, cited in Kinkel (2020).

De Backer, Koen and Dorothee Flaig (2017): The future of global value chains – business as usual or "a new normal?" OECD science, technology and innovation policy papers, July 2017, No. 41, cited in Kinkel (2020).

Deloitte (2017): Digital sustainability. How do companies prepare for the opportunities and challenges of digitalisation? Deloitte, Johannesburg

German Foundation for World Population (DSW) (2020): Social and demographic data worldwide. DSW DATA REPORT 2020. DSW Hanover

Felber, Christian (2014): Gemeinwohl-Ökonomie. Vienna, ISBN 978-3-552-06299-3

Fortunato, Piergiuseppe (2020): How COVID-19 is changing global value chains. UNCTAD Geneva

Frey, Carl Benedikt, Osborne, Michael A. (2013): The Future of Employment: How Susceptible are Jobs to Computerisation, Oxford University

Frey, Carl Benedikt, Berger, Thor (2016): Structural Transformation in the OECD – digitalisation, deindustrialisation and the future of work. OECD, Paris

Gabler Wirtschaftslexikon (n.d.): https://wirtschaftslexikon.gabler.de/definition/ wirtschaftspolitisches-ziel-50933, Wiesbaden, [2.2.2021].

GIZ (n.d.): FAIR Forward – Artificial Intelligence for All. https://gizonline.sharepoint.com/sites/Digital-Gateway/SitePages/FAIR-Forward---Kuenstliche-Intelligenz-fuer-alle.aspx?web=1, last 19.4.2021

Göpel, Maja (2020): Rethinking Our World. Berlin, ISBN 978-3-8437-2311-4

Graham, Hjorth & Lehdonvirta (2017): Digital labour and development: impacts of global digital labour platforms and the gig economy on worker livelihoods. SAGE journals

GTAI (2018): Work Programme 2018: Opportunities through Change. Germany Trade&Invest, Berlin

GTAI (2019a): Overview of selected countries' strengths and prospects in digitisation. https://www.gtai.de/gtai-de/meta/ueber-uns/was-wir-tun/schwerpunkte/digitalisierung/ueberblick-zu-den-staerken-und-perspektiven-ausgewaehlter-117756 [2.2.2021]

GTAI (2019b): Singapore aims to become Southeast Asia's artificial intelligence hub. https://www.gtai.de/gtai-de/trade/branchen/special/singapur/singapur-moechte-zum-suedostasiatischen-hub-fuer-kuenstliche-22638 [2.2.2021]

Hallward-Driemeier, Mary, Nayyar, Gaurav (2018): Trouble in the Making? The Future of Manufacturing-Led Development. The World Bank Group. Washington D.C.

Hornweg, Pope (2015) Population predictions for the world's largest cities in the 21st century. Environment and Urbanization vol. 29, issue 1

ILO (2016): World Employment Social Outlook 2016: Trends for Youth; UNICEF (2017)

IMD (2020): IMD world digital competitiveness ranking 2020. IMD, Lausanne

IMF (2018): Measuring the digital economy. Staff report. International Monetary Fund, Washington, D.C.

Institute for Employment Research of the Federal Employment Agency (IAB) (2017): Research Report 13/2017, p. 60ff.

International Bank for Reconstruction and Development / The World Bank (2020): World Bank Outlook 2050: Strategic Directions Note. Supporting Countries to Meet Long-Term Goals of Decarbonization. Washington D.C.

IPCC – Intergovernmental Panel on Climate Change (2018): Global warming of 1.5°C. An IPCC Special Report on the Impacts of Global Warming of 1.5°C Above Pre-Industrial Levels and related Global Greenhouse Gas Emission Pathways, in the Context of Strengthening the Global Response to the Threat of Climate Change, Sustainable Development, and Efforts to Eradicate Poverty. Summary for Policymakers. IPCC Geneva

Itterman, Peter; Niehaus, Jonathan, Hirsch-Kreinsen, Hartmut (2015): Arbeiten in der Industrie 4.0: Trendbestimmungen und arbeitspolitische Handlungsfelder. Hans Böckler Foundation, Düsseldorf

Jäger, A., Moll, C., Som, O., Zanker, C., Kinkel, S., Lichtner, R. (2015): Analysis of the impact of robotic systems on employment in the European Union. Unpublished final report of the Fraunhofer Institute for Systems and Innovation Research ISI. Commissioned by: European Commission, Directorate-General of Communications Networks, Content & Technology.

Kagermann, Henning, Reiner Anderl, Jürgen Gausemeier, Günther Schuh, Wolfgang Wahlster (2016): Industrie 4.0 im globalen Kontext – Strategien der Zusammenarbeit mit internationalen Partnern. Acatech, Munich

Kinkel, Steffen (2019): Zusammenhang von Industrie 4.0 und Rückverlagerungen von Produktionsaktivitäten aus dem Ausland. Volume 20 of FGW Study Digitalisation of Work. Research Institute for Social Development (e. V.)

Kinkel, Steffen (2020): Digitalisierung und Reshoring – Weniger globale und mehr regionale Wertschöpfungsketten in Deutschland und der EU. In: sustainable digital – digital sustainable / Virtual Conference 4/5 December 2020, University of Göttingen.

Kleinhans, Jan-Peter & Dr Nurzat Baisakova (2020): The global semiconductor value chain. A technology primer for policy makers. Stiftung Neue Verantwortung Berlin

Korotayev, Andrey, Malkov Artemy, Khaltourina Daria (2006): Introduction to Social Macrodynamics: Compact Macromodels of the WorldSystem Growth. Moscow

Lall, Somik Vinay et al (2017): Africa's Cities: Opening Doors to the World. Worldbank Group and UK Aid. Washington D.C.

Lenz, Fulko (2018): Digitalisierung und Beschäftigung – Ein Ende ohne Arbeit oder Arbeit ohne Ende. Stiftung Marktwirtschaft – Argumente zu Marktwirtschaft und Politik, No. 141, April 2018;

Manchanda, Sumit, Hassan Kaleem and Sabine Schlorke (2020): AI Investments Allow Emerging Markets to Develop and Expand Sophisticated Manufacturing Capabilities. EM Compass, IFC

McKinsey (McKinsey Global Institute), (2017): Jobs Lost, Jobs Gained: Workforce Transitions in a Time of Automation

McKinsey (McKinsey Global Institute) (2020). Risk, resilience, and rebalancing in global value chains

Meyer, Laurin (2015): Fair gehandelte Smartphones: Schmerzfrei telefonieren. In: Die Tageszeitung: taz. 8 September 2015, ISSN 0931-9085. https://taz.de/Fair-gehandelte-Smartphones/!5226988/ [27.3.2021])

Sustainable Economic Development in the Mining Sector, Democratic Republic of Congo. GIZ project brief. https://www.giz.de/de/weltweit/19891.html, [10.1.2021].

Norasatya, Erik, Prasad Sahuivan Peterson (2020): COVID-19 highlights need for digitizing and automating trade in South Asia. World Bank blogs

OECD Economic Outlook 2019. https://www.oecd-ilibrary.org/sites/70409513-de/index.html?item Id=/content/component/70409513-de [2.2.2021]

OECD (2020a): Trustworthy AI in health. Background paper for the G20 AI Dialogue, Digital Economy Task Force, SAUDI ARABIA, 1–2 APRIL 2020

OECD (2020b): A roadmap towards a common framework for measuring the digital economy. Report for the G20 Digital Economy Task Force, SAUDI ARABIA, 2020

Osman, Maddy (2020): E-commerce statistics for 2021 – chatbots, voice, omni-channel marketing. THE KINSTA BLOG. https://kinsta.com/de/blog/e-commerce-statistik/#:~:text=There%20 are%%2020more%20global%20Märks,to%2016%2C9%25%20sink [2.2.2021].

Otte, Ralf (2019): Künstliche Intelligenz für Dummies. Weinheim, ISBN 978-3-527-71494-0

Piketty, Thomas (2020): Capital and Ideology. Munich

Plattform Industrie 4.0 (2020): Progress Report 2020. Shaping Industry 4.0. Sovereign. Interoperable. Sustainable. BMWi Berlin

Raworth, Kate (2017): Doughnut economics. London, ISBN 978-1-847-94137-4

Saslow, Kate (2020): Memorandum Foreign Policy Engagement with African Artificial Intelligence. Foundation for New Responsibility Berlin

Schmiedchen, F. et al (2018): Statement on the Asilomar Principles on Artificial Intelligence. Ed. by the Federation of German Scientists e.V., Berlin

Statista (2021): Trends in global export volume of trade in goods from 1950 to 2019. https://www.statista.com/statistics/264682/worldwide-export-volume-in-the-trade-since-1950/ [2.2.2021].

Tetzlaff, Rainer (2018): Africa. An introduction to history, politics and society. Wiesbaden

UNCTAD (2017): World Investment Report (p. 156ff.). Geneva

UNCTAD (2018): World Investment Report. Geneva

UNCTAD (2020): World Investment Report. Geneva

UNICEF (2017): Generation 2030 – Africa 2.0: Prioritizing investments in children to reap the demographic dividend. New York

United Nations General Assembly (2015): Transforming our world: the 2030 Agenda for Sustainable Development. UN, Washington D.C.

UN (2019): World Population Prospects 2019. https://esa.un.org/unpd/wpp/Download/Standard/ Population [27.3.2021]

WBGU (2019): Our Common Digital Future. Main Report on Digitalisation 2019. ISBN 978-3-946830-20-7

WEF (World Economic Forum in collaboration with The Boston Consulting Group), (2018a): Towards a Reskilling Revolution: A Future of Jobs for All, Cologne/Geneva

WEF (World Economic Forum in collaboration with The Boston Consulting Group), (2018b): Eight futures of Work – Scenarios and their Implications. Cologne/Geneva

World Bank (2019): World Development Report 2019: The Changing Nature of Work

WTO (2021): WTO Chairs Programme. Adapting to the digital trade era: challenges and opportunities. Ed. Maarten Smeets. Geneva

Automobile Production magazine (25.3.2019): BMW stops buying cobalt from Congo. https://www.automobil-produktion.de/hersteller/wirtschaft/bmw-stoppt-kobalt-einkauf-aus-dem-kongo-270.html [27.3.2021]

Zhan, James, Richard Bolwijn, Bruno Casella, Amelia U. Santos-Paulino (2020): Global value chain transformation to 2030: Overall direction and policy implications. VoxEU.org, Centre for Economic Policy Research London

Chapter 15: Future of the Digital Working Society

Albrecht, T. / Kellermann, C. (2020): *Artificial Intelligence and the Future of the Digital Work-Oriented Society. An Outline for a Holistic Technology Impact Assessement.* Friedrich Ebert Foundation Publishing. Cited from: (http://library.fes.de/pdf-files/bueros/bratislava/16955.pdf)

Acemoglu, D. (2019): *Elizabeth Warren's Bold Ideas Don't Go Far Enough.* Project Syndicate. Cited from: (https://www.project-syndicate.org/commentary/good-jobs-agenda-us-by-daron-acemoglu-2019-12).

Acemoglu, D. / Restrepo, P. (2018). *Artificial Intelligence, Automation and Work.* NBER Working Paper No. 24196. Cited from: (https://www.nber.org/papers/w24196.pdf).

Albrecht, T. (2020): Regulation, Transparenz, Beteiligung und Befähigung – Voraussetzungen für den erfolgreichen Einsatz Künstlicher Intelligenz am Arbeitsplatz. (Regulation, Transparency, Participation, and Competence: Prerequisites for the Successful Implementation of AI in the Workplace). In: Jens Nachtwei & Antonia Sureth (2020). *Sonderband "Zukunft der Arbeit", HR Consulting Band 12*, Berlin: Veröffentlichungsreihe für Qualitätssicherung in Personalauswahl und -entwicklung (in Vorbereitung).

Arntz, M. / Gregory, T. / Zierahn, U. (2017). Revisiting the Risk of Automation. In: *Economics Letters*, No. 159, pp. 157–160.

Arntz, M. / Gregory, T. / Zierahn, U. (2018). *Digitalisierung und die Zukunft der Arbeit: Makroökonomische Auswirkungen auf Beschäftigung, Arbeitslosigkeit und Löhne von morgen.* (Digitisation and the Future of Labour: Macroeconomic Impacts on Employment, Unemployment, and the Wages of Tomorrow). Bundesministerium für Forschung und Entwicklung (BMBF), Mannheim. Cited from: (http://ftp.zew.de/pub/zew-docs/gutachten/DigitalisierungundZukunftderArbeit2018.pdf).

Autor, D. (2015). Why Are There Still So Many Jobs? The History and Future of Workplace Automation. In: *Journal of Economic Perspectives*, No. 29(3), pp. 3–30. Cited from: (https://economics.mit.edu/files/11563).

Autor, D. / Dorn, D. / Katz L. F. / Patterson, C. / Van Reenen, J. (2017). Concentrating on the Fall of the Labor Share. In: *American Economic Review Papers & Proceedings*, No. 107(5). pp. 180–185. Cited from: (https://economics.mit.edu/files/12544).

Autor, D. / Salomons, A. (2017). Robocalypse Now – Does Productivity Growth Threaten Employment? In: *ECB Forum on Central Banking – Investment and Growth in Advanced Economies – Conference Proceedings.* Frankfurt am Main: European Central Bank, pp. 119–128. Cited from: (https://www.ecb.europa.eu/pub/pdf/other/ecb.ecbforumcentralbanking2017.en.pdf).

Bessen, J. (2018). AI and Jobs: The Role of Demand. In: *NBER Working Paper Series*, No. 24235. Cited from: (https://www.nber.org/papers/w24235.pdf).

Dauth, W. / Findeisen, S. / Südekum, J. / Wößner, N. (2017). German Robots – The Impact of Industrial Robots on Workers. In: *IAB-Discussion Paper*, 30/2017. Cited from: (http://doku.iab.de/discussionpapers/2017/dp3017.pdf).

Dispan, J. / Schwarz-Kocher, M. (2018). Digitalisierung im Maschinenbau. Entwicklungstrends, Herausforderungen, Beschäftigungswirkungen, Gestaltungsfelder im Maschinen- und Anlagenbau. (Digitisation in Machine Engineering: Development Trends, Challenges, Employment Impact, and Design Fields in Machine and Equipment Construction). In: *Working Paper Forschungsförderung*, No. 094. Cited from: (https://www.boeckler.de/pdf/p_fofoe_WP_094_2018.pdf).

Frey, C. B. / Osborne, M. A. (2013). *The Future of Employment: How Susceptible are Jobs to Computerisation?* Oxford Martin School (OMS) Working Paper. University of Oxford. Cited from: https://www.oxfordmartin.ox.ac.uk/downloads/academic/ The_Future_of_Employment.pdf

Frey, C. B. / Osborne, M. A. (2017). The Future of Employment: How Susceptible are Jobs to Computerisation? In: *Technological Forecasting & Social Change.* No. 114, pp. 254–280.

Fuchs et al. (2019): Zuwanderung und Digitalisierung. Wie viel Migration aus Drittstaaten benötigt der deutsche Arbeitsmarkt künftig? (Immigration and Digitisation. How Much Migration from Third Countries will the German Labour Market Need in the Future?). Bertelsmann Stiftung. Cited from: (https://www.bertelsmann-stiftung.de/fileadmin/files/Projekte/Migration_fair_gestalten/IB_Studie_Zuwanderung_und_Digitalisierung_2019.pdf).

IBM (2019): KI-Studie über die Folgen für Arbeitnehmende und Arbeit. (An AI Study on the Consequences for Employees and Labour). Cited from: (https://www.ibm.com/de-de/blogs/think/2019/09/17/watson-ki-studie/).

Kellermann, C. / Obermauer, R. (2020). Von der Würde der Arbeit in digitaler und klimaneutraler Zukunft. (On the Dignity of Work in a Digital and Climate-neutral Future). spw 238

Korinek, A. / Stiglitz, J. E. (2017). Artificial Intelligence and Its Implications for Income Distribution and Unemployment. In: *National Bureau of Economic Research*. Cited from: (https://www.nber.org/papers/w24174.pdf).

McKinsey Global Institute [MGI] (2018). *Notes from the AI Frontier: Modeling the Impact of AI on the World Economy*. Cited from: (https://www.mckinsey.com/featured-insights/artificial-intelligence/notes-from-the-ai-frontier-modeling-the-impact-of-ai-on-the-world-economy).

Muro, M. / Maxim, R. / Whiton, J. (2019). Automation and Artificial Intelligence: How Machines are Affecting People and Places. Metropolitical Policy Program. Cited from: (https://www.brookings.edu/wp-content/uploads/2019/01/2019.01_BrookingsMetro_Automation-AI_Report_Muro-Maxim-Whiton-FINAL-version.pdf).

Nilsson, N. (2005). Human-level Artificial Intelligence? Be Serious! American Association for Artificial Intelligence. Cited from: (http://ai.stanford.edu/~nilsson/OnlinePubs-Nils/General%20Essays/AIMag26-04-HLAI.pdf).

Pfeiffer, S., / Suphan, A. (2015). *Der AV-Index. Lebendiges Arbeitsvermögen und Erfahrung als Ressourcen auf dem Weg zu Industrie 4.0* (The AV Index: Living Work Capacity and Experience as Resources on the Way to Industry 4.0). Working Paper 2015 No. 1 (final version of a draft originally published on 13 April 2015). Cited from: (https://www.sabine-pfeiffer.de/files/downloads/2015-Pfeiffer-Suphan-final.pdf).

Webb, M. (2019). *The Impact of Artificial Intelligence on the Labour Market*. Stanford University Working Paper.

World Economic Forum (2018). *The Future of Jobs Report* (2018). Cited from: (http://www3.weforum.org/docs/WEF_Future_of_Jobs_2018.pdf).

Chapter 16: New Social Question and the Future of Social Security

Bäcker, Gerhard/ Kistler, Ernst/ Stapf-Finé, Heinz (2011): Erwerbsminderungsrente – Reformnotwendigkeit und Reformoptionen. In: Friedrich-Ebert-Stiftung (ed.): WISO-Diskurs May 2011)

Bertelsmann Foundation (2019): Platform Work in Germany. Free and flexible work without social security, Gütersloh

Bundesanstalt für Arbeitsschutz und Arbeitsmedizin (BAUA) (2017): Alterns- und altersgerechte Arbeitsgestaltung. Grundlagen und Handlungsfelder für die Praxis, 2nd edition, Dortmund.

Bruckmeier, Kerstin/ Konle-Seidl, Regina (2019): Reforms of basic income support in international comparison: new ways yes, system change no, In: IAB Forum 10 July 2019, https://www.iab-forum.de/reformen-der-grundsicherung-im-internationalen-vergleich-neue-wege-ja-systemwechsel-nein/ [10 March 2020].

Federal Ministry of Labour and Social Affairs (BMAS) (2017): Weissbuch Arbeiten 4.0. Berlin

Federal Ministry of Labour and Social Affairs (BMAS) (2015): Grünbuch Arbeiten 4.0. Berlin

Federal Ministry of Labour and Social Affairs (BMAS) (2013): Fortschrittsreport Altersgerechte Arbeitswelt. Issue 2: "Age-appropriate work design", Bonn.

German Institute for Economic Research (DIW) (2014): Wochenbericht 37/2014

Dullien, Sebastian (2014): Eine Europäische Arbeitslosenversicherung als Stabilisator für die Euro-Zone. In: Friedrich-Ebert-Stiftung (ed.): WISO direkt, June 2014

European Commission (2019): Clarification on alleged call by Commission President Juncker for European unemployment insurance. In: Press Portal, 3 January 2019.

Fröhler, Norbert/ Fehmel, Thilo/ Klammer, Ute (2013): Flexibel in die Rente. Gesetzliche, tarifliche und betriebliche Perspektiven, Berlin.

Gesellschaft für Versicherungswissenschaft und -gestaltung (GVG) (2015): Soziale Sicherung in einer modernen Arbeitswelt. Position of the GVG on the Green Paper "Work 4.0", Cologne

Giegold, Sven (2006): Globalisation. In: Urban, Hans-Jürgen (ed.): ABC zum Neoliberalismus. Hamburg.

Hans, Jan Philipp/ Hofmann, Sandra/ Sesselmeier, Werner/ Yollu-Tok, Aysel (2017)

Labour Insurance – Costs and Benefits. In: Friedrich-Ebert-Stiftung # 2017 plus, Bonn

HDI (2019): Professionals fear job losses due to digitalisation. In Insurance Industry Today, 28 November.

IHK für München und Oberbayern (2018): Auswirkungen der Digitalisierung auf den Arbeitsmarkt. Munich.

Institute for Employment Research (IAB) (2015): In hardly any occupation is a person completely replaceable: In: IAB-Kurzbericht 24/2015.

IGZA author team (Institute for the History and Future of Work) (2018): Zeitsouveränität, Neues Normalarbeitsverhältnis und Sozialstaat 4.0 – Plädoyer für ein Lebensarbeitszeitkonto. Working Paper #4, Berlin

Muro, Mark/ Whiton, Jacob/ Maxim, Robert (2019): What jobs are affected by AI? Better-paid, better-educated workers face the most exposure. Metropolitan Policy Program at Brookings, Washington.

OECD (2014): Focus on Inequality and growth. December 2014, Paris.

Oxford Economics (2019): How robots change the world. What automation really means for jobs and productivity. Oxford, June 2019

Pfaff, Martin/ Stapf-Finé, Heinz (2004): Bürgerversicherung – solidarisch und sicher. The role of SHI and PHI, contribution bases, benefits catalogue, legal implementation, Hamburg

Ramge, Thomas (2018): Humans ask, machines answer. How artificial intelligence is changing the economy, work and life. In: APuZ 6-8/2018, 15–21

Schmidt, Florian A. (2016): Arbeitsmärkte in der Plattformökonomie. On the Functioning and Challenges of Crowdwork and Gigwork, Bonn

Simon, Dieta/ Heger, Günther/ Reszies, Sabine (eds.): Praxishandbuch betriebliche Gesundheitsförderung. A guide for small and medium-sized enterprises, Stuttgart

Sommer, S./ Kerschek, R./ Lehnhardt, U. (2018): Risk assessment in company practice: results of the GDA company surveys 2011 and 2015. In BAUA (ed.): Fokus September 2018

Stapf-Finé, Heinz (2018): Kita(sozial)politik – Politische und gesellschaftliche Entwicklungstrends in der Kindertagesbetreuung. In: Brodowski, Michael (ed.): Das große Handbuch für die Kita-Leitung. Cologne, 834–859

Stapf-Finé, Heinz (2009): Nein zum Grundeinkommen, ja zum Grundanliegen. In: Neuendorff H, Peter G, Wolf F (eds.) Arbeit und Freiheit im Widerspruch. Unconditional Basic Income – A Model in the Controversy of Opinion, Hamburg.

UN Refugee Agency (2020): Facts & Figures on People on the Move. Available at: https://www.uno-fluechtlingshilfe.de/informieren/fluechtlingszahlen/?donation_custom_field_1628=J102&gclid=EAIaIQobChMI157I7JOG6AIVWeDtCh3VYQHwEAAYASABEgJKEfD_BwE [6.3.2020].

Weber, Enzo (2019): Digital Social Security. Drafting a concept for the 21st century. HBs Working Paper Research Funding Number 137, May 2019

Wintermantel, Vanessa (2017): Research Report IV. Results of the Legacy Study on Social Cohesion and the Welfare State. WZB-discussion paper 2017-009, Berlin

IV. Responsibility of Science

Snodgrass, Melinda (1989): Star Trek – The Next Generation. 35th episode: The Measure of a Man.

Kling, Marc-Uwe (2020): Qualityland 2.0. Hamburg.

McKinley, Roger (2016): Technology: Use or lose our navigation skills. In: Nature, 531, March 30, pp. 573–575.

Editors

Federation of German Scientists e. V. (VDW)
The Federation of German Scientists (VDW) was founded in 1959 as the German section of the Pugwash Conferences on Science and World Affairs, which was awarded the Nobel Peace Prize in 1995. While Pugwash, founded by Albert Einstein and Bertrand Russell, still focuses on peace and security issues and responsibility in science, the VDW, founded by Carl-Friedrich von Weizsäcker, Werner Heisenberg, Otto Hahn, Max Born, and other leading scientists, has gradually expanded its range of tasks since the 1960s to include other significant human issues, above all in the development of technology.

VDW Study Group on Technology Assessment of Digitalisation
The Digitalisation Technology Assessment (TA) study group was established in 2017 by resolution of the Executive Board, and Frank Schmiedchen was entrusted with its leadership. It deals in a transdisciplinary manner with research and application issues of digitalisation, networking, and artificial intelligence, as well as their social prerequisites, applications, limits, and consequences. In 2018, the study group presented its first statement on ethical questions about artificial intelligence.

Frank Schmiedchen is an economist and MBA. He has taught at different universities in Germany. From 1996 to 1999 he was Dean of the Faculty of Management of SME at the Catholic University of Ecuador (Ambato). Since 1999, he has been entrusted as a German civil servant and diplomate with multilateral negotiations in the European Union and the United Nations. For eleven years, he headed a Programme to Foster Local Pharmaceutical Production and on Intellectual Property Rights in the Development Ministry. Frank Schmiedchen is a member of the VDW Advisory Board (2002–2009; since 2016) and has been leading the VDW Study Group TA Digitalisation since 2017.

Prof. Dr. Klaus Peter Kratzer is Professor of Computer Science at the Ulm University of Technology. His teaching and research areas are programming, databases, and intelligent systems. In addition, he has held and continues to hold various positions in university administration as Dean of Studies, Dean, and Prorector. In addition to his university activities, he supports projects to accompany and promote European study reform at the national and European level. He is a member of the VDW and of the VDW Study Group TA Digitalisation since 2017.

Dr. Jasmin S. A. Link is mathematician and sociologist. From 1999 until 2009, she worked in a research and education project for mathematically talented pupils at Universität Hamburg. Since 2010, she is associated scientist at the Research Group Climate

Change and Security (CLISEC) at Universität Hamburg. 2012/2013 she worked on the technological impact assessment of climate engineering for the Max-Planck-Institute for Meteorology in Hamburg. 2013–2015, Dr. Link co-founded an international young researchers network in complex systems sciences and was member of its advisory board (yrCSS). She joined VDW in 2015 and is part of the VDW Study Group TA Digitalisation since 2020.

Prof. Dr. Heinz Stapf-Finé (sociologist, economist) is Professor of Social Policy at the Alice Salomon University of Applied Sciences Berlin and Academic Director of the Paritätische Akademie Berlin. He wrote his doctoral thesis on "Old Age Security in Spain" and is an international expert in the field of labour and social policy. Previously, he was Head of Social Policy at the Federal Executive Board of the German Trade Union Confederation (DGB). Before the DGB, he worked as Operations Manager of the Luxembourg Income Study, as a researcher at the Institute for Health and Social Research (IGES) Berlin, and as a policy officer at the German Hospital Association. He is a member of the VDW and of the VDW Study Group TA Digitalisation.

Authors

Michael Barth is Head of Corporate Affairs and Head of the Berlin branch of genua GmbH. Previously, he was Division Manager for Public Security and Defence at the IT industry association BITKOM e. V. He completed a 12-year officer career in the German Armed Forces, where he was primarily deployed in the area of Operational Information and the Armed Forces Support Command. Barth studied history and social sciences, as well as communications. He is involved as chairman of the BITKOM working group on security policy, on the board of the Future Forum for Public Security, and on the board of the working group on cyber security and economic protection of the Federation of German Industries. He is a member of the VDW and of the VDW Study Group TA Digitalisation.

Prof. Dr. Stefan Bauberger, SJ is a physicist, philosopher, and theologian. He is a member of the Jesuit Order and teaches as Professor of Natural Philosophy and Philosophy of Science at the Munich School of Philosophy. He worked for several years in theoretical elementary particle physics. He researches and teaches on borderline issues between philosophy and natural science, especially physics, in the field of dialogue between natural science and religion as well as on the philosophy of Buddhism, and in the fields of philosophy of technology and philosophy of science. Previously he was head of education for the Jesuit Order in Germany, Austria, and Switzerland. He is a ZEN Master and runs a meditation centre. Prof. Bauberger is a member of the VDW and of the VDW Study Group TA Digitalisation.

Prof. Dr. Johann Behrens has as a sociologist and health economist conducted numerous research projects on social and health policy issues. In 1998 he founded the Center for Evidence-Based Nursing and in 1999 in Halle-Wittenberg as founding director of the first Institute for Health including Therapy and Nursing Sciences. at a German-speaking public medical faculty. As a university lecturer, he teaches at Halle and previously at universities in North America and Europe. Prof. Behrens is a long-standing board member of the Dean's Conference on Nursing Science and the Commission for Evaluation of the ICOH/WHO. He is a member of the VDW and since 2020 has been the spokesperson for the VDW Study Group "Health as Self-Determined Participation".

Prof. Dr. Paula Bleckmann is a media educator and computer game addiction expert. She has been a professor of media education at Alanus University in Alfter near Bonn since 2015. She completed her habilitation in health education and her doctorate in media education. She is a member of the advisory board of the Federation of German Scientists (VDW), co-initiator of the network initiative UNBLACK THE BOX and researches and publishes on media (addiction) prevention, digital education policy, and parent counseling. Prof. Bleckmann has led the VDW Study Group on digitalisation and education since 2016.

Dr. Rainer Engels is an agricultural economist and has been working for the Gesellschaft für Internationale Entwicklung (GIZ) since 2003; currently as Project Manager of the Sustainable Economic Development Sector Project. Before that, he worked for eight years at Germanwatch e. V., among other things as Managing Director. His academic focus is on industrial policy, with an emphasis on SMEs, services, and quality infrastructure. At GIZ, he has also been supporting the area of digitalisation and networking of the economy (e.g. Industry 4.0, machine learning) since 2016. He is a member of the VDW and of the VDW Study Group TA Digitalisation.

Alexander von Gernler is the Head of *Research and Innovation* at genua GmbH. He is interested not only professionally, but also in all new technological developments in IT security. Alexander von Gernler is Vice President of the Gesellschaft für Informatik e. V. (GI). (GI). The development of his subject as well as its reception in the population, above all the topic of responsibility of computer science, are a matter of concern to him. From 2005 to 2010, he was a developer in the OpenBSD free software project *OpenBSD*. Alexander von Gernler is a member of the VDW and of the VDW Study Group TA Digitalisation.

Dr. Christian Kellermann teaches at the University of Applied Sciences in Berlin in the field of digitalisation and business. Previously, he was managing director of the Institute for the History and Future of Work (IGZA), where he researched and published, in particular, on artificial intelligence and the future of the labour society. He is the managing director of Denkwerk Demokratie. He also worked for the Friedrich-Ebert-Stiftung as head of the regional office for the Nordic countries based in Stockholm. Dr. Kellermann is a member of the VDW and of the TA Digitalisation VDW study group.

Dr. Reinhard Messerschmidt is a philosopher and sociologist and works in the Helmholtz Open Science Office at the GermanGeoForschungsZentrum (Potsdam). Previously, he was the Digitalisation Officer at the office of the German Advisory Council on Global Change (WBGU) and the HighTech Forum Officer at the Fraunhofer-Gesellschaft. He has been scientifically active and has published in various functions on different digitalisation topics. Dr. Messerschmidt is a member of the VDW and of the VDW Study Group TA Digitalisation.

Prof. Dr. Götz Neuneck is Senior Research Fellow at the Institute for Peace Research and Security Policy (IFSH) and Professor at the MIN Faculty of the University of Hamburg. Until 2018, he was Deputy Scientific Director of IFSH and directed the Master's programme "Peace and Security Studies" at the University of Hamburg. His research focuses on the areas of arms (control) and security. He is spokesperson for the Physics and Disarmament Working Group of the German Physical Society (DPG) and a member of the Council of the Pugwash Conferences on Science and World Affairs as well as Pugwash representative of the VDW.

Brigitte Pemberger is a media educator, teacher (1st–9th grade) with many years of professional experience and a research assistant in the research groups of Prof. Dr. Bleckmann at the Alanus University of Arts and Social Sciences in Alfter/Germany. She is a lecturer in several Teacher Education and Media Prevention courses and a founding member of "Analog-Digidaktik". Publications on conceptions that combine media literacy with health promotion.

Oliver Ponsold is an officer in the German Armed Forces and, after scientific and lecturing assignments in the field of computer science and project management, has been working in the field of cyberdefence since 2017, currently as Information Security Officer for the Berlin Service Headquarters at the Federal Ministry of Defence.

Prof. Dr. Eberhard Seifert (bioeconomist) works on sustainability, ecological economics, and environmental management. He is a founding member of the "New Models of Prosperity" working group at the Wuppertal Institute for Climate, Environment and Energy, co-founder of the Institute for Ecological Economy Research (iöw) Berlin, and the European Association for Bioeconomic Studies (EABS). Since 1992 he has been active in international environmental management standardisation in various functions, including as chairman of the German Industrial Standards (DIN) Committee for "climate change". He has taught at various universities. He is a member of the VDW and of the VDW Study Group TA Digitalisation.

Christoph Spennemann is a lawyer and works at the Swiss Patent Office, for which he conducts negotiations at the UN Intellectual Property Organisation (WIPO). Previously, he led a programme at UNCTAD since 2006 to support emerging and developing countries in intellectual property issues and also advised the UN Secretary-General on these issues. Christoph Spennemann is a member of the VDW and of the VDW Study Group TA Digitalisation.

Dr. Stefan Ullrich is a computer scientist and philosopher. He heads the research group "Responsibility and the Internet of Things" at the Weizenbaum Institute for the Networked Society, Berlin, and was previously a long-time member of the working group "Informatics in Education and Society" at Humboldt University Berlin. Since 2019, Stefan Ullrich has been a member of the Expert Commission for the Third Equality Report of the Federal Government. He is the deputy spokesperson for the "Informatics and Ethics" specialist group of the German Informatics Society. Since 2019, he has been on the advisory board of the International Federation for Information Processing (IFIP), Chapter TC 9. Dr. Ullrich is a member of the VDW and of the VDW Study Group TA Digitalisation.